Low-Dimensional Cooperative Phenomena

Cooperative Phenomena

The Possibility of High-Temperature
Superconductivity

NATO ADVANCED STUDY INSTITUTES SERIES

A series of edited volumes comprising multifaceted studies of contemporary scientific issues by some of the best scientific minds in the world, assembled in cooperation with NATO Scientific Affairs Division.

Series B: Physics

The series is published by an international board of publishers in conjunction with NATO Scientific Affairs Division

A	Life Sciences	Plenum Publishing Corporation
B	Physics	New York and London
C	Mathematical and Physical Sciences	D. Reidel Publishing Company Dordrecht and Boston
D	Behavioral and Social Sciences	Sijthoff International Publishing Company Leiden
E.	Applied Sciences	Noordhoff International Publishing Leiden

Low-Dimensional Cooperative Phenomena

The Possibility of High-Temperature Superconductivity

Edited by

H. J. Keller

Anorganisch-Chemisches Institut der Universität
Heidelberg, Germany

SPRINGER SCIENCE+BUSINESS MEDIA, LLC

Library of Congress Cataloging in Publication Data

Nato Advanced Study Institute on Low-dimensional Cooperative Phenomena
   Starnberg, Ger., 1974.
   Low-dimensional cooperative phenomena; the possibility of high-temperature
superconductivity.

   (Nato advanced study institutes series: Series B, Physics; v. 7)
   Includes bibliographical references.
   1. One-dimensional conductors — Congresses. 2. Solids — Magnetic proper-
ties — Congresses. I. Keller, Heimo J., 1935-   ed. II. North Atlantic
Treaty Organization. III. Title. IV. Series.
QC176.8.E4N37 1974   530.4'1   74-34312
ISBN 978-1-4757-1401-2   ISBN 978-1-4757-1399-2 (eBook)
DOI 10.1007/978-1-4757-1399-2

Lectures presented at the 1974 NATO Advanced Study Institute on
Low-Dimensional Cooperative Phenomena and the Possibility of High-Temperature
Superconductivity, held in Starnberg, Germany, September 3-15, 1974

© 1975 Springer Science+Business Media New York
Originally published by Plenum Press, New York in 1975
Softcover reprint of the hardcover 1st edition 1975

United Kingdom edition published by Springer Science+Business Media, LLC
A Division of Plenum Publishing Company, Ltd.
4a Lower John Street, London W1R 3PD, England

Preface

Theoretical and experimental work on solids with low-dimensional cooperative phenomena has been rather explosively expanded in the last few years, and it seems to be quite fashionable to contribute to this field, especially to the problem of one-dimensional metals. On the whole, one could divide the huge amount of recent investigations into two parts although there is much overlap between these regimes, namely investigations on magnetic exchange interactions constrained to mainly one or two dimensions and, secondly, work done on 1d metallic solids or linear chain compounds with 1d delocalized electrons. There is, of course, overlap from one extreme case to the other with these solids and in some rare cases both phenomena are studied on one and the same crystal. In fact, however, most of the scientific groups in this area could be associated roughly with one of these categories and, in addition, a separation between theoreticians and experimentalists in each of these groups leads to a further splitting of interests although many theories about these solids have been tested by experimentalists. Nevertheless, more cooperation and understanding between scientists working on low-dimensional cooperative phenomena should appreciably stimulate further development. With a better interdisciplinary understanding, new ideas could possibly help chemists in synthesizing tailor-cut solids. This would in return give experimentalists new phenomena to examine and finally would stimulate new theoretical work.

It was the purpose of the NATO ASI held in Starnberg to stimulate this cooperation and to develop a common "language", which means that the theoretical background of solid state physics of extremely anisotropic solids was taught from the beginning to chemists and experimental physicists. Furthermore, the experimental physicists gave a detailed account of what they had done on "magnetic" and "metallic" 1d solids and what type of experiments could be done in the future, together with a mention of technical application of these ideas. Finally, the preparative chemists gave an up-to-date summary of the progress made in this field regarding the preparation of tailor-cut molecules and pointed out what kind of compounds could be prepared in the near future.

In several evening sessions some participants presented sum-
maries of their recent work and these and other new results were
discussed. A draft of these discussion could not be added in a
printed form because of the limitations set by the total page number of this volume, but to give at least an idea of the problems
touched upon during these sessions, a list of the main contributors together with the title of the discussed contribution is
given as an appendix.

I hope that the participants have profited from the meeting
and, furthermore, that at least some of the readers of the following papers are stimulated to high-dimensional cooperative efforts
on low-dimensional cooperative solids.

I would like to thank NATO who made this project possible
through generous financial support. The Advanced Study Institute
could not have taken place without the efforts of Mrs. G. Egerland
who did all the correspondence and typing of particulars and without the organizational work of the Co-directors, Dr. D. Nöthe and
Dr. H. H. Rupp. The help of these people and of H. Endres who did
most of the book-keeping is thankfully acknowledged. The generosity
of IBM, Germany, contributed to the completeness of the program.

Heimo J. Keller

Heidelberg
October 1974

Contents

BASIC PRINCIPLES AND CONCEPTS IN THE PHYSICS OF LOW DIMENSIONAL

COOPERATIVE SYSTEMS

Philip Pincus

Department of Physics, University of California

Los Angeles, California, 90024

A. INTRODUCTION

This NATO institute is devoted to a study of quasi- one and two dimensional systems which exhibit strong electronic correlations. Low dimensionality and strong interactions may lead to many fascinating quasi- ordered states in solids, e.g. magnetism, super conductivity, Peierls distortions, charge or spin density waves, etc. This presentation shall attempt to introduce to students or other non-solid state scientists some of the basic ideas which the author has found to be useful and relevant for thinking and visualizing the systems under investigation. We shall usually <u>not</u> reproduce detailed calculations which we already have available in the literature, but rather attempt to give a feeling for the physical arguments which lead to the predicted results. While our emphasis will lie with theory, the models considered are direct manifestations of the experimentally known nature of the systems under study. We shall present very little detailed experimental results; these have been reviewed by Shchegolev,[1] for the highly conducting 1d systems, and by DeJongh and Miedema[2] and Hone and Richards[3] for the lower dimensionality magnetic systems. Of course, at this institute I expect that much of the most recent work will be presented.

The Physics of quasi- one dimensional solids involves several types of interesting and unusual phenomena which are well exemplified by 1) the "1d" magnetic systems and 2) the organic charge transfer salts based on the tetracyanoquinodimethan (TCNQ) anion.

The quasi- 1d magnetic insulators [2,3] are beautiful

examples of the special situation in 1d that short range forces cannot generate true long range order; i.e. ferro- or antiferro-magnetism. On the contrary, in 3d these interactions lead to those well known infinitely correlated states. In section B, we shall discuss the role of fluctuations in supressing phase transitions in 1d and present some results on the incipient order that develops. We shall then show how weak coupling between 1d strands can reestablish a temperature region where long range order exists.

Section C, will be devoted to a discussion of the basic physics associated with the TCNQ type charge transfer salts. Some of these systems exhibit metallic properties but are inherently of the narrow band type. The tight binding (LCAO) method is natural to treat the electronic motion of such systems but, because of the strongly localized nature of the wavefunctions, Coulomb interactions. play an even greater role than in the more common metals, e.g. alkalis. We shall discuss some of the consequences of the strong electron-electron interactions and the correspondingly important coupling of the electrons to both intra- and intermolecular vibrations.

B. COOPERATIVE PHENOMENA IN QUASI 1d SYSTEMS

As was mentioned in the introduction, strictly speaking, at any finite temperature there does not exist long range order in a 1d system. The basic topological reason for this can be understood by considering the simple case of an Ising chain,[4] which is a 1d lattice where at each site there exists a two level system (e.g. a spin $S = 1/2$) which may be in either of the two states. Let us denote the state of the ith site (Fig.1) by the integer n_i which takes on the values +1 or -1 respectively when the site is in the state labelled by ↑ or ↓. One assumes to be an interaction between nearest neighbors of the form $-J n_i n_{i+1}$ ($J > 0$ is the energy of interaction) leading to a total energy

$$E = -J \sum_i n_i n_{i+1} \qquad (B.1)$$

The effect of the interaction is to attempt to push all the sites into the same (i.e. ferromagnetic) state either ↑ or ↓ (Fig. 1a), with a total energy -NJ for a chain of N+1 sites. This ordered state is indeed the ground state of the Ising ferromagnet; but let us consider the situation at any finite (non-zero) temperature T. It is the free energy, $F = E - TS$ (S is the entropy) whose mininum determines the state of the system. Let us imagine one mistake in the chain (Fig. 1a) where there is a reversal of orientation. Relative to the ground state energy (-NJ) such a break

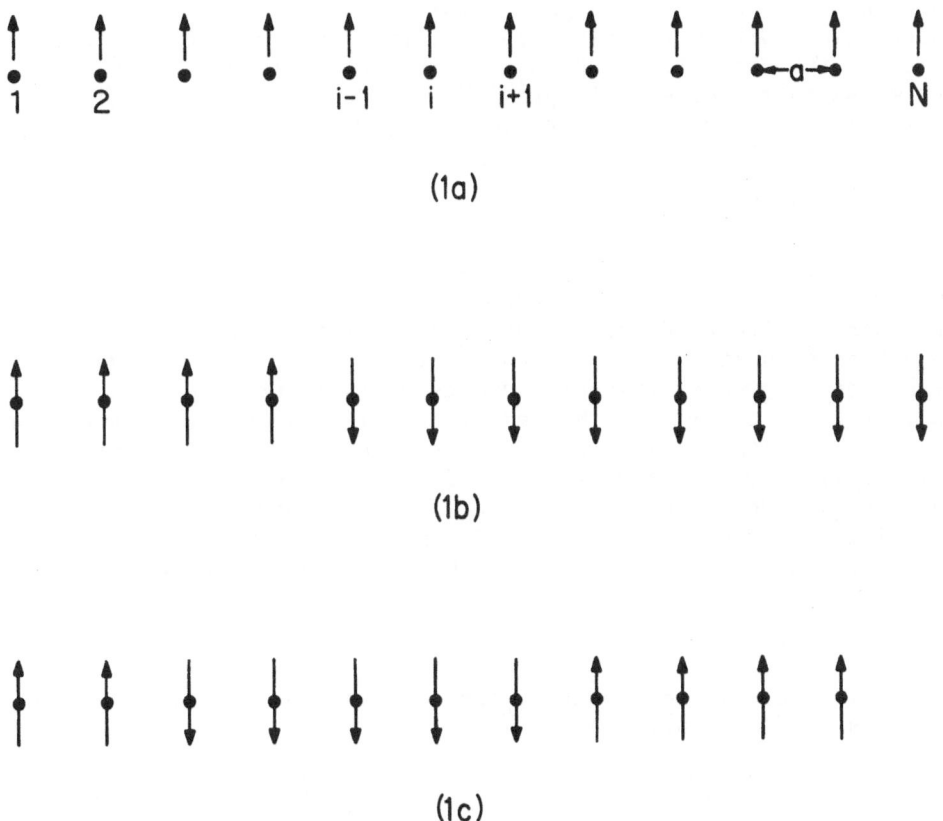

Fig. 1. A one dimensional Ising lattice (interatomic spacing a).
        a) Ferromagnetic State (T=0).
        b) One domain boundary.
        c) Part of a long chain showing one complete ferromagnetic
           cluster.

costs an energy 2J. On the other hand, the break may occur at any
of N sites, generating an entropy $k_B \ln N$, resulting in a change in
free energy

$$\Delta F = 2J - k_B T \ln N \qquad\qquad (B.2)$$

which at any finite T is negative for sufficiently long chains.
Indeed proceeding with this argument, it is energetically favorable for the chain to break up into ferromagnetic domains (Fig.
1c) whose characteristic length $\xi(T)$ (the correlation length) is
approximately given by setting $\Delta F = 0$ in (B.2);

$$\xi(T) \overset{\sim}{=} a\, e^{2J/k_B T}$$

where a is the interatomic spacing. Thus at finite temperature, we
encounter a situation where there are domains of length $\xi(T)$ within
which there is strong "short range" order. However spins separated
by distances greater than $\xi(T)$ are essentially uncorrelated - hence,
no long range order. As the temperature is lowered the domains increase in length until finally at absolute zero we attain the
completely ferromagnetic state. This cluster formation with no
long range order seems to be a general phenomenon in 1d systems; it
is only the correlation length $\xi(T)$ which is apparently strongly
model dependent. For example, in the classical Heisenberg model[5]

$$E = - J \sum_i \vec{S}_i \cdot \vec{S}_{i+1} \qquad\qquad (B.4)$$

where $\vec{S}_i$ is a classical vector, the correlation length, at low temperatures $(k_B T < J)$, varies as

$$\xi(T) \sim a(J/k_B T) \qquad\qquad (B.5)$$

This temperature dependence for the correlation length has been
beautifully demonstrated by Hutchings et al[6] in the quasi 1d anti-
ferromagnet $(CD_3)_4 NMnCl_3$ (TMMC), (Fig.2). The T^{-1} behavior of the
correlation length is also characteristic of 1d superconductors[7].
It is then essentially $\xi(T)$ which governs the detailed thermodynamic
behavior of the system.

Let us contrast this situation with the case of higher dimensionality where phase transitions often occur. In particular, the
Ising model has a phase transition in 2d. Let us consider a
topological argument to see how 2d differs from 1d in this respect.
In Fig. 3, we depict a square lattice with a ferromagnetic domain.
We estimate the free energy by noting that lattice points on the
perimeter have one near neighbor whose spin is reversed relative

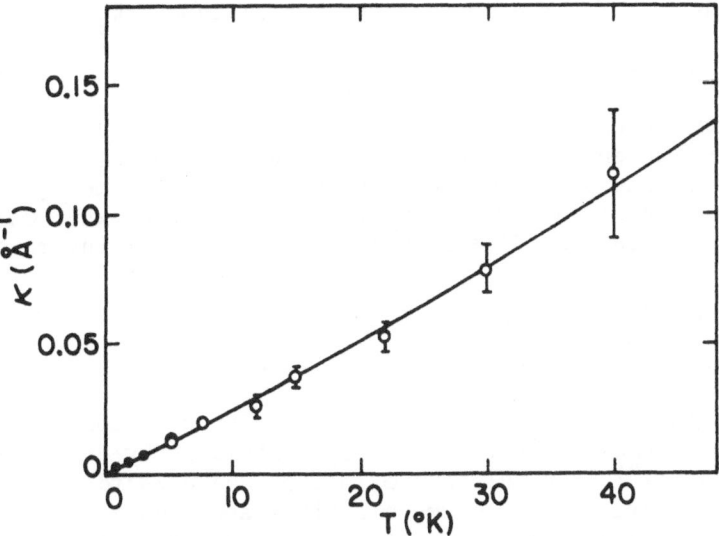

Fig. 2.   The inverse static correlation length $\xi^{-1}(T)$ for TMMC[6].
The solid line corresponds to Fisher's theory[5].

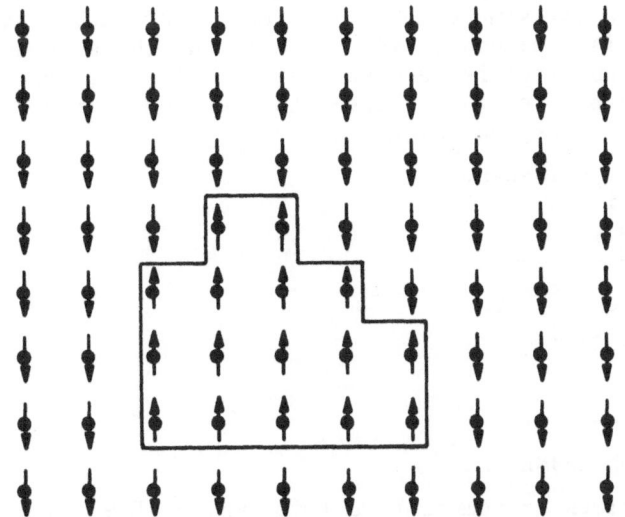

Fig. 3.   A reversed spin domain in a square Ising lattice.

to the ground state; then if there are P perimeter sites this results in an increase in energy JP. The entropy may be estimated
by noting that each perimeter site has one emerging line that may
connect to any of three nearest neighbors; thus $S \simeq k_B \ln 3^P$, or

$$\Delta F \simeq 2JP - k_B T \ln 3^P = P(2J - k_B T \ln 3) ; \tag{B.6}$$

there seems to exists a critical temperature $T_c \simeq (2J/k_B) \ln 3$ below
which the free energy increases upon the creation of reversed spin
domains, leading us to the conclusion that below some T_c there
exists a state of long range order. In fact, the exact Onsager
solution[9] to the two dimensional square lattice Ising model gives

$$k_B T_c = 2J/\ln(1+\sqrt{2}) \tag{B.7}$$

for the phase transition to the ordered state.

The traditional elementary method to study second order (no
latent heat) phase transitions in 3d systems is by use of mean
field theory[10]. In this approximation one uses the (usual) translational invariance to assume all sites are identical. Then one
confines his attention to a particular site and replaces its interactions with its neighbors by a "mean field", i.e. one replaces the
variables associated with the z neighbors by their average values
which are then determined self consistently   As an example
consider the Ising model (B.1) . In general the ith site interacts
with its z nearest neighbors (z = 2. for 1d; z = 4 for 2 d square
lattice; z = 6 for simple cubic lattice). For each of these
neighbors n_i is replaced by $<n>$ which is the thermodynamic value
of $<n_g>$ (no subscript is necessary since all sites are identical).
The many body problem then reduces to a one body problem and using
Boltzman statistics, we calculate $<n>$ as

$$<n> = \frac{\sum\limits_{n_i=1,-1} n_i e^{+\beta z J <n> n_i}}{\sum\limits_{n_i=1,-1} e^{\beta z J <n> n_i}} \tag{B.8}$$

or

$$<n> = \tanh z\beta J <n> \tag{B.9}$$

The lowest values of temperature T_c for which (B.9) has a non-zero
value of $<n>$ as a solution is taken as the transition temperature,
which is $k_B T_c = zJ$. Notice that this results depends on geometry
only through the number of nearest neighbors. In fact, mean field
theory predicts incorrectly a phase transition in 1d. The error

lies in the neglect of the fluctuations about the average value of the magnetization. Mean field theory always gives a discontinuous jump in the specific heat at T_c which for the Ising model is (3/2) k_B per spin. In Fig. 4, this mean field approximation to the specific heat is contrasted with the exact 1d Ising model result. Notice the strong effect of the fluctuations in smearing out the phase transition. In higher dimensional systems, however, the mean field theory does a reasonable job in describing many phase transtions, if the region (critical region) in the immediate vicinity of the phase transition is avoided.

The real systems with which we are concerned in the laboratory are, however, never completely 1d. At best, many of the magnetic and TCNQ salts which will be discussed here may be considered as two dimensional arrays of weakly coupled strands. Using a magnetic example, in addition to the exchange coupling J between spins on the same chain, there must surely exist an interaction J' (<<J) between neighboring spins on different strands. This coupling between chains may eventually, at a sufficiently low temperature, lead to a real 3d phase transition[2]. Roughly speaking, such a system will exhibit three temperature regions, demarked by the transition T_c which occurs at some intermediate temperature J' < $k_B T_c$ < J: (a) For T>>T_c, the system behaves essentially as an incoherent assembly of independent 1d strands; (b) For T<<T_c the behavior is essentially that of an ordinary anisotropic 3d system. (c) In the neighborhood of T_c there is an intermediate "cross-over" regime which is the transition region between 1d and 3d characteristics.

We shall briefly indicate a theoretical method[11] to study these effects which is an extension of mean field theory. The approach is based on the idea that the non-existence of phase transition in 1d arises from the strong effect of the 1d fluctuations. These fluctuations must be treated rather accurately. On the other hand, the weak interchain coupling may be considered as giving rise to a mean field on entities which are not insolated spins (as is the previous mean field theory) but entire strands. The results of such calculations for T_c may be described as follows: we have already seen for a strictly 1d system that exist regions of length $\xi(T)$>>a ($k_B T$>>J) over which there is short range order. This implies that in response to a weak external perturbation the (ξ/a) spins of these strongly correlated segments must behave as a rigid unit. Therefore, the energy required to turn a spin relative to a spin or a neighboring chain is $\sim(\xi/a)$ J' rather than $\sim$ J' for an ordinary 3d system with exchange energy J'. The mean field argument will then predict a transition temperature at

$$k_B T_c \sim \left[\xi(T_c)/a\right] J' \qquad (B.10)$$

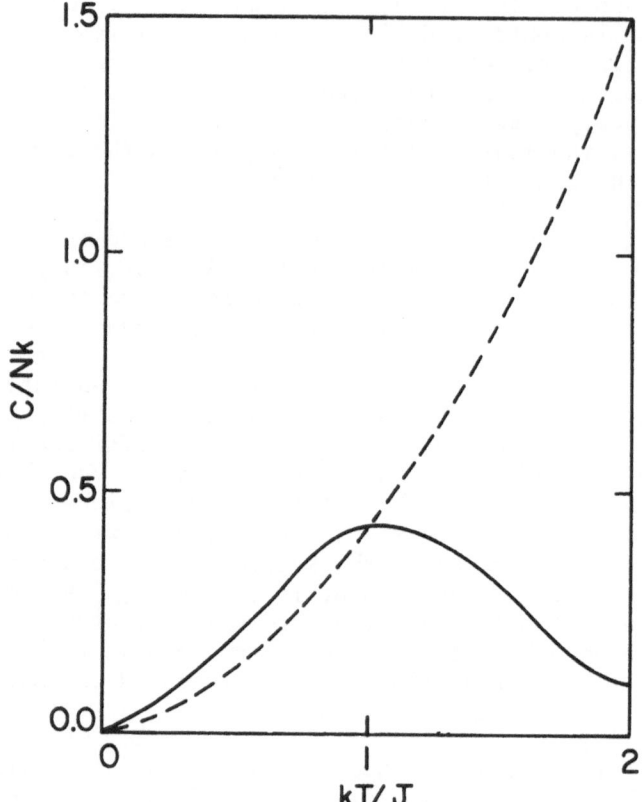

Fig. 4. Comparison of the exact (solid) and mean field theory
(dashed) results for the specific heat of the Ising chain.
The effect of fluctuations in smearing out the phase transition is apparent.

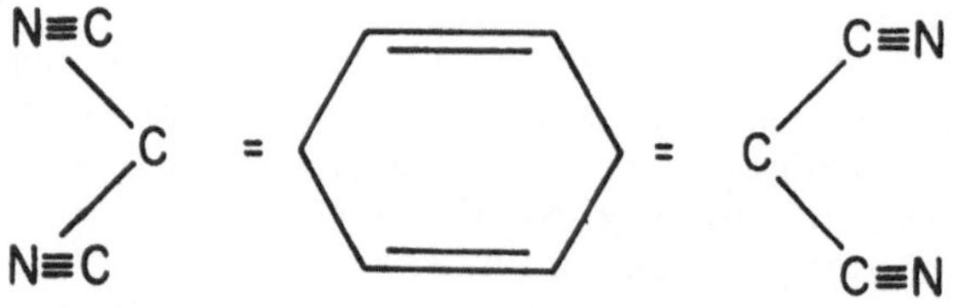

Fig. 5. The neutral TCNQ molecule.

which for the Ising model (B.3) leads to

$$k_B T_c \sim 2J/\ln(2J/J') \qquad \qquad \text{(B.11)}$$

and for the classical Heisenberg chain (B.4,5) to

$$K_B T_c \sim (JJ')^{1/2} \qquad \qquad \text{(B.12)}$$

In the next section we shall see a further example of a physical effect special to 1d.

C. PHYSICAL INTERACTIONS IN THE ORGANIC CHARGE TRANSFER SALTS

One of the major themes of this institute is the properties of the organic charge transfer salts based on the TCNQ anion. The TCNQ molecule (Fig. 5) is a rigid planar structure whose most important property for us is its high electron affinity. Indeed it easily forms salts[1] of the types X^+ TCNQ$^-$ where X = (Na, K, Li, NH$_4$ or any of several organic doners) and R^+ TCNQ$^-$ TCNQ^0 where R = (Quinolinum, Acridinium, etc.). The TCNQ$^-$ anion contains one unpaired electron whose wave function is substantially[12] pulled out into the region of the CN groups.[12] The salts often[13] crystallize into a two dimensional array of separated cation and anion stacks. The planes of the TCNQ$^-$ molecular ion, are typically not perpendicular to the symmetry axis but tilted relative to it (Fig. 6). Most of the cations are closed shell systems; e.g. alkali ions, N-methyl phenazinium. In those cases we expect most of the interesting electronic properties to be associated with the TCNQ$^-$ chains. Sometimes [e.g. tetrathiofulvalene (TTF)$^+$] the cations also possess an unpaired electron leading to more complex electronic structures.

Our zeroth order picture of the various charge states of the TCNQ molecule is to consider only the lowest unfilled molecular orbital of TCNQ^0. In TCNQ$^-$, this state becomes occupied by one electron. The next higher molecular orbital probably lies a few electron volts higher in energy and is usually neglected - however the effects of these states should be considered at some point. In the metallic systems, the configuration TCNQ^{--} sometimes must occur; in this case, we assume that the second electron goes into the same orbital but at the expense of some addition Coulomb energy, U_0, arising from the extra Coulomb repulsion. This "bare" Coulomb energy may be estimated by assuming that the two "extra" electrons correlate to be on opposite sites of the molecule leading to $e^2/r \sim 1eV$. Various bond lengths are known to stretch with increased valence leading to a reduction of U_0 by several tenths of an electron volt and possible small polaron formation[15]. In the crystalline environment U_0 is further reduced by dielectric screening. This is most effecient when the cations are molecules having

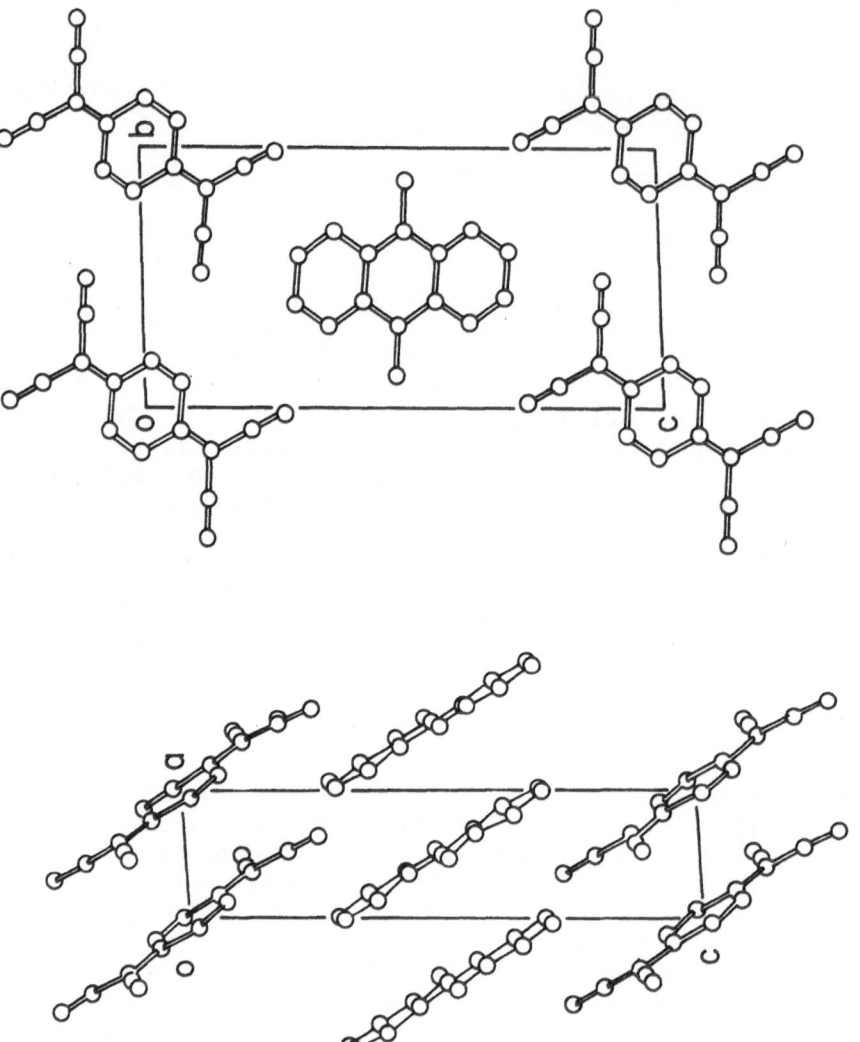

Fig. 6. Crystal structure of N-methyl phenazinium (NMP) TCNQ[16].

low lying excited states. In such cases, Chaikin et al [15] have estimated dielectric screening reductions of the bare U_O by an amount of the order of U_O itself, again leading to possible excitonic polaron formation.

Let us begin to formulate the solid state electronic structure by considering a face-to-face TCNQ anion dimer (Fig. 7). If we let ϕ be the molecular wavefunction corresponding to the lowest un-occupied orbital of $TCNQ^O$, a first guess to the low lying dimer wavefunctions might be to construct bonding and antibonding combination i.e. a linear combination of molecular orbitals

$$\psi^{\pm} = c \ (\phi_a \pm \phi_b) \qquad\qquad (C.1)$$

where + and - refer to the bonding and antibonding states respect-fully; c is a normalization constant. As is customary, the bonding orbital ψ^+ has lower energy because it gives a finite amplitude to the wavefunction on the nodal plane, i.e. it allows the electron to be shared more effectively by the two attracting centers. The $TCNQ_2$ anion dimer would then have two electrons occupying the bond-ing orbital in a spin singlet configuration; i.e.

$$\psi_{MO} = \psi^+(\uparrow) \ \psi^+ \ (\downarrow) \qquad\qquad (C.2)$$

A model Hamiltonian which is convenient to describe this situation is, in second quantized notation, [17],

$$H = - t \sum (a_\sigma^+ b_\sigma + b_\sigma^+ a_\sigma) + U \ (n_{a\uparrow}n_{a\downarrow} + n_{b\uparrow}n_{b\downarrow}); \qquad (C.3)$$

where the operator a_σ removes an electron with spin σ (= $\uparrow$ or $\downarrow$) from the a molecule; in the orbital ϕ; a_σ^+ places an electron with σ on the molecule; $n_{a\sigma} = a_\sigma^+ a_\sigma$ is the number of electrons with spin σ ($n_{a\sigma} = 1,0$) on the molecule a in the orbital ϕ. The energy associated with the orbital ϕ is taken as the zero of energy; the transfer (or resonance) integral [18] t measures the attraction of an electron on molecule a toward molecule b and vice-versa; U re-presents the relative extra Coulomb energy when a given molecular orbital is doubly occupied. This model Hamiltonian when extended to an antire crystalline array of sites was introduced by Hubbard [19] and Kanamori [20] to discuss correlation effects in magnetic transition metals, alloys, and compounds. In this notation, the bonding and antibonding orbitals (c.1) are described by operators $\psi_\sigma^{\pm}$ which create an electron with spin σ shared equally between both members of the dimer with respectively similar and opposite phases,

$$\psi_\sigma^{\pm} = \frac{1}{\sqrt{2}} \ (a_\sigma^+ \pm b_\sigma^+) \qquad\qquad (C.4)$$

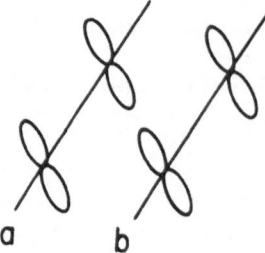

Fig. 7.   Schematic representation of a TCNQ dimer with the associated
          π orbitals.

Fig. 8.   A linear chain of dimers where t and t' are respectively the
          intra- and interdimer nearest neighbor resonance integrals.

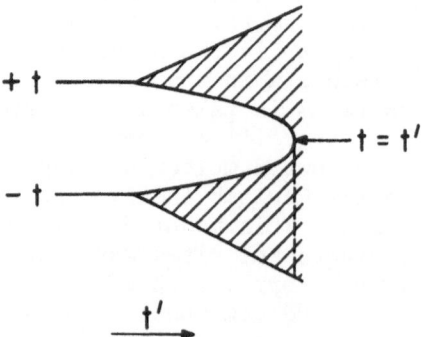

Fig. 9.   Schematic representation of the increasing bandwidths of the
          bonding and antibonding states with increasing t'; i.e., as
          the dimers approach each other. When b becomes equal to a,
          i.e., a chain of uniformly spaced molecules, the two bands
          merge.

The bonding orbital gains an energy t relative to the zero of energy
while the antibonding state is higher in energy by the same amount,
i.e. the bonding orbital has energy -t and the antibonding orbital
+t. The anion dimer with both electrons in the bonding state (C.2)
will then gain a resonance energy (-2t). In this state, each
orbital(ϕ_a or ϕ_b) has a probability of 1/2 of being occupied by an
electron of a given spin; this then leads to a Coulomb energy (C.3)
of $\frac{1}{2}$U. Thus while the linear combination of molecular orbital wavefunctions lowers the total energy via the resonance term, the
Couloumb energy is increased because of the configurations where
the electrons get near one another; i.e. on the same molecule. If
t>>U/4, we might expect the MO states to be a good approximation;
let us momentarily make this assumption and proceed to discuss how
a crystalline enviroment modifies the bonding and antibonding states.
Consider a linear chain of dimer anions (Fig. 8) which we imagine
to be constructed by pushing dimers in one dimension, towards one
another uniformly from infinity. As the interdimer distance
diminishes, the interdimer resonance integral t' (<t) (t' is the
matrix element which takes an electron from the right hand molecule
of one dimer to the left hand member of the next dimer and visa
versa) increases; this leads to spreading of the bonding and antibonding dimer wavefunctions over many dimers with relative phases
and amplitudes resulting in a broadening of the states into bands
(Fig. 9). At the point where the chain becomes uniform (b=a and
t'=t) the bonding and antibonding bands merge. Here symmetry
dictates that the wavefunctions at all sites have equal amplitudes,
i.e. completely delocalized states, charactistic of a metal. Indeed for the ground state of a uniform chain of TCNQ⁻ anions all
the bonding orbitals must be occupied by singlet pairs and all the
antibonding states empty. Thus for a long chain (N → ∞), it takes
arbitrarily little energy to excite an electron from an occupied
bonding into an unoccupied antibonding states, i.e. we have a metal.
Let us now retreat somewhat from the uniform situation back to a
dimerized chain (Fig. 8). Then t'<t, and there is still an energy
gap between the bonding and antibonding bands (Fig. 9). A finite
energy of order 2|t-t'| is required to excite an electron from the
filled bonding (valence) band into the empty antibonding (conduction)
band; this is the energetics characteristic of a direct band gap
semiconductor. In fact, most of the one-to-one (R⁺ TCNQ⁻) salts
involve dimerized TCNQ chains[1] and are indeed semiconductors.
Kommandeur and his coworkers[21] have discussed many of the experimental properties of the alkali salts in terms of just this type
of semiconducting band structure.

It is of interest to pose the question of why dimerized structures seem to be so prevalent relative to the uniform chains which,
at first glance, might be favored by symmetry. It is likely that
the Peierls instability[22] of uniform one dimensional metallic
chains is playing a role. A clue to the understanding of this

lattice instability toward dimerized structure may be obtained from
(Fig. 9). As one distorts the chains (i.e. moves to the left from
the t=t' point), the bands split with a net lowering a energy of
the filled bonding states; this gain in resonance energy is balanced by an increased in the elastic energy. The special feature of
one dimension is that such a distortion is always favorable at
sufficiently low temperatures [23]. The Groningen group [24] has
suggested that the Peierls effect is at the heart of the dimerization
of the alkali TCNQ salts. Since several of the subsequent pre-
sentations at this conference will be concerned with the dynamic
and thermodynamic properties of the Peierls state, I will cut short
the discussion of this fascinating subject at this point.

Some of the TCNQ salts with organic cations seem to remain
uniform down to reasonably low temperatures; e.g. TTF-TCNQ,
NMP-TCNQ. These systems, indeed, seem to be metallic, at least
above some critical temperature region below which they also
become semiconducting. In TTF- TCNQ, just above the metal- semiconductor transition ($\sim 60^{\circ}K$), the Pennsylvania group has reported
an extraordinary [25] increase in the low frequency electrical conductivity which they associate with 1d fluctuations of a collective
superconducting state. They describe the transition itself as a
Peierls transition; the temperature may have been supressed to
this relatively low value by the Beni effect [23] arising from the
relatively high transverse resonance integrals associated with the
open shell structure of the TTF^{+} cation. NMP-TCNQ has also been
intensively studied by the Garito-Heeger [26] group; in that case
they conclude that the metal-semiconductor transition is an example of a Mott transition [27] driven by Coulomb correlation
effects. We shall discuss theseideas later in the framework of
the Hubbard model hamiltonian (C.3) Bloch et al [28] have offered
an alternative explanation for the transport properties of quasi -
1d materials such as NMP-TCNQ, where there is intrinsic disorder.
This disorder is associated with the asymmetric structure of the
cation (Fig. 10). In NMP-TCNQ crystals, the methyl group randomly
occurs attached to either of the two nitrogens, leading to a random
dipolar potential. In 1d, it is known [29] that any random
potential leads to localized states which can transport current
only by electron hopping. Using this idea, Mott's [30] variable
range hopping theory for transport between localized states was
generalized to 1d and fit to the conductivity of several systems [28].
However the applicability of this model to NMP-TCNQ is not yet
unambiguously established. [31]

Let us return now to consider the Coulomb forces which lead
to correlations between the electrons. As we saw previously for
the Hubbard model dimer problem, the MO wavefunction (C.2)
contained ionic configurations (i.e. $TCNQ^{0}-TCNQ^{--}$) which resulted
in a Coulomb repulsion U/2. If U>4t, it would be energetically

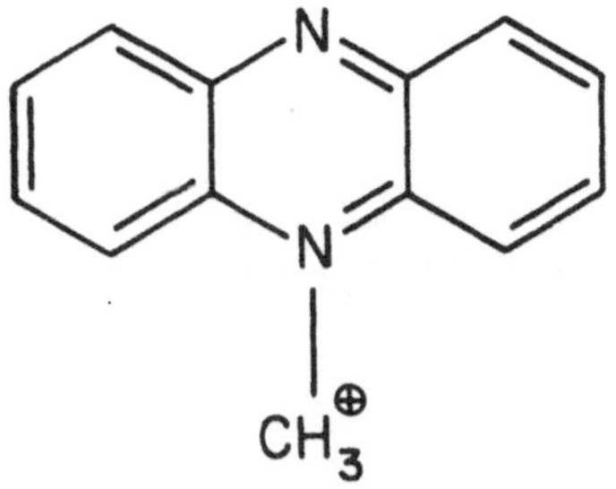

Fig. 10.   The molecular structure of N-methyl phenazinium (NMP).

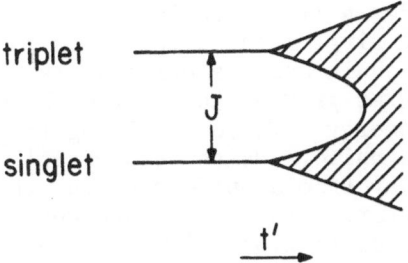

Fig. 11.   Spreading of the singlet and triplet Heitler - London bands
           as interdimer spacing decreases. When b→a, the two bands
           merge.

more favorable to localize the electrons, i.e. to <u>correlate</u> the two
electrons so that there are no ionic configurations. This is the
traditional valence bond (or Heitler-London) approach, where the
four states are written as a singlet

$$\psi_s = \frac{1}{\sqrt{2}} \left[\phi_a(\uparrow)\phi_b(\downarrow) - \phi_a(\downarrow)\phi_b(\uparrow) \right] \tag{C.5}$$

and three triplets

$$\psi_1 = \phi_a(\uparrow)\phi_b(\uparrow) ;$$

$$\psi_0 = \frac{1}{\sqrt{2}} \left[\phi_a(\uparrow)\phi_b(\downarrow) + \phi_a(\uparrow)\phi_b(\downarrow) \right] ; \tag{C.6}$$

$$\psi_{-1} = \phi_a(\downarrow)\phi_b(\downarrow)$$

These purely covalent states have zero expectation values for the
energy (C.3) because of the complete absence of ionicity. For large
Coulomb energies, these correlated states are then a better first
approximation than the uncorrelated electron molecular orbitals.
In this limit, we might expect to improve the approximation by
allowing the resonance term to mix pertubatively ψ_s with some of the
high energy ionic configurations[32] of the type $\phi_a(\uparrow)\phi_a(\downarrow)$. This
leads to a lowering in energy of the singlet state relative
to the triplet (which is not admixed with $\phi_a(\uparrow)\phi_a(\downarrow)$ by conservation
of spin angular momentum) by $-4t^2/U$. This singlet-triplet splitting
may be rewritten as an effective antiferromagnetic Heisenberg
exchange interaction between the electron spins localized on each
of the molecules, $J \vec{S}_a \cdot \vec{S}_b$ where $J = 4t^2/U$.

If we now consider a linear chain of these highly correlated
dimers, the low lying singlet and triplet states spread into bands
(Fig. 11). When the chain becomes uniform, the two bands merge,
giving rise to a singlet ground state but with some triplet
excitations at zero energy. These low lying triplets are the well
known spin waves of a uniform Heisenberg chain[33]. As one
dimerizes slightly away from a uniform chain, the singlet energies
are reduced thus suggesting that a uniform Heisenberg chain should
also suffer a Peierls distortion[34]. Thus an alternate possible
description of the dimerized alkali TCNQ salts is in terms of
highly correlated localized electrons experiencing an antiferromagnetic Heisenberg interaction[35,36]. As for the electronic
conductivity of the uniform chain, the low lying excitations are
triplets which do not carry a current. Indeed in order to move
an electron one must create polar (i.e. doubly occupied) sites
which cost an energy of $\sim U$ [32] i.e. the uniform system is a semiconductor. This type of behavior is called a Mott insulator
because the nonmetallic nature arises from the many electron
correlations rather then the interplay of crystalline symmetry with

one electron theory. This is the model utilized by the Pennsylvania group[26] to interpret the behavior of NMP-TCNQ.

In the complex salts, e.g. Quinolinium-TCNQ, where approximately one half of the TCNQ sites are neutral, these arguments would predict that the uniform chains are metallic, because electrons are able to move down the chain without creating doubly charged configurations. Coll[27] has, in fact, shown that the Hubbard model for uniform chains has zero energy electronic excitations which should be conducting. Certainly in such cases where the mean spacing between electrons exceeds a lattice constant, the use of the Hubbard model is somewhat suspect. One should at least allow electrons on nearest neighbor sites to experience Coulomb repulsions. Such an "extended" Hubbard model has been considered [39,40] and for sufficiently large repulsions relative to the resonance integral a non-conducting ground state is recovered.

The truncation of the full r^{-1} Coulomb interaction must also be reconsidered when studying the optical properties of solids. It is well known [17] that the long range character of this force in a metal gives rise to a plasma resonance frequency above which the material becomes transparent to electromagnetic radiation. Such effects have been reported in some highly conducting TCNQ salts[41, 42]. The author believes that somewhat more theoretical [43] work in this direction is needed (in the context of quasi-one dimensional strongly correlated systems) before a complete interpretation of the experiments is possible.

In the beginning of this section, we alluded to the possible formation of small polarons in the TCNQ salts [15,16]. These entities arise when an electron is strongly coupled to some other degree of freedom of the system, e.g. intra-molecular vibrations[15] or Frenkel excitons[16]. Let us illustrate the physical ideas involved by first considering one single electron moving on a uniform TCNQ chain. The Coulomb forces that come into play when an extra electron is added to the molecule may cause some intra-molecular bonds to stretch (e.g. C$\equiv$N bond[44]). (Fig. 12). Then as the transfer integral drives the electron to move from site to site along the chain, the accompanying distortion must also be dragged along [45]. This additional load effectively reduces the resonance integral[46] i.e. $t \rightarrow te^{-S}$ where the factor S = $E_B/\hbar\omega_0$; E_B the local energy gain by the distortion and ω_0 is the vibrational frequency associated with the extra degree of freedom. The fact that one electron distorts the molecule makes it relatively easier for a second electron to come onto the same site. This leads to an effective screening of the Coulomb repulsion between the electrons which may even result in a locally

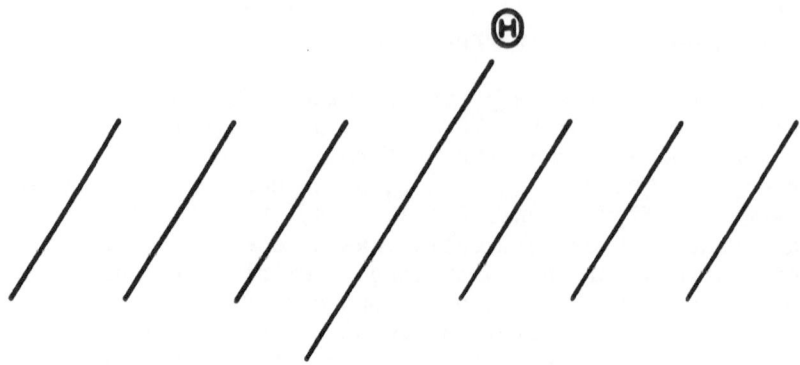

Fig. 12. A snapshot of an intramolecular small polaron. The site
          where the electron resides is distorted by the strong elec-
          tron-lattice coupling.

attractive interaction which in turn may lead to various types
of correlated ground states some of which are reminiscent of superconductivity[47]. The effects of polaron formation in the highly
correlated limit of the Hubbard model have also begun to come under
investigation [48,49]. The author believes that polaronic effects
probably do play a vital role in the TCNQ salts and much more
theoretical development in this direction is necessary.

    In this section we have attempted to present a descriptive
survey of what appears to be the dominant physical mechanism
responsible for the behavior of the TCNQ salts. In the future,
we expect to see the interplay between theory and experiment lead
to a sufficiently detailed understanding of the physical phenomena
to enable the solid state scientists and organic chemists to
collaborate in order to design materials with a wide variety of
desired properties[50]. I hope that the bare bones presented here
will give a common basic physical language so that the most recent
aspects of this exciting field can be fully appreciated by all of
us.

References

1. I.F. Shchegolev, Phys. Stat. Sol. 12, 9 (1972).
2. L.J. DeJongh and A.R. Midema, Adv. in Physics 23, 1 (1974).
3. D.W. Hone and P.M. Richards, To be Published in Ann. Rev. of Material Science, 4.
4. A good discussion of the Ising model may be found in K. Huang, Statistical Mechanics, (J. Wiley and Sons, New York, 1963), Chapters 16, 17.
5. M.E. Fisher, Am. J. Phys. 32, 343 (1964).
6. M.T. Hutchings, G. Shirane, R.J. Birgeneau, and S.L. Holt, Phys. Rev. B5, 1999 (1972).
7. D.J. Scalapino, M. Sears, and R.A. Ferrell, Phys. Rev. B6, 3409 (1972).
8. This estimate neglects (a) the fact that the domain must have a closed perimeter, and (b) that some perimeter points have 2 reversed near neighbors. For sufficiently large P these effects are small.
9. L. Onsager, Phys. Rev. 65, 117 (1944).
10. For a good discussion see Daniel C. Mattis, The Theory of Magnetism, (Harper and Row, New York, 1965) Chapter 8.
11. J.W. Stout and R.C. Chisholm, J.Chem. Phys. 36, 979 (1962) have treated the Ising Model. The method has been elaborated and extended to other magnetic systems as well as coupled superconducting strands by D.Scalapino, Y. Imry, P. Pincus, To Be Published
12. This is established by nuclear quadrapole resonance experiments in Li^+ $TCNQ^-$ by J. Murgich and S. Pissanetzky, Chem. Phys. Letters 18, 420 (1973).
13. There also exist stacked structures of the type cation-anion-cation-anion; e.g $TMPD^+$ - $TCNQ^-$. We shall not discuss these further here becuase they are intrinsfcally semiconducting not metallic.
14. C.J. Fritchie, Acta Cryst. 20, 892 (1966).
15. P. Pincus, Solid State Comm. 11, 51 (1972).
16. P. Chaikin, A.F. Garito, and A.J. Heeger, Phys. Rev. B5, 4966 (1972) P. Chaikin, A.F. Garito, and A.J. Heeger, J. Chem. Phys. 58, 2336 (1973).
17. See for example C. Kittel, Quantum Theory of Solids, (J. Wiley and Sons, Inc., New York, 1964) Chapt. 5.
18. For the case of TTF-TCNQ, t has been calculated by A.J. Berlinsky, J.F. Carolan, and L.Weiler, [Solid State Comm., To be Published], to be about 0.1ev.
19. J. Hubbard, Proc. Roy. Soc. (London) A276, 238 (1963); A237 (1963); A281, 401 (1964).
20. J. Kanamori, Prog. Theo. Physics (Kyoto) 30, 275 (1963).
21. J.G. Vegter, J. Kommandeur, and P.A. Fedders, Phys. Rev. B7, 2929 (1973); J.G. Vegter and J. Kommandeur, Phys. Rev. B9, 5150 (1974).

22. R.E. Peierls, Quantum Theory of Solids, (Oxford Univ. Press, London, 1955), Chapter V.

23. Indeed G. Beni, Solid State Comm., To Be Published, has shown that if d>1, the electron bandwidth arising from transfer between strands tends very strongly to supress the instability.

24. J.G. Vegter, P.I. Kuindersma, and J. Kommadandeur, in N. Klein, D.S. Tannhauser, and M. Pollak, eds. Conduction in Low Mobility Materials, (Taylor and Francis Ltd., London, 1971 p. 213.

25. L.B. Coleman, M.J. Cohen, D.J. Sandman, F.G. Yamagishi, A.F. Garito, and A.J. Heeger, Solid State Comm. 12, 1125 (1973); D.B. Tanner, C.S. Jacobsen, A.F. Garito, and A.J. Heeger, Phys. Rev. Letters 32, 1301 (1974); M.J. Cohen, L.B. Coleman, A.F. Garito, and A.J. Heeger, Phys. Rev. B. To Be Published.

26. A.J. Epstein, S. Etemad, A.F. Garito, and A.J. Heeger, Phys. Rev. B5, 952 (1972).

27. N.F. Mott, Proc. Phys. Soc. (London) A62, 416 (1949).

28. A. Bloch, R.B. Wiseman, and C.M. Varma, Phys. Rev. Letters 28, 753 (1972).

29. R.E. Borland, Proc. Phys. Soc. London 78, 926 (1901) N.F. Mott and W.D. Twose, Advan. Phys. 10, 107 (1961).

30. N.F. Mott, Phil. Mag. 19, 835 (1969); V. Ambegaokar, B.I. B.I. Halperin, and J.S. Langer, Phys. Rev. B4, 2612 (1971).

31. E. Ehrenfreund, S. Etemad, L.B. Coleman, E.F. Rybaczewski, A.F. Garito, and A.J. Heeger, Phys. Rev. Letters 29, 269 (1972).

32. Of course the dimer Hubbard model (C.3) is exactly soluble. See article by P. Pincus in de Laredo and Jurisic, (eds) Selected Topics in Physics, Astrophysics, and Biophysics (D. Reidel, Dordecht, 1973).

33. J. des Cloizeaux and J. Pearson, Phys. Rev. 128, 2131 (1962).

34. P. Pincus, Solid State Comm. 9, 1971 (1971); G. Beni and P. Pincus, J. Chem. Phys. 57, 3531 (1972); G. Beni, J. Chem. Phys. 58, 3200 (1973).

35. Experimental arguments for $U \gtrsim t$ in the alkali salts is given by S. K. Khanna, A.A.Bright, A.F. Garito and A.J. Heeger, Phys. Rev. To Be Published.

36. An excellent general review of the electronic properties and spin dynamics in the TCNQ salts will appear in Z. Soos, Ann. Rev. Phys. Chem. 25, (1974) To Be Published.

37. A. Ovchinnikov, Soviet Physics JETP 30, 1160 (1970) has calculated exactly the low lying excitations in this model of both the conducting and triplet types. Indeed he finds that so long as U>0, the uniform chain is a semiconductor.

38. C.F. Coll III, Phys. Rev. B9, 2150 (1974).

39. A.A. Ovchinnikov, I.I. Ukrainskii, and G.V. Kventsel, Soviet Physics Uspekhi 15, 575 (1973).

40. G. Beni and P. Pincus, Phys. Rev. B9, 2963 (1974).

41. A.A. Bright, A.F. Garito, and A.J. Heeger, Solid State Comm. 13, 943 (1973). A.A.Bright, A.F. Garito, and A.J. Heeger, Phys. Rev. B. To Be Published.

42. P.M. Grant, R.L. Greene, G.C. Wrighton, and G. Castro, Phys. Rev. Letters 31, 1311 (1973).
43. P. F. Williams and A.N. Bloch, Phys. Rev. B10, 1097 (1974).
44. H. Gutfreund, B. Horovitz and M. Weger, Solid State Comm. To Be Published.
45. The detailed study of small polaron motion has been carried out by T. Holstein and his coworkers in precisely this model. See for example: T. Holstein, Ann. Phys. 8, 325 (1959); T. Holstein, Ann. Phys. 8, 343 (1959); L. Freedman and T. Holstein, Ann. Phys. 21, 494 (1963); D. Emin and T. Holstein, Ann. Phys. 53, 439 (1969).
46. The details depend somewhat on the nature of the extra degree of freedom to which the electron is coupled. The result given here assumes that this is a simple harmonic oscillator. R. Bari, Phys. Rev. Letters 30, 790 (1973) has considered the opposite limit of a two level system which he feels is more suited for the excitonic polarons.
47. G. Beni, P. Pincus, and J. Kanamori, Phys. Rev. To Be Published.
48. R. Bari, Phys. Rev. B9, 4329 (1974).
49. C.F. Coll III, and G. Beni, Solid State Comm., To Be Published.
50 As a first step in this direction see A.F. Garito and A.J. Heeger, Accounts of Chem. Res. 7, 232 (1974).

THEORY OF THE QUASI ONE-DIMENSIONAL BAND CONDUCTOR

M. J. Rice

Xerox Webster Research Center

Webster, N.Y. 14580, U.S.A.

LECTURE ONE

Chiefly simplified discussion of the (meanfield) Peierls transition; nearly-divergent density response of the 1-d conduction electron system at $2k_F$; its consequences for the stability of a hypothetical 1-d metal; derivation of T_c; calculation of distortion amplitude and energy gap below T_c.

LECTURE TWO

Chiefly the dynamics of the pinned charge density wave (CDW) below T_c; phenomenological theory of the pinned CDW involving n_s (condensed electron no. density) and m^* (its effective inertial mass). Frequency dependent dielectric constant and conductivity. Relation between m^* and n_s for given order parameter Δ. Simple derivation of paraconductivity above T_c.

INTRODUCTION

Consider a one-dimensional (1-d) system of non-interacting conduction electrons that are coupled to the lattice vibrations of a linear chain of ions. For such a model of a hypothetical 1-d metal, and within mean-field theory, a critical temperature T_c can always be found below which the linear lattice will develop a permanent periodic distortion that will open up an energy gap about the Fermi energy of the conduction electrons. This is the so-called Peierls-Fröhlich (P-F) transition of a 1-d metal. The basic physical ideas behind this model, which were initiated

23

independently by Fröhlich[1] and by Peierls[2] some twenty years ago, have become of considerable relevance to some of the current experimental and theoretical research[3] on the 1-d conductors.

The purpose of my first lecture is to present a simplified discussion of the P-F transition. This discussion will emphasize the important role played by the fundamental instability of a 1-d system of conduction electrons.[4] The topic of my second lecture will be one of current research interest, namely, the dynamics of the charge density wave (CDW) that accompanies the periodic lattice distortion below T_c.[5,6]

THE PEIERLS-FRÖHLICH TRANSITION

Let us first consider the question of the density response of a 1-d system of non-interacting conduction electrons. The energies ε_k of the single-electron states may be assumed to be those appropriate to a linear chain of N atoms and lattice spacing a, and may depend on the wavevector k as indicated in Fig. 1. In the following we shall measure energies with respect to the Fermi energy ε_F.

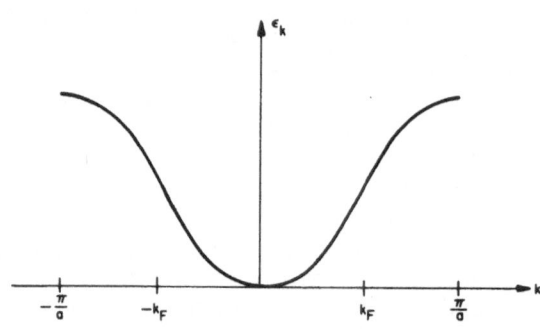

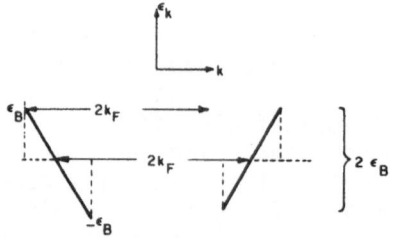

Fig. 1. The single electron energies and the geometric construction that generates the property [3] of the text.

Suppose we ask for the density response that results from the application of some external force $F(x)$ directed along the length of the system. For a translationally invariant system[7] the change in density at the point x will, in general, be of the form

$$\delta\rho(x) = -\int dx' \; F(x') \; \chi(x'-x)$$

or, on taking Fourier components,

$$\delta\rho(q) = -F(q) \; \chi(q) \qquad\qquad [1]$$

where $\chi(x)$ is the density response function[8] characteristic of the system of conduction electrons. The function $\chi(q)$ which determines the magnitude of the "density-fluctuation" $\delta\rho(q) \; e^{-iqx}$ that results from the periodic force $F(q) \; e^{-iqx}$ may be calculated for the non-interacting electron system to be[8]

$$\chi(q) = (2/N) \sum_{k} (f_k - f_{k-q})/(\epsilon_{k-q} - \epsilon_k) \qquad\qquad [2]$$

where $f_k = 1/(\exp(\epsilon_k/T)+1)$ is the Fermi-Dirac equilibrium occupation of the electronic state ϵ_k ($K_{Boltzmann} = 1$). In terms of the k states $\delta\rho(q)$ consists of a linear superposition of all possible electron-hole pair excitations (k, k-q) of total momentum $-\hbar q$. Each term in [2] represents an "oscillator strength" for these excitations. The function [2] is analogous to the dielectric function of an atom.

In one-dimensional [2] has a remarkable resonant property at the particular wavevectors $q_0 = \pm 2k_F$. The key to the recognition of this is to note that for these particular wavevectors it will be possible to excite a special series of electron-hole excitations with the property

$$\epsilon_{k-q_0} = -\epsilon_k \qquad\qquad \text{for} \qquad\qquad |\epsilon_k| < \epsilon_B \qquad\qquad [3]$$

where ϵ_B is some constant (assumed $\gg k_B T$) which will depend on the actual details of the band structure. To see this use the expansion $\epsilon_k = v_F \; (|k|-k_F)$ in the vicinity of $k = q_0$ in order to make the geometric construction indicated in the lower portion of Fig. 1. The contribution to $\chi(q_0)$ arising from these excitations is

$$\chi(q_0) = 2 \cdot (a/2\pi) \int_{k(-\epsilon_B)}^{k(\epsilon_B)} dk \; (f(\epsilon_k)-f(-\epsilon_k))/(-2\epsilon_k)$$

or, on introducing the density of states (per atom per both spin states) at the Fermi energy $N(o) = 2a/\pi h v_F$, where v_F denotes the Fermi velocity,

$$\chi(q_0,T) = \frac{1}{2} N(o) \int_0^{\varepsilon_B} \frac{d\varepsilon}{\varepsilon} \tanh (\varepsilon/2T) \ .$$ [4]

$$= \frac{1}{2} N(o) \, \ell n \, (1.14 \ \varepsilon_B/T)$$ [5]

which diverges as $T \to 0$. The contribution to $\chi(q_0)$ arising from electron-hole pair excitations other than those of [3] will be found not to be divergent so that [5] will dominate $\chi(q_0,T)$. For $q \neq q_0$ the property [3] will be lost ("off-resonance") and the logarithmic divergence [5] will not arise. As a function of q, therefore, $\chi(q_0)$ will exhibit a strong peaking at $q = q_0$. The divergence at $T = 0$ implies that at absolute zero the electrons will <u>tend</u> to arrange themselves in such a way that there is a net density fluctuation of period q_0. We stress that the pairing [3], and subsequent logarithmic divergence [5], is a unique property of the <u>one-dimensional</u> system.

This resonant property of $\chi(q)$ is extremely relevant to the question of a hypothetical 1-d metal because an important factor which will determine the <u>stability</u> of the latter will be the magnitude of the conduction electron density response to the vibronic motion of the ions.[9] Thus, for example, if we assume that the conduction electrons are coupled to the ionic lattice vibrations via

$$H_{e-p} = 1/\sqrt{N} \sum_q g(q) \ Q(q) \ \rho_{-q}$$ [6]

where $Q(q)$ denotes the normal mode displacement operator (in units of $\sqrt{\hbar/2M\omega_0(q)}$, M = ionic mass) for the lattice vibrations,

$$u_j = \frac{1}{2} \frac{1}{\sqrt{N}} \sum Q(q) \ e^{iqR_j} + h.c. \ ,$$

and $g(q)$ denotes a coupling constant, we shall see later that the resulting normal mode frequences of the "metallic" linear chain will be just

$$\omega^2(q) = \omega_0^2 (q) \ [1-(2g^2(q)/\hbar\omega_0(q)) \ \chi(q,T)]$$ [7]

where $\omega_0(q)$ is the normal mode frequency that would be obtained if we had neglected the response of the conduction electrons. Since $\omega^2(q)$ measures the restoring force for the $Q(q)$ normal mode, it will be necessary to have $\omega^2(q) > 0$ (for all q) for stability of the metallic lattice. The second term in [7] arises from the density response of the conduction electrons and acts to weaken the net restoring force: the electrons move in an attempt to "screen-out" the ionic displacements. But according to [5] the latter response at $q = q_0$ can become arbitrarily large at sufficiently low T, so that $\omega^2(q_0)$ can become negative. There occurs therefore a lattice instability at the temperature T_c at which $\omega^2(q_0)$ vanishes, or from [7] and [5], at the temperature

$$T_c = 1.14 \ \varepsilon_B \ \exp \ (-1/\lambda) \tag{8}$$

where we have introduced the so-called electron-phonon coupling constant (dimensionless) $\lambda = g^2 N(o)/\hbar\omega_0$ ($g = g(q_0)$, $\omega_0 = \omega_0(q_0)$). Before discussing what must happen below T_c let us first indicate the derivation of [7].

In the absence of the coupling [6] the equation of motion for $Q(q)$ is simply

$$\ddot{Q}(q) + \omega_0^{\ 2}(q) \ Q(q) = 0 \tag{9}$$

If we include the coupling [6] we must add to the right hand side of [9] the extra driving term $-\alpha(q)\rho_q$, where $\alpha(q) = 2g(q)\omega_0(q)/\hbar\sqrt{N}$. We now adopt a mean-field approach and average this term over the conduction electron system using the linear response formula [1]:

$$<\rho_q>_e = - \sqrt{N} \ g(q) \ Q(q) \ \chi(q) \tag{10}$$

The coefficient of $\chi(q)$ in [10] follows from the form of [6]. The meanfield equation of motion for $Q(q)$ is therefore

$$\ddot{Q}(q) + \omega^2(q) \ Q(q) = 0$$

with the new normal mode frequency given by [7].

What happens below T_c? Since $\omega^2(q_0) < 0$, the zero equilibrium position (as a function of time) that we have assumed for $Q(q_0,t)$ is no longer tenable. We will have to look for a new equilibrium position, i.e.

$$Q(q_0,t) = Q_0 + \delta Q(q_0,t) \tag{11}$$

where the mean value of Q_0 is non-vanishing. According to [10], the latter will also imply a non-vanishing mean charge density fluctuation

$$\rho_{q_0} = \Delta\rho + \delta\rho_{q_0}(t) \tag{12}$$

and the coupled equations of motion for δQ and $\delta\rho$ will determine the new normal mode frequencies. The stability of the system below T_C will depend on whether or not we can find consistent solutions for Q_0 and $\Delta\rho$. Now $Q_0 \neq 0$ implies a <u>permanent lattice distortion</u> of wavevector q_0 ($u_j \propto Q_0 \cos(Rjq_0 + \text{phase factor})$). Such a distortion will have to be balanced against the elastic restoring forces (associated with ω_0^2) by the non-vanishing electronic density fluctuation $\Delta\rho$, or charge density wave (CDW). Thus the condition for stability will be

$$\omega_0^2 Q_0 = -\alpha(q_0) \Delta\rho$$

or by [10],

$$\omega_0^2 Q_0 = (2g^2\omega_0/\hbar)Q_0 \, \chi(q_0,T;Q_0)$$

that is,

$$1/\lambda = (2/N(o)) \, \chi(q_0,T;Q_0) \tag{13}$$

The solution (if any) of [13] determines Q_0 as a function of temperature. Note that in view of [8] we have $Q_0(T_c) = 0$ as will be necessary. In order to solve [13] we will require to evaluate $\chi(q_0,T;Q_0)$ which in view of [2] will require the evaluation of the single electron states, $E(k) = E(k,Q_0)$, of the <u>periodically distorted lattice</u>. It follows from the form of [6] that the latter are determined by the electronic Hamiltonian

$$H = \begin{array}{l}\text{original} \\ \text{band states}\end{array} + \begin{array}{l}\text{interaction with periodic} \\ \text{potential } V(q_0) = \dfrac{1}{\sqrt{N}} Q_0 g \equiv \Delta\end{array}$$

As is well known, this Hamiltonian may be diagonalized to give the new band states

$$E_k = (\varepsilon_k/|\varepsilon_k|) \sqrt{\Delta^2 + \varepsilon_k^2} \qquad (|\varepsilon_k| < \varepsilon_B) \tag{14}$$

which exhibit an <u>energy gap</u> 2Δ at the wavevector $q_0/2$, i.e. about the Fermi energy. We may now evaluate χ as we did before and find that [13] reduces to

$$1/\lambda = \int_0^{\varepsilon_B} \frac{d\varepsilon_k}{E_k} \tanh(E_k/2T) \tag{15}$$

which is precisely the gap equation which appears in the BCS theory of superconductivity.[4] Δ or Q_0 therefore follows a BCS-type of temperature dependence (see Fig. 2).[10] At T = 0, [15] yields

$$\Delta(0) = 2\varepsilon_B \exp(-1/\lambda) = 1.76\ T_c$$

where we have assumed $\Delta(0)/\varepsilon_B$ small. The picture we have obtained of the 1-d metal is shown schematically in Fig. 2.

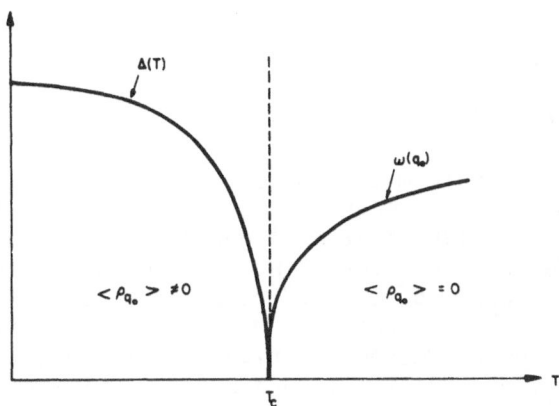

Fig. 2. Mean-field behavior of the one-dimensional metal.

Although there is an energy gap $2\Delta(T)$ below T_c, the lower temperature phase may not necessarily be an insulator since -- as was stressed in the brilliant paper of Fröhlich[1] -- movement of the lattice distortion-CDW condensate relative to the laboratory frame (LF) will constitute an electrical current. For a perfectly translationally invariant system the phase of the condensate relative to the LF is arbitrary and, consequently, Fröhlich was able to demonstrate that the condensate would be able to move <u>unattenuated</u> with a uniform velocity v_s ($v_s < \Delta/\hbar k_F$), i.e. super-conductivity would result below T_c. The actual discreteness of the underlying linear lattice, and in practice, the presence of impurity centres and neighboring linear chains, will remove the translational invariance of the condensate (the condensate energy becomes phase dependent), causing it to become "pinned-down" to some equilibrium position.[5] The d.c. electrical conductivity then behaves in an insulating manner. Nevertheless as we shall see in the following lecture, there will be some interesting electrodynamic properties associated with the pinned CDW.

Electrodynamics of the CDW

In this lecture I will outline a phenomenological theory of the pinned CDW, due to M. J. Rice, S. Strässler and W. R. Schneider.[6] The theory was suggested by the microscopic results of an elegant paper by P. A. Lee, T. M. Rice and P. W. Anderson[5] and its phenomenological approach is similar to that adopted by D. Allender, J. W. Bray and J. Bardeen[11] in discussing paraconductivity above T_c.

Below T_c, the induced CDW will have the form

$$\delta\rho(x) = (eN_e/L) A \cos (q_o x + \phi)$$

where eN_e/L is the uniform charge density of the normal metal (N_e = no. of conduction electrons, L = length of the linear conductor). A is a constant that will depend upon the details of a microscopic model of the interacting electron-phonon (e-p) system, and ϕ denotes the phase of the CDW relative to the laboratory frame. When the CDW is pinned the mean value of ϕ will assume some fixed value and the corresponding fixed spatial position of the CDW, which we may denote by a co-ordinate X_0, will be such that there will exist no net dipole moment between the CDW and the rest of the system (ions, neighbouring linear chains).

The first step of the phenomenological theory is to assume that the equation of motion governing the displacement X of the CDW from its equilibrium position X_0 is of the form

$$M\ddot{X} + \gamma\dot{X} + kX = e^*E \qquad\qquad [16]$$

where M denotes an effective inertia for the CDW, e* its total charge, k a harmonic restoring force constant and γ a friction constant. $E = E_\omega e^{-i\omega t}$ denotes an external field applied along the X-direction of the linear conductor.

Next we note that since e* must arise from an integer number of elementary charges we can identify a number of "condensed" or "super-fluid" electrons, N_s, according to the relation

$$e^* = N_s e \qquad\qquad [17]$$

The current density induced by the motion of the CDW is therefore

$$i = (N_s/L)e \ \dot{X} \equiv n_s e \dot{X} \qquad\qquad [18]$$

so that the N_s electrons (from the total of N_e electrons) are carried along with the CDW. We now define an effective mass m^* for the superfluid electrons according to

$$M = N_s m^* \qquad\qquad [19]$$

We assume that both n_s and m^* are <u>intrinsic</u> properties of the <u>unpinned</u> CDW in the absence of the <u>applied</u> electric field. This will necessitate restricting the phenomenological discussion to situations of weak pinning. In general we expect n_s and m^* to be functions of temperature. If we now introduce the harmonic frequency ω_F and damping constant Γ according to $k = m^* n_s \omega_F^2$ and $\gamma = m^* n_s \Gamma$, eq. [16] can be expressed as

$$\ddot{X} + \Gamma \dot{X} + \omega_F^2 X = eE/m^* \qquad\qquad [20]$$

and may be interpreted as describing the mean motion of a super-fluid electron in the P-F condensate. We will regard Γ and ω_F as phenomenological parameters that can be determined from experiment. m^* and n_s can be calculated from an underlying microscopic theory of the CDW. We shall see later that for a given value of the order parameter Δ, n_s and m^* are simply related to each other, so that a microscopic theory for only one of these two quantities will be required.

Since the dipole moment arising from X is $p = n_s e X$ it follows from [20] that the ω-dependent dielectric constant due to the motion of the CDW is

$$\epsilon_F(\omega) = \frac{\omega_p^2 (n_s/n)(m/m^*)}{\omega_F^2 - \omega^2 - i\Gamma\omega} \qquad\qquad [21]$$

while the associated ω-dependent electrical conductivity is

$$\sigma_F(\omega) = (n_s e^2/i\omega m^*)\omega^2/(\omega_F^2 - \omega^2 - i\omega\Gamma) \qquad\qquad [22]$$

where $\omega_p^2 = 4\pi n e^2/m$ is the conduction electron plasma frequency and $n = N_e/L$.

[21] is of the conventional form for an optical mode and shows that the oscillator strength of the CDW is just

$$f_F = (n_s/n)(m/m^*) \qquad\qquad [23]$$

Since n_s/n cannot exceed unity we obtain the inequality

$$(m^*/m) < f_F^{-1}$$

indicating that an experimental measurement of f_F will provide an upper limit to the value of the effective mass m^*. We note that in general $n_S \neq n$ at $T = 0$. The contribution to the static dielectric constant coming from the CDW is

$$\varepsilon_F(0) = (\omega_p/\omega_F)^2 f_F \qquad [24]$$

and can be expected to be very large for the weakly pinned CDW (e.g., if $\omega_F \sim 10^{-3}$ eV, $\omega_p \sim 1$ eV, $f_F \sim m/m^* \sim 10^{-3}$ at $0°K$ (see later) then $\varepsilon_F(0) \sim 10^3$).

   Since the CDW is pinned the collective or Fröhlich conductivity $\sigma_F(\omega)$ is, of course, finite only at finite frequencies. In the absence of pinning ($\omega_F = 0$) a finite d.c. conductivity will result and according to [22] this will be given by

$$\sigma_F(0) = n_S e^2 \Gamma^{-1}/m^* \qquad [25]$$

Fröhlich's superconductivity arises from the fact that for $\omega < 2\Delta$ the e-p interaction alone cannot lead to an attenuation of the CDW ($\Gamma = o$). In this case it follows from [22] that [25] is to be replaced by

$$\sigma_F(\omega) = -n_S e^2/i\omega m^* \qquad [26]$$

which is the London equation expressed as a formula for $\sigma(\omega)$. Anharmonicties in the underlying linear lattice are likely to be a major factor in determining Γ.

   The relationship between m^* and n_S may be derived by noting that the kinetic energy of the moving CDW, $\frac{1}{2} m^* N_S \dot{X}^2$, must consist of that required to translate the N_S participating electrons of actual (band) mass m and that required to move the accompanying lattice distortion with $\dot{\phi} = q_0 \dot{X}$:

$$\frac{1}{2} m^* N_S \dot{X}^2 = \frac{1}{2} m N_S \dot{X}^2 + 2 \cdot \frac{1}{2} M <|\dot{Q}(q_0)|^2> (\hbar/2M\omega_0) \qquad [27]$$

where $\dot{Q}(q_0) = i\dot{\phi}Q(q_0)$ for phase motion. Introducing $\Delta = (g/\sqrt{N}) \times <Q(q_0)>$ into [27] we obtain the result

$$m^*/m = 1 + (4|\Delta|^2/\lambda\omega_0^2)(n/n_S) \qquad [28]$$

connecting m^* and n_S. Note if $n_S \sim n$ and $\lambda = 0.4$ and $\Delta \sim 10 \omega_0$ (say), then $m^*/m \sim 10^3$, i.e. $m^* \gg m$.

Within the framework of the simple mean-field theory of Lecture 1 it can be shown[6] that $n_s/n = 1$ at $T = 0$ while

$$n_s(T)/n = 7 \, \xi(3) \, (|\Delta(T)|/2\pi T_c)^2 \tag{29}$$

for $T \sim T_c$. According to [28] these results imply $m^*/m = (4|\Delta(0)|^2/\lambda\omega_0^2)$ for $T = 0$, and

$$m^*/m = 1.52 \, m^* \, (T = 0)/m \tag{30}$$

for $T \sim T_c$, where in obtaining the latter result we have employed $\Delta(0)/T_c = 1.76$. Note that within this model the oscillator strength is just m/m^* at $T = 0$.

Above T_c the mean value of Δ will be zero. The mean value of $|\Delta|^2$ will be finite although small ($\sim 1/N$), and according to [29] such a "fluctuation" will imply a finite value, δn_s (say), for n_s. In the absence of pinning, and if the life-time of the fluctuation is long compared to $1/\Gamma$ (i.e. $T \sim T_c$), we may, in view of [25], expect the "paraconductivity" $\delta\sigma_F = (\delta n_s \, e^2/m^*\Gamma)$. Fluctuations with wavevectors close to q_0 will also contribute similar para-conductivities and if all these are summed we obtain, within the mean field theory, $\sigma_F = \tilde{n}_s e^2/m^*\Gamma$ where

$$\tilde{n}_s = n \, \sqrt{7\xi(3)/4\epsilon(T)} \tag{31}$$

with $\epsilon(T) = (T-T_c)/T_c$. This is the paraconductivity first proposed by Bardeen,[11] and should be large close to T_c. The divergence of [31] at $T = T_c$ is an artifact of the mean field theory which must break down at values of $T - T_c$ for which [31] predicts $\tilde{n}_s \lesssim n$.

REFERENCES AND FOOTNOTES

[1] H. Fröhlich, Proc. Roy. Soc. A223, 296 (1954); see also C. G. Kuper, Proc. Roy. Soc. A227, 214 (1955).

[2] R. E. Peierls, "Quantum Theory of Solids", (Clarendon Press, Oxford, 1955) p. 108.

[3] For a review see H. R. Zeller, Festkörperprobleme 13, 31-58 (1973).

[4] M. J. Rice and S. Strässler, Solid State Commun. 13, 125 (1973).

[5] P. A. Lee, T. M. Rice and P. W. Anderson, Solid State Commun. 14, 703 (1974).

[6] M. J. Rice, S. Strässler and W. R. Schneider, "Some Fluctuation and Electrodynamic Properties of the Peierls-Fröhlich Conductor", to be published in the Proceedings of the German Physical Society Conference on "One-Dimensional Conductors", University of Saarbrücken, 10-12 July, 1974. (Editor H. G. Schuster).

[7] We neglect here the periodicity of the underlying linear lattice.

[8] See, for example, D. Pines and P. Nozieres, "The Theory of Quantum Liquids, 1: Normal Fermi Liquids", (Benjamin, New York, 1966).

[9] Ref. 8, p. 237.

[10] The mathematical structure of the mean field theory of the Peierls-Fröhlich transition is in fact identical to that of the BCS theory of the pairing superconductor (J. Bardeen, L. N. Cooper and J. R. Schrieffer, Phys. Rev. 108, 1175 (1957)).

[11] D. Allender, J. W. Bray and J. Bardeen, Phys. Rev. B9, 119 (1974); J. Bardeen, Solid State Commun. 13, 1389 (1973).

EXCITON MECHANISM OF SUPERCONDUCTIVITY

W. A. Little

Physics Department

Stanford University, Stanford, California 94305

BACKGROUND

A large number of metals and alloys become superconducting when cooled below a certain critical temperature, T_c which is typically of the order of a few degrees Kelvin.[1] In this state the electrical resistivity is immeasurably small and currents can be carried without dissipation of energy. However, a magnetic field which is normally excluded from the interior of a superconductor, penetrates when its value exceeds a certain critical value H_c driving the metal into the normal state. Two types of superconductors occur: the type I, in which the superconductivity is quenched at a field H_c and the type II in which field penetration occurs at H_{c1} but superconductivity remains up to a higher critical field H_{c2}. Practical applications of superconductivity are based on the use of type II superconductors because, while H_c is typically less than about 1000 gauss, H_{c2} can be much higher, exceeding 200 Kgauss. Superconducting magnets have been built using the alloys Nb_3Sn and $NbTi$ which produce fields of the order of 150 Kgauss and 80 Kgauss, respectively. These magnets offer something totally new to the area of electromotive machinery for the field energy and electromagnetic forces produced are orders of magnitude greater than those attainable with iron cored magnets. Large scale application of superconducting technology in high speed transportation, motors, generators and ship propulsion is expected in the next few decades. However, in all these applications sophisticated refrigeration is needed to maintain the metal in the superconducting state. We shall examine therefore, those factors which limit the value of T_c, the superconducting transition temperature and what prospects exist for substantially raising T_c.

The characteristic changes which occur when a metal enters the superconducting state are the result of the formation of a condensate of electron pairs. Electrons in the metal become bound in pairs and each pair is constrained to be in the same quantum mechanical state as all the others. The binding energy for the pairs comes from the interaction of the electrons with the lattice. As one electron moves in the lattice the interaction of its field with the positive ions of the lattice distorts the lattice. A second electron is attracted to this distorted region and thus indirectly is attracted to the first electron. This is the phonon-induced electron-electron attraction. With all the pairs in the same state a coherent interaction occurs which strengthens the overall binding force. At low temperatures most of the conduction electrons condense into this paired structure. As one raises the temperature some of the pairs are thermally dissociated and as these are lost from the condensate their absence weakens the interaction of the remaining pairs. Above a critical temperature T_c no more pairs remain bound and the metal reverts to the normal state. The detailed theory of Bardeen, Cooper, and Schrieffer (BCS) shows that the transition temperature is determined by the expression,

$$k_B T_c \simeq \hbar\omega_o \, \exp\left\{\frac{-1}{N\,V}\right\} \, , \tag{1}$$

where ω_o is the Debye frequency of the lattice, N is the density of states at the Fermi surface and V is the effective electron-electron attraction.

To understand the limitations on T_c it is instructive to model the lattice as a set of point masses, M attached to springs of spring constant, k. Then ω_o is given by

$$\omega_o = \sqrt{k/M} \, . \tag{2}$$

If we make the lattice excessively rigid an electron moving through it will generate a negligible distortion and thus the attractive interaction V also will be negligible. From (1) then, we would find $T_c \approx 0$. As we make the lattice less rigid by decreasing k, V increases, resulting at first in an increase in T_c. But as the lattice gets softer with further decrease of k, ω_o begins to get rather small through (2) and eventually this causes T_c to flatten out and then to decrease. Thus we see how we could expect a limit to exist on the maximum attainable value of T_c. Actually we expect an even stronger limit on T_c. Let us examine this simple model a little closer. As we reduce the value of k, eventually the interaction of the ions with the electrons themselves will begin to make a contribution to the effective spring constant. This results in a renormalization of the phonon frequencies so that

$$\omega^2 = \omega_o^2 \Big(1 - g(V)\Big) \, ,$$

where ω_0 is the original Debye frequency and $g(V)$ is a quantity which depends on the interaction, V .

If V gets too large, the renormalization parameter $g(V)$ causes the lattice frequencies to become imaginary, resulting in a lattice instability. The resultant phase change thus prevents the attainment of anomalously large values of T_c .

In the exciton model of a superconductor one attempts to replace the ionically polarizable phonon system with an electron-ically polarizable "exciton" system. By so doing one attempts to replace ω_0 , the frequency of the order of the Debye frequency in (1) with a much higher frequency associated with an electronic transition and at the same time one attempts to keep V , the attractive term, at approximately the same value. This gives one hope of attaining superconductivity at substantially higher tempera-tures than with the phonon mechanism but imposes some severe elec-tronic and structural requirements on the design of any such mate-rial. Before considering in detail these requirements we will derive first the expression for T_c given in (1) so one can under-stand how each factor plays its role in the expression.

SUPERCONDUCTIVITY THEORY

The electrons in the metal can be described by the Hamiltonian

$$\mathcal{H} = \sum_k \epsilon_k c_k^\dagger c_k + \sum_{k,\bar{k},q} V_{(q)} c_{k+q}^\dagger c_{\bar{k}-q}^\dagger c_{\bar{k}} c_k \quad (3)$$

where ϵ_k is the energy of an electron in state k , and $V_{(q)}$ is the net interaction between electrons for a momentum transfer, q . We are using the notation of second quantization[2] where the $c_k^\dagger$ and c_k are the creation and annihilation operators for a particle in state k . One does not have to understand the full meaning of these operators to appreciate the essence of their behavior. This can be seen as follows. Consider the first term in (3). If this term operates on a state in which an electron is in the state k then c_k destroys this electron but $c_k^\dagger$ recreates it leaving the state unchanged. This operation then leaves us with a contribution to the energy ϵ_k . On the other hand, if there is no electron in state k then we cannot destroy it so c_k acting on such a state gives us zero. Also the interaction term may be viewed pictorially as shown in Fig. 1 where electrons k and $\bar{k}$ are annihilated but reappear in states k + q and k - q or in other words, they col-lide with one another and change their momentum states as shown.

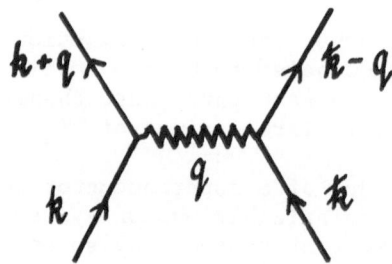

Fig. 1.   Scattering of two electrons   k   and   k̄   with momentum
          transfer   q.

Consider now a state described by

$$\psi = \prod_k (u_k + v_k c_k^\dagger c_{-k}^\dagger) \phi_o \tag{4}$$

where

$$u_k^2 + v_k^2 = 1 . \tag{5}$$

Here ϕ_o is the "vacuum" i.e., the absence of all electrons. If
we expand the product we will obtain a series of terms, the first
from the product of the u's corresponds to no electrons, the next
from all u's except one v_k corresponds to a pair of particles in
k and -k , the next to two pairs of particles in k, -k , k'
and -k' etc. In fact ψ corresponds to a linear combination of
Slater determinants in each of which the electrons are paired as
above. This state, for particular values of u_k and v_k , gives
a good description of the BCS superconducting ground state. Using
(3) and (4) we can calculate the expectation value of the energy to
obtain

$$w = w_{KE} + w_I = \sum_k 2\epsilon_k v_k^2 + \sum_{k,k'} V_{(q)} u_k v_k u_{k'} v_{k'} \tag{6}$$

where k' - k = q . Now we may find v_k and u_k by minimizing
w with respect to variations in v_k and u_k subject to the
constraint (5). We then find

$$\frac{v_k(1 - v_k^2)^{1/2}}{1 - 2v_k^2} = \frac{- \sum_{k'} V_{(q)} u_{k'} v_{k'}}{2\epsilon_k} \tag{7}$$

The quantity

$$\Delta_k \equiv - \sum_{k'} V_{(q)} u_{k'} v_{k'} \tag{8}$$

plays an important role in the theory and defines the superconducting energy gap. In terms of it, u_k and v_k may be expressed from (7) as

$$v_k^2 = \frac{1}{2} \left\{ 1 + \frac{\epsilon_k}{\sqrt{\epsilon_k^2 + \Delta_k^2}} \right\}$$

$$u_k^2 = \frac{1}{2} \left\{ 1 - \frac{\epsilon_k}{\sqrt{\epsilon_k^2 + \Delta_k^2}} \right\} \tag{9}$$

Thus v_k and u_k both depend upon Δ_k, so (7) is an integral equation for the gap parameter, Δ_k. Using (8) and (9) we find

$$\Delta_k = - \sum_{k'} \frac{V_{(q)} \Delta_{k'}}{2 \sqrt{\epsilon_{k'}^2 + \Delta_{k'}^2}} \tag{10}$$

The solution of this equation together with (9) and (4) gives a description of the superconducting state at $T = 0°K$. A similar calculation made at finite temperatures, where some of the pairs are broken with the individual electrons in states which are not paired as in (4), gives an analogous expression to (10) of the form

$$\Delta_k = - \sum_{k'} \frac{V_{(q)} \Delta_{k'} \tanh \left\{ \frac{\beta}{2} \sqrt{\epsilon_k^2 + \Delta_k^2} \right\}}{2 \sqrt{\epsilon_k^2 + \Delta_k^2}} \tag{11}$$

where $\beta = 1/k_B T$. The factor, $\tanh\{...\}$, results from the loss of the electrons from the pair condensate through thermal excitation. This weakens the effect of the potential in (11) as we discussed qualitatively earlier.

To proceed further we must look at the detailed behavior of the interaction term $V_{(q)}$. This consists of two parts, the Coulomb repulsion between the electrons together with the phonon induced attraction.   It takes the form

$$V_{(q)} = V_c(q) + \frac{2\,|M_{(q)}|^2\,\hbar\omega_q}{(\epsilon_{k+q} - \epsilon_k)^2 - (\hbar\omega_q)^2} \tag{12}$$

$M_{(q)}$ is the matrix element describing the interaction between the electron and the phonon, and $\hbar\omega_q$ is the energy of the phonon. For electron energies such that $(\epsilon_{k+q} - \epsilon_k)$ is less than $\hbar\omega_q$ the second term will be negative, whereas for larger energy differences it is positive. It is illustrated in Fig. 2(a). The approximation used by BCS was to treat V as a constant for $|\epsilon_k|$, $|\epsilon_{k'}|$ less than the Debye energy, ω_0, and zero outside this range [Fig. 2(b)]. Using this approximation and assuming that the density of states is a constant, N near the Fermi surface, we may change the sum in (10) to an integral over the energy and obtain

$$\Delta_k = \int_{-\hbar\omega_0}^{\hbar\omega_0} \frac{N\,V\,\Delta_{k'}\,\tanh\frac{\beta}{2}\sqrt{\epsilon_{k'}^2 + \Delta_{k'}^2}}{2\sqrt{\epsilon_{k'}^2 + \Delta_{k'}^2}}\,d\epsilon_{k'} \quad . \tag{13}$$

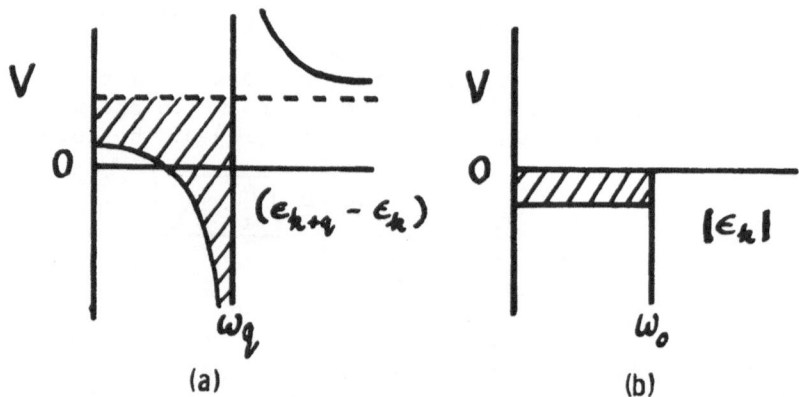

(a)   (b)

Fig. 2.   (a) Form of net electron-electron interaction due to
              screened Coulomb interaction plus phonon-mediated
              attraction.
          (b) BCS approximation to the net interaction.

Now Δ_k no longer depends on k .

The transition temperature can then be found by finding from (13) the largest value of T which gives a finite solution to Δ . One finds then

$$k_B T_c = 1.14 \, \hbar\omega_o \exp \left\{ \frac{-1}{N \, V} \right\} , \qquad (14)$$

the result quoted earlier.

We see then that the essential factors which determine T_c are contained in the expression (12) for the effective electron-electron interaction. These are the Coulomb interaction $V_c(q)$ which is not the pure Coulomb interaction but rather the screened interaction, the electron phonon coupling term, $M_{(q)}$, the phonon energy $\hbar\omega_q$ for given momentum, q , and the band structure of the electrons which determines ϵ_k , etc. One should note that the pre-exponential term, $\hbar\omega_o$ in (14) arises from the range of integration over which the interaction is attractive. It is thus a measure of the number of electrons which can interact coherently with one another. One does not need to make the BCS approximation to solve (11) and various numerical or approximate analytic methods have been developed to obtain better expressions for T_c for particular systems.

In the next section we will discuss a particular excitonic system and show how each of the elements in (13) were calculated and how this predicts the possibility of obtaining a relatively high temperature superconducting transition temperature.

A MODEL EXCITONIC SUPERCONDUCTOR

Most of what has been discussed earlier has been known for over a decade but nevertheless no high temperature superconductors have been designed or built on these principles. The problem lies in translating this information into a useable material. Our original suggestion[3] was based on the use of an organic polymeric conductor with polarizable dye-like side chains to provide the attractive excitonic interaction. The preparation of such a material presented some formidable synthetic difficulties. Since that time a large number of other systems have been considered as potential model excitonic superconductors[4] and the current interest in one-dimensional metals has largely arisen from these considerations. At present we believe that compounds related to the Krogmann salts[5] such as $K_2Pt(CN)_4Br_{0.3} \cdot 3H_2O$ offer the most promising approach. Following a suggestion of Collman we have studied certain metal-organic systems involving a polarizable dye-like structure complexed to platinum where the metal atoms form a linear conductive chain. We have examined in detail the structure shown in Fig. 3. This is

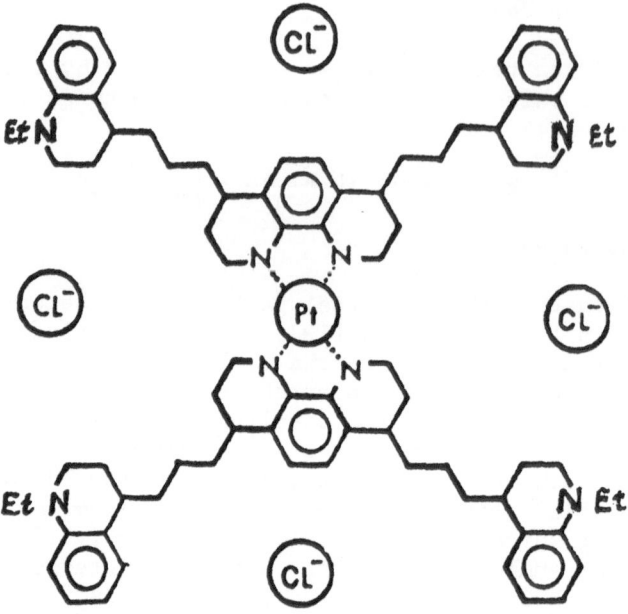

Fig. 3.   Monomer of proposed excitonic superconductor.

a Pt-complex of phenanthroline in which the phenanthroline moities
have been substituted at the 4 and 7 positions to yield two cyanine-
dye-like structures as shown. It is expected that such a complex
would stack so as to form a linear metal chain. Synthetic work on
this compound is underway but it has not been completed as yet.
For purposes of calculation, we assume that the complex can be
formed and that it would stack with a metal-metal spacing determined by the van der Waals' contacts between the organic ligands.
Prior to complexation, each half of the phenanthroline acts like a
cyanine-dye base. Upon complexation the base is converted to the
dye.

     The main calculations which have to be made are the following.
First we must calculate the transition density for the excited
states of the dye-like substituents. Next we must take into account
the interaction between each of the polarizable substituents on the
square planar complex and between those of different complexes in
the stack. This interaction causes each state of the individual
dye to form an exciton band and the calculation gives the dispersion
characteristics of these bands.

Next we must calculate the band structure of the platinum chain spine. As we do not know the exact metal-metal spacing, we have calculated this band structure for a range of such spacings.

With this band structure we can determine the extent to which the Coulomb field is screened by the movement of electrons in the platinum chain and in adjacent chains. We must also determine the effective dielectric constant of the neighborhood of the spine and then find the net Coulomb repulsion between electrons on the chain.

The attractive component (11) which results from the inter-action between the electrons and the excitons $|M_{(q)}|$ is now obtained from the exciton band structure and wave functions. The sum of this term and the screened Coulomb repulsion gives the net interaction.

This then is all the information which is needed to determine T_c and other properties of the superconducting state. The transition temperature is given by an integral equation whose kernel contains the net interaction and the density of states from the electron band structure. It may be solved by iteration.

A calculation of the above form has recently been completed. We start with a LCAO calculation of the polarizability of the phenanthroline structure using a modified Pariser-Parr-Pople technique using certain elements of the random phase approximation.[6] A configuration interaction (C.I.) calculation is made to determine the low lying excited states and their transition densities. Next we use these results to calculate the exciton band and exciton wave functions using a second C.I. calculation.

In a separate calculation we determine the structure of the 5-d and 6-p bands of the platinum chain. This is done using a multiple scattering procedure.[7]

Previously[8] we had solved the problems of the screening of the Coulomb field in these filamentary type materials and we thus find the net Coulomb interaction between electrons in the spine and between electrons in the spine and on the side chains. The latter allows us to calculate the attractive component resulting from the interaction with the exciton system.

Using all the above information we can set up and solve the integral equations for the superconducting gap parameter and T_c The result of this calculation was that, assuming no Peierls distortion, superconducting transition temperatures of order 3000K should result.[9] While one should not take the numerical value too seriously, it does indicate that strong coupling effects can be expected in such structures. The most striking features of the calculations though were the effects of small changes in the ligand system. If the polarizable ligands were moved $2\overset{\circ}{A}$ further from the Pt atom, T_c was driven to zero. Also if a ligand structure with only two polarizable side chains (instead of four) was used, again superconductivity was not found. Variations of the metal-metal distance and the Fermi energy had a weak effect upon T_c .

We conclude then that high temperature superconductivity may be a possibility in these systems but that a high density, close packing of the polarizable substituents is absolutely essential for its attainment.

REFERENCES

1. Several excellent books describe the superconducting state, C. Kittel, "Intro to Solid State Physics," J. Wiley, New York, 1966; P. G. de Gennes, "Superconductivity of Metals and Alloys," Benjamin, 1964.

2. A. L. Fetter and J. D. Walecka, "Quantum Theory of Many Particle Systems," McGraw-Hill, New York , 1971, Chapter 1.

3. W. A. Little, Phys. Rev. A134, 1416 (1964).

4. E. B. Yagubskii and M. L. Khidekel, Russian Chem. Reviews 41, 1011 (1972), and I. F. Shchegolev, Phys. Stat. Sol. (A) 12, 9 (1972).

5. K. Krogmann, Angew. Chem. Int. Ed. 8, 35 (1969).

6. H. Gutfreund and W. A. Little, J. Chem. Phys. 50, 4468 (1969).

7. K. H. Johnson, Advances in Quantum Chemistry 7, ed. P. O. Löwdin.

8. D. Davis, Phys. Rev. B7, 129 (1973).

9. D. Davis, Ph.D. thesis, Stanford University (1974).

MAGNETIC INTERACTIONS IN LINEAR CHAIN CRYSTALS

Zoltán G. Soos

Department of Chemistry, Princeton University
Princeton, New Jersey, 08540, U.S.A.

I. INTRODUCTION

Several recent reviews[1-5] summarize the physical properties of one-dimensional crystals based on either organic molecules or inorganic complexes. We focus here on magnetic properties and specifically exclude magnetic conductors, or one-dimensional "metals," in which the approximation of localized moments fails. The crystals of interest are then magnetic insulators or semiconductors that contain linear arrays of localized moments, $\vec{S}_n$. The linear alternating Heisenberg antiferromagnetic chain

$$\mathcal{H}_{ex} = \sum_n \left\{ J\,(1+\delta)\,\vec{S}_{2n} \cdot \vec{S}_{2n+1} + J\,(1-\delta)\,\vec{S}_{2n} \cdot \vec{S}_{2n-1} \right\} \tag{1}$$

describes either exchange or superexchange interactions among neighboring paramagnetic sites. $\mathcal{H}_{ex}$ contains structural information through the alternation parameter δ, which we take to be positive, with $0 \leq \delta \leq 1$. Regular chains are defined to have $\delta = 0$ and thus identical exchange constants J along the chain. Alternating chains have $0 < \delta < 1$ and consequently a stronger exchange $J\,(1+\delta)$ between sites 2n and 2n+1 than the exchange $J\,(1-\delta)$ between sites 2n and 2n-1. Noninteracting magnetic dimers occur in the limit $\delta \to 1$. All three possibilities of $\delta = 0$, $0 < \delta < 1$, and $\delta \to 1$ are illustrated below for magnetic insulators based on transition-metal ions and for magnetic semiconductors based on planar organic ion-radicals.

Static magnetic susceptibility, specific heat, and neutron diffraction studies have, in ideal cases,[5] been correlated with exchange interactions. Electron paramagnetic resonance (EPR) in single crystals is emphasized here. The characteristic effects of

exchange in linear chain crystals provides important insights into
their electronic structure. For example, the <u>angular</u> dependence of
the spectra reflects elementary interactions between magnetic
moments and establishes the triplet nature of the excited state of
the magnetic dimer considered in Section II and of the alternating
chain in Section III. The <u>temperature</u> dependence of the spectra is
another convenient variable, especially in systems with $J \sim kT$ in
which the number of spin excitations depends sensitively on temperature. Single crystal EPR is shown in Section V to provide a simple
demonstration of the one-dimensionality of $\mathcal{H}_{ex}$ in many representative
crystals. As Richards will discuss in greater detail, EPR can also
be exploited to measure small exchange interactions J' between
chains.

Single crystal EPR studies yield information about electronic
properties even in cases where a quantitative analysis has yet to be
achieved. Indeed, this may be the most important role for magnetic
studies in novel systems. Ionic charge-transfer (CT) crystals of
organic ion-radicals are illustrated in Section IV as a qualitative
application of exchange in regular chains. In contrast to such
qualitative applications, the fact that numerical solutions of $\mathcal{H}_{ex}$
are now available[6] makes it possible to test quantitatively the
magnetic properties of proposed theoretical models for novel compounds.

The connection between exchange and optical or electrical
properties is particularly striking in one-dimensional organic semiconductors.[1,6] The characteristic electronic feature in these
π-molecular crystals is a CT excitation around $\Delta E_{CT} \sim 1.0$ eV that
occurs below any single-molecule excitation. The transfer matrix
element for antiparallel electrons at the free radicals $\dot{R}_1$ and $\dot{R}_2$,

$$t = \langle R_1 \ddot{R}_2 | \mathcal{H} | \dot{R}_1 \dot{R}_2 \rangle \qquad (2)$$

controls electron motion along the chain, and, as discussed by
Pincus, is the Hubbard hopping integral;[7] t is also the Mulliken CT
integral[8] when R_1 is an electron acceptor (A) and R_2 is a donor (D).
We have suppressed the charges associated with the radicals $\dot{R}_1$ and
$\dot{R}_2$, since either π-electron anion radicals A^- or cation radicals D^+
can stack separately (segregated stacks) or alternately (mixed
stack).[1,6] When $t \ll \Delta E_{CT}$, virtual transfer of antiparallel electrons leads to an antiferromagnetic exchange constant

$$J \sim 4t^2/\Delta E_{CT} \qquad (3)$$

which can be related to $\mathcal{H}_{ex}$ for different values of δ. The occurrence of a low-lying CT excitation in organic semiconductors thus

leads naturally to various types of Hubbard models[6,1] and, in the limit of small t, to $\mathcal{H}_{ex}$ in Eq. (1). More complicated exchange Hamiltonians can also be found when the ratio of $t/\Delta E_{CT}$ is small but finite.[9]

The analysis of different contributions to J is far more complex in inorganic systems.[10,11] There are generally d-d or f-f electronic transitions below the lowest CT state and, in addition, the orbitals of bridging ligands must be considered together with a detailed analysis of the superexchange pathway.[11,12] Small polynuclear clusters[12-16] of transition-metal ions provide many interesting examples, some of which will be elaborated on by Kokoszka. Since EPR does not probe the contributions to J, $\mathcal{H}_{ex}$ is equally appropriate for any linear chain crystal with isotropic exchange. Thus magnetic properties demonstrate close similarities between materials with very different chemical, optical, and electric properties.

II. THE DIMER LIMIT

In the limit $\delta \rightarrow 1$, $\mathcal{H}_{ex}$ reduces to noninteracting dimers containing sites 2n and 2n+1. For $S_n = 1/2$ and antiferromagnetic exchange (J > 0), the ground state of each dimer is a singlet. There is an excited triplet state, at energy 2J, whose thermal equilibrium population is

$$\rho = \{1 + 1/3 \exp (2J/kT)\}^{-1} . \tag{4}$$

The triplet sublevels are split by electron dipolar, Zeeman, and hyperfine interactions, as described[17] by the spin Hamiltonian for any triplet state

$$\mathcal{H}_T = \mu_B \vec{\mathcal{S}} \cdot g \cdot \vec{H}_0 + \vec{\mathcal{S}} \cdot \mathcal{D} \cdot \vec{\mathcal{S}} + \vec{\mathcal{S}} \cdot \mathcal{A} \cdot \vec{I} . \tag{5}$$

Here μ_B is the Bohr magneton, $\vec{\mathcal{S}} = 1$ is the triplet spin associated with an excited dimer, g is the g-tensor, $\vec{H}_0$ is the applied magnetic field, $\mathcal{D}$ is the fine structure tensor, $\mathcal{A}$ is the hyperfine tensor, and $\vec{I}$ is the nuclear spin associated with a dimer. There is a vast literature[17-19,12,13] associated with the measurement and interpretation of single-ion magnetic parameters like g, $\mathcal{D}$, or $\mathcal{A}$ tensors. Such magnetic studies are important for theories of electronic structure and of bonding, but will not be discussed here.

When J is large compared to the magnetic terms in $\mathcal{H}_T$, the static paramagnetic susceptibility χ_p is simply given by the thermal equilibrium of triplet states,

$$\chi_p = N\rho \, \mathcal{S}(\mathcal{S}+1) \, \mu_B^2 \, g^2 / 6 \, kT \qquad\qquad (6)$$

where N is the number of sites (N/2 dimers). The total <u>intensity</u> of the EPR spectrum is another measure of χ_p. The <u>structure</u> of the spectrum reflects the small splitting of the triplet sublevels by $\mathcal{H}_T$.

Bleaney and Bowers[20] first used this type of EPR analysis to deduce the dimeric nature of copper acetate monohydrate. Each Cu^{II} ion has a $(3d)^9$ electronic structure and a net spin 1/2. Subsequent crystallographic studies[21] showed that the four bridging acetate ligands lead to a Cu_2Ac_4 unit that is common to many complexes.[12,13] The pyrazine complex[22,23] shown in Fig. 1 is the first linear chain of such dimers.

The singlet-triplet separation of $2J \sim 300 \text{ cm}^{-1}$ in Cu_2 Ac_4 complexes can be obtained from fitting absolute χ_p data, which also yields g values in Eq. (6), or from EPR intensity data. The singlet-triplet model is very nearly quantitative over a wide range of temperature. The density ρ of excited triplet states is therefore accurately known.

Both the $\mathscr{D}$ and $\mathscr{A}$ tensors in Cu_2 Ac_4 pyrazine are axially symmetric[23] along the chain axis shown in Fig. 1. When the Zeeman splitting μ_B g H_0 is much larger than $\mathscr{D}$ or $\mathscr{A}$, the fine structure leads to a splitting whose angular dependence is $\mathscr{D}(3 \cos^2\theta - 1)$, with θ the angle between $\vec{H}_0$ and the Cu-Cu axis. The center of the spectrum determines the g-tensor. $\mathscr{D}$ contains[24] both spin-orbit and electron dipolar interactions for a copper dimer. We have recently separated[23] the two contributions by taking advantage of anisotropy

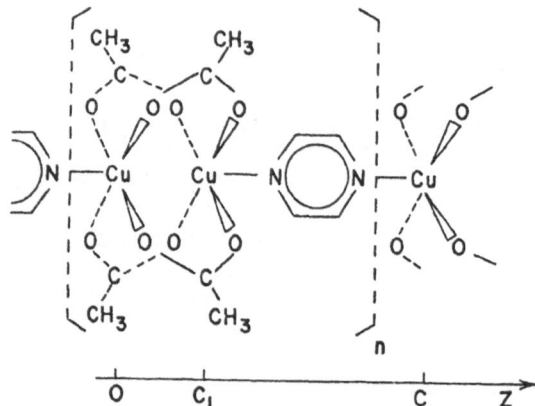

Fig. 1. Schematic representation of the pyrazine-bridged Cu_2 Ac_4 dimers in Cu_2 Ac_4 pyrazine. (Reproduced from Refs. 22 and 23)

of the g-tensor in Cu_2Ac_4 pyrazine. In organic ion-radicals, the spin-orbit contribution is negligible and $\mathscr{D}$ contains only electron-dipolar interactions.[25]

When H_0 is along the chain axis, the EPR transitions in Cu_2 Ac_4 pyrazine occur at g μ_B H_0 $\pm\mathscr{D}$ and each fine-structure line shows the hyperfine structure in Fig. 2. Either ^{63}Cu or ^{65}Cu has I = 3/2 and both have comparable hyperfine coupling constants. The total nuclear spin of a dimer can therefore be any integer between -3 and +3 with relative weights of 1:2:3:4:3:2:1 deduced from the different ways of adding I_1 and I_2. The simulation shown in Fig. 2 consists of seven equally-spaced Lorentzian lines with the theoretical relative intensities. The observation of resolved hyperfine structure proves that a triplet state remains on a dimer

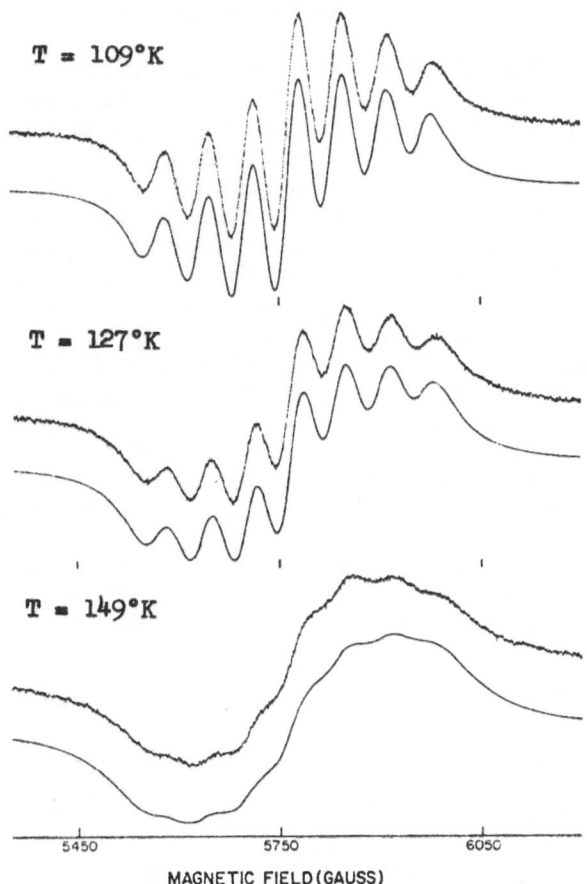

Fig. 2. Experimental and simulated hyperfine spectra for Cu_2 Ac_4 pyrazine with $\vec{H}_0$ along the chain axis. (Reproduced from Ref. 22)

for a time longer than 10^{-8} sec. Indeed, there is no evidence for
triplet motion even as the temperature is raised,[22] although each
hyperfine component in Fig. 2 is seen to broaden with increasing T.
The temperature-dependent line broadening is evidence for a small
exchange of $J(1-\delta) \sim 0.1$ cm^{-1} between triplets on adjacent
dimers.[22,23]

Cu$_2$ Ac$_4$ pyrazine thus corresponds to a linear chain in the
dimer limit, with $J(1+\delta) = 325$ cm^{-1}, $J(1-\delta) = 0.1$ cm^{-1} and thus
$\delta = 0.9994$. There is no evidence for triplet motion, in spite of
interdimer exchange, and the triplet state is consequently most
easily described[23] as a self-trapped spin polaron. It is interesting that mobile triplet states (triplet spin excitons) have yet
to be found unambiguously in inorganic systems.

III.   ALTERNATING CHAINS:TRIPLET SPIN EXCITONS

Figure 3 shows the strong π-electron donors (D), NNN'N' tetra-
methyl-p-phenylenediamine (TMPD) and tetrathiafulvalene (TTF), and
the strong π-electron acceptors (A), tetracyanoquinodimethane
(TCNQ) and chloranil. These planar, neutral, diamagnetic molecules
form many crystals[26,1] containing paramagnetic ion-radicals,
D^+ or A^-, with $S = 1/2$. Wurster's blue perchlorate (TMPD-ClO$_4$) is
dimerized[27] below 186°K and its low-temperature EPR illustrates[28]
the alternating case of $\mathcal{H}_{ex}$ for an ion-radical stack of the type
$\cdots(D^+D^+)\,(D^+D^+)\cdots$. There is a far more extensive[1,29,30] series
of TCNQ salts based on face-to-face stacking[26] of TCNQ$^-$ ion-radicals. Both simple (or 1:1) and complex (1:2, 2:3, or 3:4) salts
occur, with the latter containing formally neutral TCNQ sites in
the stack. Although the optical and electric properties of complex and single salts differ,[1,6] the central feature of alternating strong and weak exchange along the stack is common to both.
The alternating case, $0 < \delta < 1$, of $\mathcal{H}_{ex}$ is therefore appropriate.

The temperature dependence of the EPR of two complex TCNQ
salts,[29] $(\phi_3 \text{ As CH}_3^+) \text{ (TCNQ)}_2^-$ and $(\phi_3 \text{ P CH}_3^+) \text{ (TCNQ)}_2^-$, is shown in
Fig. 4. The spectra are identical up to 315°K, where the

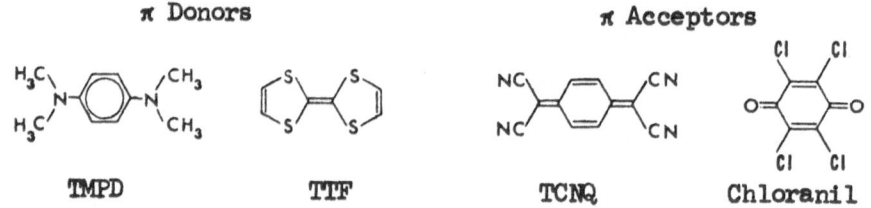

π Donors π Acceptors

TMPD   TTF   TCNQ   Chloranil

Fig. 3. Representative planar π-electron donors (D) and
        acceptors (A).

phosphorous complex undergoes a phase transition.[31,32] The susceptibility for noninteracting dimers is a fair approximation for the
EPR intensity, or χ_p, of these alternating chains.[31,25,1] A perturbation expansion in $(1-\delta)/(1+\delta) \leq 1$ leads to an activated susceptibility at low temperature[6,33,34] and thus retains the principal feature of the singlet-triplet model. The activation energy[35]
is 520 cm^{-1} for $(\phi_3 \text{ As CH}_3^+) (\text{TCNQ})_2^-$ and 250 cm^{-1} for TMPD-ClO$_4$.[28]
The thermal equilibrium density ρ of triplets is then very low,
($\sim 0.1\%$) below 100°K. Room temperature, by contrast, often corresponds to a dense ($\sim .1$ or more) triplet-exciton system.

The angular dependence of the representative dipolar splitting
in Fig. 4 establishes the triplet nature of the excitation. The
occurrence of a dipolar splitting, which decreases[17] as $\langle r_{12}^{-3} \rangle$,
establishes that the two electrons are on adjacent sites. This can
be demonstrated quantitatively[6,36-38] by actually computing the
fine-structure splitting from the known crystal structure and the
π-electron spin distribution for TMPD$^+$ or TCNQ$^-$ deduced from
solution hyperfine or from approximate quantum mechanical computations. Very good agreement is found in general.[6] This type of
data supports the picture[1,6] of weakly overlapping, essentially
unperturbed molecular ion-radicals in these π-molecular crystals.
The smaller[37,25] dipolar splittings in complex TCNQ salts indicates
additional delocalization of the unpaired electrons.

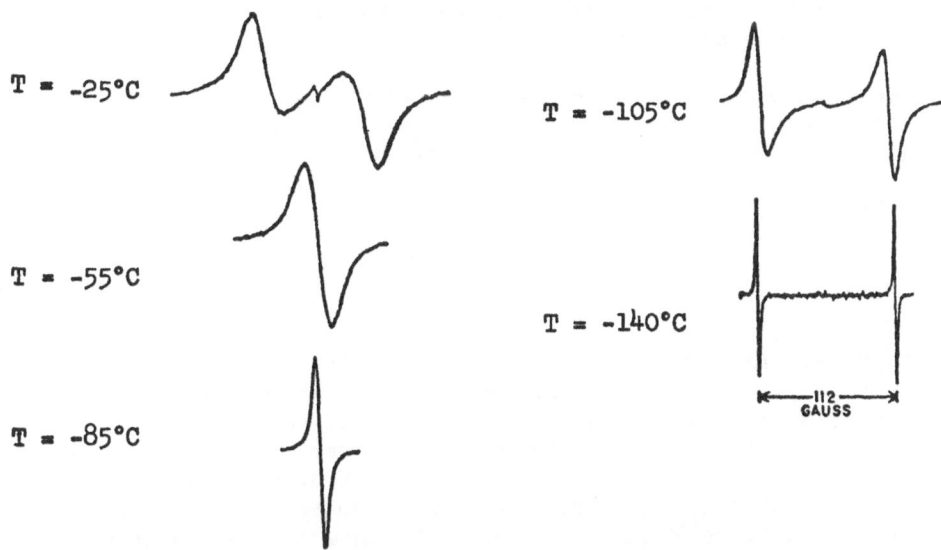

Fig. 4. Temperature dependence of the fine structure splitting for
   triplet (Frenkel) spin excitons. The orientation is fixed,
   but the spectrometer gain settings are different at each
   T. (From Ref. 29)

The most striking difference between the triplet exciton spectra in Fig. 4 and the Cu_2Ac_4 pyrazine spectra in Fig. 2 is the complete absence of hyperfine in the former. If $J(1-\delta)$ exceeds the width of the hyperfine splitting, or about 10-200 G for these ion-radicals, then triplet motion along the chain averages the hyperfine splittings and leads to a sharp, Lorentzian line for each dipolar line.[25] The absence of hyperfine in triplet exciton spectra thus demonstrates rapid ($\gtrsim 10^{10}$ sec^{-1}) motion along the chain. In contrast to Cu_2Ac_4 pyrazine, no stacked ion-radical organic system has yet shown sufficiently slow motion to resolve the hyperfine interactions. The characteristic temperature and angular dependence of triplet spin excitons yields detailed information even in the otherwise complicated, complex TCNQ salts.

Triplet spin excitons are among the best understood excitations in organic semiconductors.[25] The motion of strongly correlated spins along the chain does <u>not</u> average the dipolar interaction. The unpaired electrons in the triplet state remain adjacent and thus form a Frenkel, or strongly correlated, spin excitation[39] for the dimer. Such strong correlations are characteristic of Hubbard models[1,6] with $t < \Delta E_{CT}$ in Eq. (3). Thus the EPR of alternating chains shows that electron correlations are important in these ion-radical salts. Since neither the CT band, ΔE_{CT}, nor the largest π-electron overlap along the chain suggested by the crystal structure[26] are different on going to a regular chain, it is likely that correlations are of central importance in paramagnetic organic insulators and semiconductors.[1,6]

The temperature dependence of the EPR spectrum shown in Fig. 4 indicates spin exchange among the triplet excitons. In the absence of either attractive or repulsive interactions, the collision frequency is proportional to the concentration of triplets, as indicated by

$$\omega_e \sim \rho \, J(1-\delta)/\hbar \tag{7}$$

where we have neglected factors of order unity. Collisions between excitations first broaden and merge the fine structure lines and eventually average the splitting. Even more frequent collisions lead to a single, exchange-narrowed line whose width is proportional to ω_e^{-1}, and hence to ρ^{-1}. This behavior is observed in many TCNQ salts,[35,25] but not with a single activation energy. Exciton-phonon interactions[40] and improved treatments[41] of exciton collisions must be introduced in a more quantitative analysis. The temperature dependence of Frenkel (triplet) spin systems, together with the angular dependence of the dipolar splitting, characterize the EPR of alternating chains.

IV. REGULAR CHAINS:IONIC CT CRYSTALS

The regular ($\delta = 0$) Heisenberg chain has been of interest as a many-body problem long before its experimental realization was seriously considered. It turns out that even such a restricted class of compounds as the halides of first-row transition metals provide many examples of linear-chain systems.[42] Hundreds of $S = 1/2$ Cu[II] linear chains have been found.[14,13] The best illustrations[5,43,44] of the quantitative application of $\mathcal{H}_{ex}$ to a wide variety of physical properties are $N(CH_3)_4 Mn Cl_3$ (TMMC) and $Cu Cl_2$ (pyridine)$_2$ (CPC). TMMC crystallizes[45] in a $\delta = 0$ array of Mn[II] ions, each with $S = 5/2$, that are triply-bridged by chloride ions as shown in Fig. 5(a). The characteristic doubly-bridged Cu Cl$_2$ structure is shown in Fig. 5(b). Lewis bases, such as pyridine, complete the coordination of the $S = 1/2$ Cu[II] ions above and below the Cu Cl$_2$ plane and account for the many possible examples.

These and other $\delta = 0$ inorganic systems are characterized by modest exchange interactions of $J \lesssim 20$ cm^{-1}. Their recent analysis in terms of Eq. (1) has afforded, in favorable cases, quantitative agreement not only for static properties but for dynamic quantities such as spin correlation functions. The general theories[46] of magnetic resonance have been developed for the high-temperature limit, with $kT \gg J$, in which the detailed spectrum of the electronic states is not required. A rather complete analysis of EPR spectra in small J inorganic insulators is therefore possible.

Ionic organic CT crystals like TMPD-TCNQ, TMPD-chloranil, or PD-chloranil crystalize in mixed stacks of the type $\cdots D^+ A^- D^+ A^- \cdots$ and are regular in the sense of equally strong exchange along the chain, in spite of having two radicals per unit cell.[1,25] These mixed stacks exhibit a strongly temperature

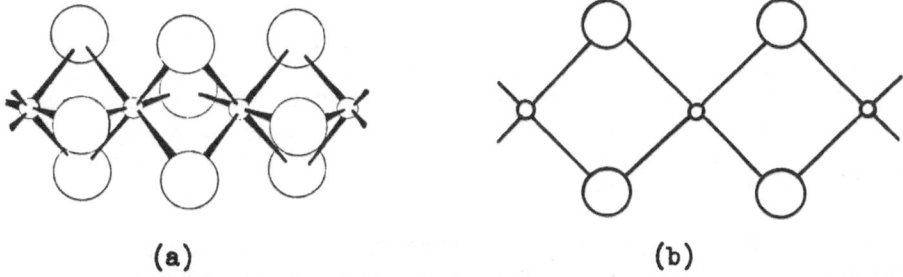

(a)   (b)

Fig. 5. Schematic representation of ligand-bridged, regular linear chains of transition-metal ions. (a) The Mn Cl$_3^-$ chain in TMMC (from Ref. 44); (b) the Cu Cl$_2$ or Cu Br$_2$ chain. Singly-bridged chains are also common.

dependent susceptibility

$$\chi_p T \quad \alpha \quad \exp\left(-\Delta E_p/kT\right) . \tag{8}$$

Values of ΔE_p as high as 1000 cm^{-1} are obtained from a plot of the logarithm of the EPR intensity and temperature against $1/T$. Since a regular chain does not have a gap in the spin states, the origin of ΔE_p remains to be found and may be associated with the slightly less than unit charges along the stack.[6]

A qualitative analysis of the single-crystal EPR data is nevertheless possible.[39,6] In a regular chain, we do not expect the spins forming a triplet state to remain on adjacent sites. The observed EPR intensity then defines the density

$$\rho(T) \equiv \chi_{obs} / \chi_C \tag{9}$$

of unpaired spins relative to the curie susceptibility, $\chi_C = Ng^2 \mu_B^2/(4kT)$ for noninteracting spins $1/2$. We obtain a model of a uniformly diluted magnetic crystal in which the degree of dilution, $\rho(T)$, can readily be controlled and in which the unpaired electrons are exchange-coupled.

Exchange-narrowing in a magnetically dilute system leads to a Lorentzian line whose width is approximately[39]

$$\Gamma \sim \rho M_2^{(0)} / \omega_e \tag{10}$$

Here $M_2^{(0)}$ is the truncated Van Vleck second moment,[17] which can be computed[47] readily for crystals with known structures. The exchange frequency ω_e naturally depends on the concentration of unpaired spins, with

$$\omega_e \sim o\omega_e^{(0)} \tag{11}$$

the simplest approximation for the collision frequency. The constant $\omega_e^{(0)}$ is, for alternating chains, given in Eq. (7). In contrast to the strong narrowing in alternating chains, the case of uniform magnetic dilution shown in Eq. (11) leads to a dipolar linewidth that is independent of ρ. Such behavior is characteristic[48,6] of ionic CT crystals.

In Fig. 6, the angular dependence of the EPR linewidth in TMPD-Chloranil is compared[47] with the calculated second moment $M_2^{(0)}$ for the spin densities of TMPD$^+$ and Chloranil$^-$ based on solution hyperfine data. The exchange rate $\omega_e^{(0)}$ and the relaxation due to other mechanisms were assumed to be isotropic, thus providing two adjustable parameters for all three principal-axis planes.

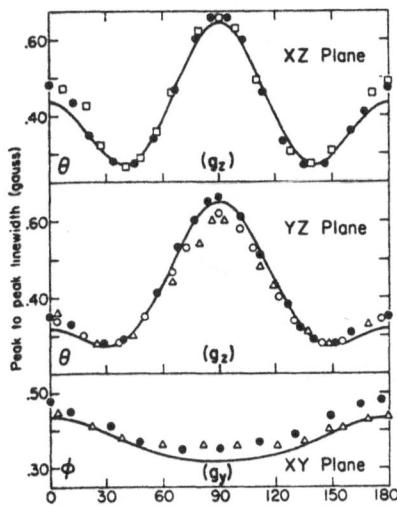

Fig. 6: EPR linewidth of several TMPD-Chloranil single crystals in
the principal-axes plane of the g-tensor. The angular
dependence of the solid line is the calculated second
moment, $M_2(0)$. (From Ref. 47)

At temperatures sufficiently high to exchange-narrow the
dipolar splitting in alternating chains, a single EPR signal near
the free-electron g value is observed in both regular and alternating systems. The ρ^{-1} dependence of the dipolar linewidth in
alternating chains and the lack of ρ dependence in Eq. (11) for
regular chains leads to characteristically different temperature
behavior for the linewidth. Any other variable, such as hydro-
static pressure, that can change the concentration $\rho(T)$ of
unpaired electrons should also show such differences. In Fig. 7,
the EPR linewidth vs pressure[49] of powdered TMPD-ClO$_4$ is shown.
The linewidth increases in the low-temperature alternating
phase, where pressure decreases ρ; the linewidth gently decreases
in the high-temperature regular phase. The constant or decreasing
linewidth vs pressure has been observed[49,48] in other regular
chains showing EPR spectra characteristic of uncorrelated spin
excitons.

V. MAGNETIC ONE-DIMENSIONALITY AND INTERCHAIN EXCHANGE

Many π-molecular ion-radical crystals contain crystallographically equivalent, but magnetically inequivalent, linear chains.[6]
The CPC system discussed by Richards also has magnetically
inequivalent chains. The structure[50] of Cs$_2$(TCNQ)$_3$ shown in
Fig. 8 clearly indicates that <u>two</u> sets of dipolar lines should

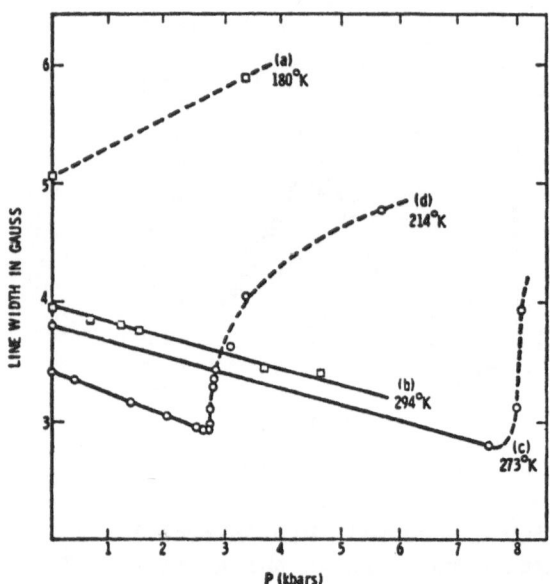

Fig. 7. The pressure dependence of the peak-to-peak EPR linewidth
in TMPD-ClO$_4$. The solid lines indicate the high-temperature, regular stack; the dashed lines correspond to the
low-temperature, alternating phase. (From Ref. 49)

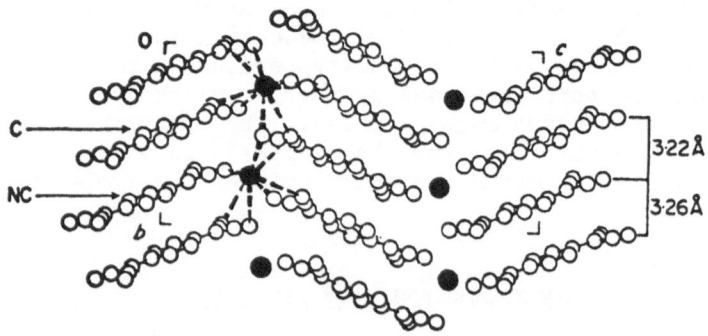

Fig. 8. The [1 0 0] projection of the Cs$_2$ (TCNQ)$_3$ structure. The
noncentric TCNQ sites are largely ionic, while the centric
sites are largely neutral. (From Ref. 50)

occur if the chains do not interact. Two sets of triplet exciton splittings were indeed observed. The structure of the ionic CT crystal PD-Chloranil, where PD is p-phenylenediamine, indicates three magnetically inequivalent chains related to each other by rotations of $2\pi/3$. The calculated positions of the EPR transitions shown in Fig. 9 are in excellent agreement with this assignment.[48] Magnetically inequivalent chains have been resolved in a complex regular TCNQ chain also.[51]

To resolve splittings of 1G ($\sim 10^{-4}$ cm^{-1}) or less, the interactions between the magnetically inequivalent sites must be less than the splitting. This immediately sets an upper bound on the interchain exchange J'. Since intrachain exchange is often in the range of 10^2 cm^{-1}, the resolved splitting of 10^{-4} cm^{-1} demonstrates a very high degree of one-dimensionality. The crystal structures indicate π-electron overlap along the stack, as does the conductivity and the polarization of the CT excitation along the stack.[1,6] The assumed one-dimensionality is thus nicely corroborated by a simple magnetic measurement.

The delocalization of the electrons in PD-Chloranil over the magnetically inequivalent chains under the influence of interchain dipolar interactions explains the very unusual linemerging[48] shown in Fig. 10. Instead of broadening into a single line, as is

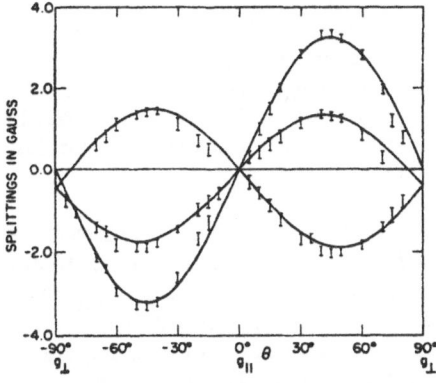

Fig. 9. Angular dependence of the 35GHz g-tensor splittings of PD-Chloranil at 263°K. The solid lines correspond to g-tensors related by $2\pi/3$ rotations about the chain axis. (From Ref. 48)

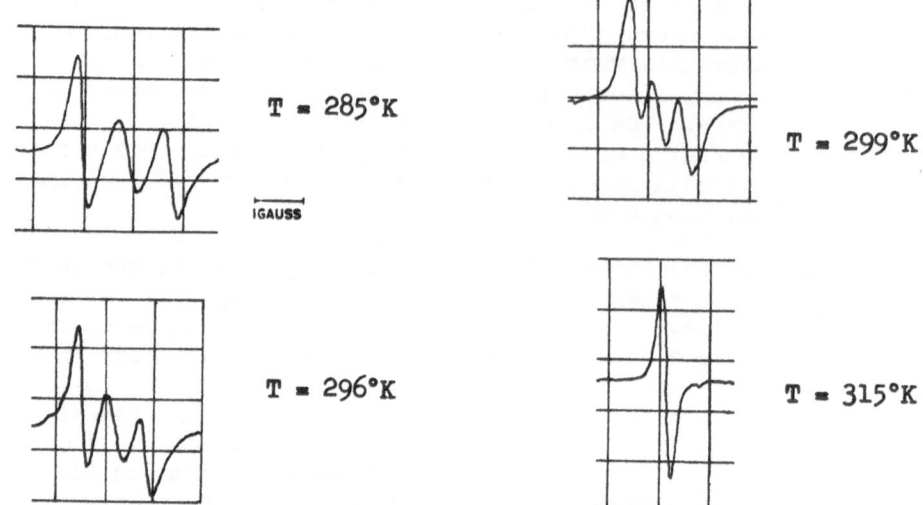

Fig. 10: Temperature dependence of the PD-Chloranil EPR spectrum
         at the maximum splitting. The sweep, but not the spectro-
         meter gain, is the same for all four spectra. (From
         Ref. 48)

required by any model[17,46] based on random jumps among inequivalent
sites, the lines are seen to merge smoothly. The interchain
dipolar field increases as $\rho^{1/2}$ with increasing temperature and is
large enough to shift the EPR signals for electrons delocalized
over the chains.[52]

    There are other unusual effects afforded by the strong,
temperature-dependent magnetic dilution in organic semiconductors.
For example, the temperature dependence of the exchange ω_e in
Eq. (11) allows[53] both $\omega_e > \omega_0$ and $\omega_e < \omega_0$ in TMPD-TCNQ, which
thus shows a "10/3" effect that is conventionally seen by con-
structing a series of EPR spectrometers whose Larmor frequencies
ω_0 span the exchange frequency ω_e.

    It is worth emphasizing that the qualitative model for
uncorrelated spin excitations provides no explanation of the
observed activation energy ΔE_p for paramagnetism. Furthermore,
the general theories of magnetic resonance[46] have yet to be
extended to the intermediate temperature regime $kT \sim J$. Thus we
have exploited the striking angular and temperature dependences
by judiciously using high-temperature results, rather than by
applying a quantitative theory. It is not clear, for example,
whether spin excitations are diffusional for $kT \sim J$. The narrow
Lorentzian lines in organic semiconductors do not show the characteristic one-dimensional lineshape identified[55] in TMMC. Indeed,

such a linewidth of

$$\Gamma \sim \left(M_2(0) \right)^{2/3} (h/J)^{1/3} \tag{12}$$

would contain a net $\rho^{1/3}$ dependence if both the second moment $M_2(0)$ and the exchange scale as ρ. No $\rho^{1/3}$ dependence has been observed in the ionic CT crystals.

Single crystal EPR provides a natural approach to evaluating small exchange constants, such as the interdimer exchange $J(1-\delta)$ in alternating chains near the dimer limit or the interchain exchange J'. The static susceptibility is usually dominated by the largest exchange interactions and is not sensitive to small perturbations. As we have seen, the intensity of the EPR signal is related to χ_p while the structure of the spectrum depends on the smaller magnetic terms in $\mathcal{H}_T$. The structure of the spectrum may be sensitive to additional small exchange constants. The angular dependence of the "10/3" effect in $Cu(NH_3)_4 Pt Cl_4$, a regular Cu^{11} chain with a $\cdots Cu\ Pt\ Cu\ Pt \cdots$ structure based on stacking the square planar complexes, yields the interchain exchange[56] $J' \sim 500$ G $(0.05$ cm$^{-1})$ and shows that $J'/J \sim 0.1$. The frequency dependence of EPR linewidth in CPC will be shown by Richards to yield J' between magnetically inequivalent chains.[44] These regular chains have a single EPR line and thus require a rather complete analysis of the linewidth.

When the small exchange is a relaxation mechanism,[22] as shown by the line broadening in Cu_2Ac_4 pyrazine, then a qualitative analysis is sufficient. It is even simpler to evaluate an exchange constant that is small enough to permit partially resolved hyperfine spectra. The $Cu(mnt)_2$ stack in $[N(n-Bu)_4]_2Cu(mnt)_2$ is such a case.[57] The paramagnetic copper ions are far enough apart to make the electron dipolar term in $\mathcal{H}_T$ negligible. A small exchange of $J \sim 120$ G $(\sim 0.01$ cm$^{-1})$ can be used to simulate the complex hyperfine spectrum observed when $\vec{H}_0$ is perpendicular to the $Cu(mnt)_2$ molecules.[57] In such a favorable case, the exchange constant is related to an observed splitting and is obtained without explicitly discussing the relaxation mechanisms responsible for the EPR linewidth. These examples illustrate some of the ways of extracting various types of exchange constants from single-crystal EPR studies.

VI. CONCLUSION

We have emphasized the EPR determination of exchange constants. The connection of J with optical and electric properties was sketched in Eq. (3) for π-molecular systems, which have a low-lying CT excitation. Exchange interactions arise from Coulombic, rather

than magnetic, effects and are consequently directly related to
electronic structure. They are usually the largest term in the
spin Hamiltonian. The resolved dipolar splitting of the two spins
of a triplet exciton in alternating chains demonstrates the importance of strong spatial correlations. The EPR of ionic CT crystals
is consistent with spatially uncorrelated spin excitations in a
magnetically dilute system. The assumed one-dimensionality of
π-electron overlap is verified in crystals with magnetically
inequivalent sites whenever resolved spectra are observed.

A quite different application of magnetic studies alluded to
in the Introduction is their role as stringent tests of theoretical
models. For example, the possibility of a direct numerical computation of the static susceptibility of a regular ($\delta = 0$) Heisenberg
chain, or a regular Hubbard chain, can be applied[6,9] to N-methyl-
phenazinium (NMP)-TCNQ. The cited values[58] of the Hubbard-model
parameters, $U = 0.14$ eV and $t = 0.021$ eV, are seen in Fig. 11 to
give a χ_p curve that is quite different from the observed susceptibility. The Heisenberg result corresponds to $J \equiv 4t^2/U$ in Eq. (1).
The static susceptibility of other regular organic chains, as well
as the $\cdots D^+ A^- D^+ A^- \cdots$ stacks in ionic CT crystals, still present

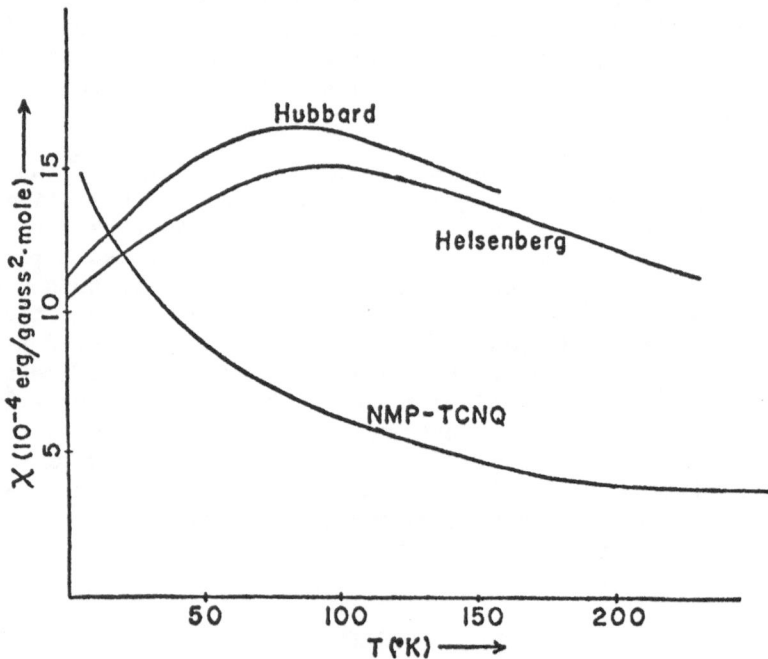

Fig. 11. The static susceptibility of the linear Hubbard model
        ($t = 0.021$ eV and $U = 0.14$ eV) and of the corresponding
        $\delta = 0$ Heisenberg chain ($J = 4t^2/U = 0.13$ eV) in Eq. (1) as
        compared with the smoothed NMP-TCNQ data from Ref. 58.

theoretical difficulties.[59] The possibility of quantitative analyses, as has largely been achieved for some inorganic insulators, is of course an important motivation for further work.

Even when the appropriate model for electronic properties is open, EPR can help sort out various hypotheses. The constant g values observed[60] in TTF-TCNQ above the metal-insulator transition at 60°K, for example, pose very severe difficulties for any model in which both the TTF and the TCNQ stacks are paramagnetic and different. Either the relative contribution of the two stacks remains the same, in which case they are both metals or both semiconductors, or only one strand contributes to the paramagnetism. The fact that the g values are close to the average of the TTF^+ and $TCNQ^-$ g values supports, but does not prove, an averaged picture. The static susceptibility as well as other features of the EPR spectrum then remain to be understood in this one-dimensional conductor. It is likely that magnetic studies will at first serve to test qualitative models for novel magnetic materials such as those discussed at this conference.

It is a pleasure to thank R. C. Hughes for a critical reading of this paper and Sandia Laboratories for their hospitality as a visiting scientist for the summer of 1974. This work was supported in part by the U.S. Atomic Energy Commission.

REFERENCES

1. Z. G. Soos, Ann. Rev. Phys. Chem., 25,   (1974).
2. I. F. Shchegolev, Phys. Stat. Sol. (a), 12, 4 (1972).
3. H. R. Zeller, Adv. Sol. State Phys., 13, 31 (1973).
4. A. F. Garito and A. J. Heeger, Accounts of Chem. Research, (to be published).
5. D. W. Hone and P. M. Richards, Ann. Rev. Mat. Sci., 4, 337 (1974), L. J. DeJongh and A. R. Miedema, Adv. in Phys., 23, 1 (1974).
6. Z. G. Soos and D. J. Klein, "Charge Transfer in Solid State Complexes," to be published in Molecular Interactions, ed. R. Foster, (Academic Press, New York, 1974). A compilation of Hubbard-model computations is given in Section IV, together with the application of Hubbard models to π-molecular CT crystals and their relation to molecular excitons.
7. J. Hubbard, Proc. R. Soc. Lond., A276, 238 (1963); A277, 237 (1964); A281, 401 (1964).
8. R. S. Mulliken, J. Am. Chem. Soc., 74, 811 (1952); R. S. Mulliken and W. B. Person, Molecular Complexes: A Lecture and Reprint Volume, (Wiley, New York, 1969).
9. W. A. Seitz and D. J. Klein, Phys. Rev., B9, 2159 (1974).
10. C. Herring, in Magnetism, Vol. II.B, (eds. G. T. Rado and

H. Suhl, (Academic, New York, 1963) p. 1-185; P. M. Levy, Phys. Rev. $\underline{117}$, 509 (1969).

11.   P. W. Anderson, Solid State Physics, Vol. 14, eds. F. Seitz and D. Turnbull, (Academic, New York, 1963) p. 99-214.

12.   R. L. Martin, in New Pathways in Inorganic Chemistry, eds. E. A. V. Ebsworth, A. G. Maddock, and A. G. Sharpe, (Cambridge University Press, 1968) Chap. 9.

13.   G. F. Kokoszka and G. Gordon, Transition Metal Chem., $\underline{5}$, 181 (1969).

14.   W. E. Hatfield and R. Whyman, Transition Metal Chem., $\underline{5}$, 47 (1969).

15.   A. P. Ginsberg, Inorg. Chim. Acta Rev., $\underline{5}$, 545 (1971).

16.   E. Sinn, Coord. Chem. Rev., $\underline{5}$, 313 (1970).

17.   A. Carrington and A. D. McLachlan, Introduction to Magnetic Resonance, (Harper and Row, New York, 1967). Any other basic texts in magnetic resonance may be used.

18.   B. R. McGarvey, Trans. Metal Chem., $\underline{3}$, 90 (1966); P. W. Atkins and M. C. R. Symons, The Structure of Inorganic Radicals, (Elsevier, Amsterdam, 1967).

19.   A. Abragam and B. Bleaney, Electron Paramagnetic Resonance of Transition Ions, (Clarendon, Oxford, 1970).

20.   B. Bleaney and K. D. Bowers, Proc. R. Soc. Lond., $\underline{A214}$, 451 (1952).

21.   J. N. Van Niekerk and F. R. L. Schoening, Acta Cryst., $\underline{6}$, 227 (1953); P. de Meester, S. R. Fletcher, and A. C. Skapske, J. Chem. Soc., 2575 (1973); G. M. Brown and R. Chidambaram, Acta Cryst., $\underline{B29}$, 2393 (1973).

22.   J. S. Valentine, A. J. Silverstein, and Z. G. Soos, J. Am. Chem. Soc., $\underline{96}$, 97 (1974).

23.   B. Morosin, R. C. Hughes, and Z. G. Soos, Acta Cryst., (submitted).

24.   B. Bleaney, Rev. Mod. Phys., $\underline{25}$, 161 (1953).

25.   P. L. Nordio, Z. G. Soos, and H. M. McConnell, Ann. Rev. Phys. Chem., $\underline{17}$, 237 (1966).

26.   F. H. Herbstein, in Perspectives in Structural Chemistry, Vol. IV, eds. J. D. Dunitz and J. A. Ibers, (New York, 1971) p. 166-395.

27.   J. L. De Boer and A. Vos, Acta Cryst., $\underline{B28}$, 835; 839 (1972).

28.   D. D. Thomas, A. W. Merkl, A. F. Hildebrandt, and H. M. McConnell, J. Chem. Phys., $\underline{40}$, 2588 (1964).

29.   D. B. Chesnut and W. D. Phillips, J. Chem. Phys., $\underline{35}$, 1002 (1963).

30.   W. J. Siemons, P. E. Bierstedt, and R. G. Kepler, J. Chem. Phys., $\underline{39}$, 3523 (1963).

31.   R. G. Kepler, J. Chem. Phys., $\underline{39}$, 3528 (1963).

32.   Y. Suzuki and Y. Iida, Bull. Chem. Soc. Japan, $\underline{46}$, 2056 (1973); Y. Iida, J. Chem. Phys., $\underline{59}$, 1607 (1972); Bull. Chem. Soc. Japan, $\underline{46}$, 320 (1972).

33.   R. Lynden-Bell and H. M. McConnell, J. Chem. Phys., $\underline{37}$, 794

(1962); A. B. Harris, Phys. Rev., B8, 2166 (1973).

34. L. N. Bulaevskii, Soviet Phys. JETP, 17, 684 (1963); Z. G. Soos, J. Chem. Phys., 43, 1121 (1965); W. Duffy, Jr., and K. P. Barr, Phys. Rev., 165, 647 (1968).

35. M. T. Jones and D. B. Chesnut, J. Chem. Phys., 38, 1311 (1963).

36. M.-A. Maréchal and H. M. McConnell, J. Chem. Phys., 43, 497 (1965).

37. T. Hibma, P. Dupuis, and J. Kommandeur, Chem. Phys. Lett. 15, 17 (1972).

38. M. J. Hove, B. M. Hoffman, and J. A. Ibers, J. Chem. Phys., 56, 3490 (1972).

39. Z. G. Soos, J. Chem. Phys., 46, 4284 (1967).

40. Z. G. Soos and H. M. McConnell, J. Chem. Phys., 43, 3780 (1965).

41. R. Lynden-Bell, Mol. Phys., 8, 71 (1964).

42. R. Colton and J. H. Canterford, Halides of the First Row Transition Metals, (Wiley-Interscience, London, 1969).

43. K. Takeda, S. Matsukawa, and T. Haseda, J. Phys. Soc. Japan, 30, 1330 (1971); W. Duffy, Jr., J. E. Venneman, O. L. Strandburg, and P. M. Richards, Phys. Rev., B9, 2220 (1974).

44. R. C. Hughes, B. Morosin, P. M. Richards, and W. Duffy, Jr., (to be published).

45. B. Morosin and E. J. Graeber, Acta. Cryst., 23, 766 (1967).

46. R. Kubo and K. Tomita, J. Phys. Soc. Japan, 9, 888 (1954); P. W. Anderson, ibid, 9, 316 (1954).

47. T. Z. Huang, R. P. Taylor, and Z. G. Soos, Phys. Rev. Lett., 28, 1054 (1972).

48. R. C. Hughes and Z. G. Soos, J. Chem. Phys., 48, 1066 (1968).

49. R. C. Hughes, PhD Thesis, Stanford (1966), (unpublished). R. C. Hughes, A. W. Merkl, and H. M. McConnell, J. Chem. Phys., 44, 1720 (1966).

50. C. J. Fritchie, Jr., and P. Arthur, Jr., Acta Cryst., 21, 139 (1966); D. B. Chesnut and P. Arthur, Jr., J. Chem. Phys., 36, 2969 (1962).

51. D. N. Fedutin, I. F. Shchegolev, V. B. Stryukov, E. Yagubskii, A. V. Zvarykina, L. O. Atovmyan, V. F. Kaminskii, and R. P. Shibaeva, Phys. Stat. Sol., (b) 48, 87 (1971).

52. Z. G. Soos, J. Chem. Phys., 49, 2493 (1968); P. J. Strebel and Z. G. Soos, ibid, 50, 2911 (1969).

53. B. M. Hoffman and R. C. Hughes, J. Chem. Phys., 52, 4011 (1970); Solid State Comm., 7, 895 (1969).

54. A. Abragam, The Principles of Nuclear Magnetism, (Clarendon, Oxford, 1961) Chap. IV and X.

55. R. E. Dietz, F. R. Merritt, R. Dingle, D. Hone, B. G. Silbernagel, and P. M. Richards, Phys. Rev. Lett., 26, 1186 (1971); M. J. Henessey, C. D. McElwee, and P. M. Richards, Phys. Rev., B7, 930 (1973).

56. Z. G. Soos, T. Z. Huang, J. S. Valentine, and R. C. Hughes, Phys. Rev., B8, 993 (1973); T. Z. Huang and Z. G. Soos,

ibid, <u>B9</u>, 4981 (1974).

57. K. W. Plumlee, B. M. Hoffman, J. A. Ibers, and Z. G. Soos, (to be published).

58. A. J. Epstein, S. Etemad, A. F. Garito, and A. J. Heeger, Phys. Rev., <u>B5</u>, 952 (1972); L. B. Coleman, J. A. Cohen, A. F. Garito, and A. J. Heeger, ibid, <u>B7</u>, 2122 (1973).

59. S. Etemad and E. Ehrenfreund, AIP Conf. Proc. No. 18, Magnetism and Magnetic Materials, eds. C. D. Graham, Jr., and J. J. Rhyne, (AIP, New York, 1972) p. 1449.

60. Y. Tomkiewicz, B. A. Scott, L. J. Tao, and R. S. Title, Phys. Rev. Lett., <u>32</u>, 1363 (1974).

ONE-ELECTRON THEORY AND THE PROPERTIES OF

SIMPLE TCNQ SALTS

Jan Kommandeur

Laboratory for Physical Chemistry

Zernikelaan, Groningen, the Netherlands

"Anyone can err, only fools persist"

Cicero

INTRODUCTION

It is in the nature of solid state physics and chemistry that no exact solutions of the equations describing the materials one is working with, can be obtained. The challenge is therefore to select from these equations the terms that are dominant in determining the properties one is considering. For a long time it sufficed in solid state physics to limit oneself to the so-called one-electron terms in the Hamiltonian, i.e. those terms that contain the coordinates of only one electron. This was more by default than by knowledge, the two-electron terms being far too hard to evaluate. In later years, however, it became clear that for a number of properties, such as magnetism and the behavior of the electrical conductivity it was necessary to at least consider the two-electron terms qualitatively. It is not unfair to say that this consideration has been one of the main occupations of theoretical solid state physics for the last decade.

It is no miracle then, when the TCNQ-salts caught the interest of solid state physicists around 1972 that they started to apply all the rather fancy theoretical machinery of recent years, without inquiring rigorously first, whether older and simpler theory could not be used just as well. To restore this balance these lectures will try to sort out how far one can get with one-electron theory for the TCNQ-salts and which modifications are necessary to account for the observations, which clearly point to the importance of two-electron terms.

First an (undoubtedly incomplete) survey will be made of the
properties that can be qualitatively understood on the basis of one-
electron theory. As an illustration we then calculate the magnetic
susceptibility of Li-TCNQ and give an illustration how the electronically induced semiconductor/semiconductor transitions come
about in the simple TCNQ-salts in the second chapter. Having thus
convinced you that one-electron theory suffices we will turn to the
triplet excitons, which occur in some of these materials, showing
that two-electron terms are at least of some importance. We will
treat these excitons in some detail in the last section, particularly also, since they appear to give a good clue as to the effective
dimensionality of the solids. We will end with some speculations
about the general validity of the results.

2.QUALITATIVE APPLICATIONS OF ONE-ELECTRON

Theory

2.1 Cohesion of the Linear Chain

The negative ions of TCNQ tend to stack in linear chains, forcing
the positive ions, which may be of almost any kind in some similar
arrangement. Only when the "counterion" is a similarly shaped open
shell species such as $TMPD^+$ does an alternating D^+A^- stack occur.
In itself, it is rather surprising that organic negative ions should
stack as closely as they do in TCNQ-salts. There must be something
highly directional that holds them together. It will be clear that
this is the one-electron transfer integral t, which depends linearly
on the overlap integral S between the π-orbitals in which the unpaired electrons are located. For the directional part of this
cohesive energy two-electron integrals can be neglected, they are
either very small (exchange-type) or non-directional (coulomb-type).
Only when the members of the chain have half-filled one-electron
orbitals is the one-electron effect considerable. Therefore, $TCNQ^-$
can be replaced by $TMPD^+$, but not by NMP^+, which has filled one --
electron orbitals.

The one-electron glue is so good that it will even permit glueing
neutral molecules into the chain, the 1:2 and 2:3 compounds, which
still stack with their TCNQ's in linear chains. In first order the
neutral TCNQ's offer the same orbitals for the extra-electron as the
ions, and the reduction in the Coulomb repulsion energy is apparently
sufficient to offset the loss in "transfer" cohesive energy arising
since two entities must now share one electron.

When the "one-electron part" of the cohesive energy is reduced to
zero, when all entities have doubly occupied one-electron orbitals,
i.e. when neutral TCNQ is considered, all tendency to stack in chains
has disappeared,neutral TCNQ shows the usual slipped fish-bone
monoclinic structure of many other aromatic crystals. We can conclude

that the one-electron integrals are very important in determining
the chain-like structure of the TCNQ-salts.

2.2.The Peierls Instability

Almost all chains appear to suffer from the Peierls instability:
1:1 salts showing dimerization, most 1:2 salts tetramerization and
one of the few 2:3 salts (Cs_2TCNQ_3) trimerization. This is exactly
what one would expect for half-filled quarter-filled and one-third
filled bands respectively, and again it should be pointed out that
these distortions are most easily understood on a one-electron
basis. In the limit of t=0 they do not occur, neither do they in the
limit of the on-site repulsive energy $U \to \infty$, which is of course a two-
electron term.

The last consideration does not apply to one of the compounds
best studied: NMP/TCNQ. Although a 1:1 solid it does not show any
dimerization along its TCNQ chain. Unfortunately, however, it has a
disordered crystal structure, the charged part of the NMP ion being
randomly distributed along the NMP^+ chain and this invalidates the
argument for a Peierls instability.

2.3. Bond Lengths

Crystallographers (1) have found three TCNQ species: $TCNQ^0$,$TCNQ^{\frac{1}{2}-}$
and $TCNQ^-$ with distinctly different bond-lengths, the core becoming
more quinone-like as the negative charge is lost. 1:1 salts contain
the $TCNQ^-$ core, 1:2 salts the $TCNQ^{\frac{1}{2}-}$, while in the Cs_2TCNQ_3 salt one
finds two $TCNQ^-$ ions and a neutral $TCNQ^0$. This is exactly what one
would expect from one-electron theory, including the surprising
result for the 2:3 salt, which comes about because only the lowest
of the three bands would be filled and it has appreciable density
only on what turn out to be the $TCNQ^-$-ions in the triad. Realizing
that the valence electrons are accomodated in π-orbitals this becomes
a symmetry argument and therefore very strong. Again, one-electron
theory suffices to adequately understand the experiments, which have
recently been confirmed by I.R.experiments (2).

2.4. Electrical Conductivity

One electron theory together with the Peierls instability applied
to the 1:1 salts yields a "valence" and a "conduction" band with a
gap=2/t_2/-2/t_1/, where /t_2/ and /t_1/ are the two occurring one-electron transfer integrals. One would expect this system to behave as a
semi-conductor.

For the 1:2 salts, in the geometry usually found: the tetrad
consisting of two equivalent dimers, both the conduction and the
valence bands as found in the 1:1 salt would be slightly split
again in the middle, yielding four bands with only the bottom one

filled . The new gaps are approximately equal to the difference in inter-dimer integrals. From geometrical considerations this difference is usually small or even zero. We expect these systems then to behave as very small gap semiconductors or metals.

In general, these one electron predictions are borne out: 1:1 complexes are semiconductors, while the metals are found amongst the 1:2 salts. NMP/TCNQ is then again an exception, but it has only one TCNQ⁻ ion per unit cell, one could thus view it as a metal. If one takes the lattice disorder into account it could be viewed as a disordered semiconductor with localized and delocalized states in the gap.

It should be realized that considerable difficulties are encountered in D.C.conductivity measurements in highly one dimensional systems. If the "strands" are occasionally broken one measures the electron hopping across a "break", and A.C.measurements should yield a different result. Such effects, particularly as regards the activation energy for conductivity have been variously reported in the literature.

2.5. Magnetic Susceptibilities

All TCNQ-salts are paramagnetic in the sense that there is some temperature where $\chi > 0$. It is useful to speak about the spinsusceptibility, which can be taken as the total susceptibility from which the diamagnetic susceptibility is subtracted. This spinsusceptibility is then either constant (in some 1:2 salts) or temperature activated (as in some other 1:2 salts and in most 1:1 salts). Very roughly the magnitude at room temperature is in the order of 1×10^{-4} e.m.u/$_{mole}$ or about 10% of the Curie Susceptibility, which one would have for a mole of non-interacting spins.

This reduction can come about in two ways:i) Through 1-electron terms leading to a Peierls gap, and thus to a semiconductor (1:1, 2:3) or to (almost) a metal (1:2, NMP/TCNQ).

ii) Through 2-electron terms yielding anti-ferromagnetism. In both cases the reduction factor must be given by something roughly like exp-E/kT, yielding an $E \approx 0,05-0,10$ eV, depending on detailed mechanisms. One-electron integrals typically have those values at distances of 3.2-3.5 Å. A direct exchange integral would be too small, but exchange would be much enhanced through configuration interaction: $J = 2t^2/u$. If we now for instance require $J \gtrsim 0,1$ eV and $t \approx 0,1$ eV we see that we have $U \gtrsim 0,1$ eV, which in its turn invalidates the approximation. At least it is clear that if these "configuration interaction terms" are at all important, so is the one-electron term. If it is possible to explain the observed phenomena in one-electron terms, it is preferred for reasons of simplicity.

2.6. Optical Spectra

Comparatively little work has so far been done on the optical spectra (3). In general their interpretation has been based on their similarity to the spectra of the so-called charge-transfer complexes with a non-ionic ground state. The lowest energy absorption (anywhere between 0,5 and 1,5 eV) is then usually called the charge-transfer band (4), which would be the absorption transfering an electron from a TCNQ$^-$-ion to its neighbour. The energy this absorption occurs at is then often taken as the on-site repulsive energy U, which is indeed correct within this interpretation. One of the major arguments used is usually the along the chain polarization of the transitions.

It should be pointed out, however, that again one-electron theory gives an alternative. In as much as the ground state is split by the Peierls distortion this will also be the case for the first excited state of the isolated ion, which from solution studies we know to occur at about 1.4 eV. On one-electron theory we therefore have a band system of the ground state and a band system of the excited state. In general one expects the splitting in the excited state to be larger. The absorptions occurring at energies from 0.5 to 2.0 eV can be interpreted as transitions from band system to band system, which would also have the correct polarizations (3). The typical shapes of these absorptions in inorganic semiconductors arising from the densities of states are obscured by the vibrational structure in these solids. High resolution spectroscopy at low temperatures may reveal more details.

One-electron theory would also predict intra-band transitions between the ground-state bands. They should lead to at least a very broad absorption from 0 to 2000 cm^{-1} in the "metallic" compounds and to a broad absorption from about 500 to 2000 cm^{-1} in the "semiconducting" materials (with gaps of about 500 cm^{-1}). These do not seem to have been reported, but their oscillator strength is relatively low, they are obscured by the molecular vibrations and most conventional apparatus has too high a resolution and too limited a range in this wavelength region to immediately see the transition. The prettiest experiment would be a low temperature infrared photo-conductivity measurement. It does not seem to have been reported either and does of course suffer from the drawbacks of a conductivity experiment as discussed above.

Another "optical" experiment would be a study of the lineshape of the intramolecular vibrational absorption. If there is any interaction between the valence electrons and the molecular vibrations, the vibrational eigenstates will mix with the electronic continuum, which will lead to a so-called Fano line-shape, which would prove the existence of the continuum as well as yield the coupling constant. Such lineshapes are indeed observed, but unfortunately they also come about through reflection effects (5). We gave up when we obtained the "Fano-lineshape" in Rb$^+$TCNQ$^-$ as well as in neutral TCNQ. The careful I.R.work of Pacelli (2) does not appear to show any of these effects on lineshapes.

Our conclusion must again be that the one-electron terms must at least not be forgotten here. The crude interpretation of the lowest "charge-transfer" absorption as being U is certainly erroneous, the one-electron terms should at least be included.

2.7. Electron Spin Resonance

All TCNQ-salts apparently show ESR-absorption and it is of some interest to speculate what this is due to. In a one-electron approximation the absorption would be due to free electrons and/or holes. If, however, U would be too high for electrons and holes to be the lowest excitation then the absorption would be due to spin-wave excitations if the system were a weakly alternating anti-ferromagnet, or to triplet states if it were a strongly alternating anti-ferromagnet. In all cases the ESR signal would be temperature activated in the "semiconductor" and (almost) temperature independent in the "metallic" salts. In all cases the ESR signal would be narrow due to exchange or motional narrowing, and in all cases one would expect some paramagnetic "impurity" contribution to dominate at low temperatures.

In some salts part of the ESR absorption is split by a spin-dipolar interaction which clearly shows that it is caused by a pair of paramagnetic species residing on neighboring ions in the solid. This fact was realized at least a decade ago by the earliest workers in the field(6). Because of the paired nature of the absorption the species received the name triplet exciton, which unfortunately has spilled over to almost any ESR absorption in organic materials without much justification.

Besides the dipolar split ESR-signal one finds a non-split central absorption, which usually has a different temperature dependence.

Now at least there is no doubt as to the reality of the two-electron terms. The "correlation" is clearly seen when eigenstates can be observed in which the spatial correlation of the electrons is evidenced by something as independent of other interactions as a spin-dipolar interaction. Nevertheless, the pair can be interpreted as a triplet state (of a dimer) when one feels the two-electron terms dominate, or alternatively as a bound hole-electron pair (Wannier exciton) if the one electron terms dominate. In the first interpretation one does not have an explanation for the central signal, in the second it is due to the unbound electrons and/or holes. Because of all the arguments given in previous paragraphs and because of the temperature activated central signal we have preferred the latter interpretation for the triplet exciton in TCNQ-salts. As we shall show in the next sections it then provides a measurement of the quantity t/U, which of course determines whether one can use one-electron theory or not. If $2t/U \gg 1$ one electron theory is fine, if $2t/U \ll 1$ one electron effects can be forgotten. As might be expected experiment yields $U \approx 2t$, which at least means that the arguments everybody gave were reasonable and that the fight can continue if one so desires.

3. ELECTRONICALLY INDUCED CRYSTALLOGRAPHIC PHASE TRANSITIONS IN THE
   SIMPLE SALTS.

In this section we will start out with a band description of a
dimerized TCNQ-chain (Peierls' state) and calculate its magnetic
susceptibility as a function of temperature, which can be well fitted to the paramagnetic susceptibilities of LiTCNQ and CuTCNQ. We
will then show that for certain values of the electronic interaction
the low temperature Peierls' state becomes unstable with respect to
the non- or less-dimerized state at higher temperatures. Paramagnetic
susceptibilities calculated from this model compare well with those
observed in the alkali-TCNQ-salts.

3.1. Magnetic Susceptibility of the Peierls State

In the dimerized TCNQ chain we will have two one-electron bands,
the energies given by

$$E(k) = \pm \left(t_1^2 + t_2^2 + 2t_1 t_2 \cos kr \right)^{\frac{1}{2}} \quad , \quad (1)$$

where t_1, t_2 are the transfer integrals, r the lattice parameter and
k the momentum quantum number. The width of the bands is given by
$E_b = 2|t_2|$, the gap by $2E_g = 2|t_1| - 2|t_2|$ and the total width of the
system by

$$E_w = 2(E_b + E_g) = 2|t_1| + 2|t_2| .$$

We further define $E_c = 2(t_1 t_2)^{\frac{1}{2}}$

and $g = \sinh\left[\frac{1}{2} \ln(|t_1|/|t_2|)\right]$, which quantities suffice to
characterize the system. The band system is depicted in fig.1.
We can calculate the paramagnetic spin susceptibility from

$$\chi = \lim_{H \to 0} \chi(H) = \lim_{H \to 0} \left\{ (\beta H)^{-1} (\partial \ln Z_{el}/\delta H) \right\} \quad (2)$$

where $\beta = (k_B T)^{-1}$ and H the applied magnetic field. We therefore
need Z_{el}, the electronic partition function

$$Z_{el} = \sum_{k,\sigma_k} \sum_{n(k,\sigma_k)} \exp\left\{ \left[-\beta E(\sigma_k, k) - \alpha \right] \right\} n(k,\sigma_k) \quad (3)$$

where $E(\sigma_H, k) = E(k) + \frac{1}{2}\sigma_k g\mu_B H$

with $\sigma_k = +1, -1$ and $\alpha \cong -\beta E_F$, other symbols having their usual signi-
ficance. From (3) we obtain

$$Z_{el} = \prod_{(k,\sigma_k)} \left\{ 1 + \exp\left[-\beta E(k,\sigma_k) - \alpha \right] \right\}$$

which together with (2) yields:

$$\frac{\chi T}{C} = \frac{1}{\Pi} \int_{-\pi/2}^{+\pi/2} \left/ \frac{e^{\alpha + \varepsilon k}}{(e^{\alpha + \varepsilon k} + 1)^2} + \frac{e^{\alpha - \varepsilon k}}{(e^{\alpha - \varepsilon k} + 1)^2} \right/ d(\tfrac{1}{2}ak) \quad (4)$$

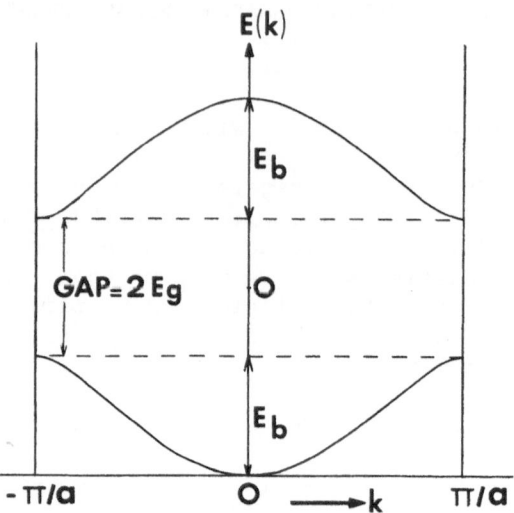

fig.1. Band system for alternating linear chain.

where the summations over k have been changed into integrations, C is the Curie constant and α can be found from $N_e = -\partial \ln z_d / \partial \alpha$, where N_{el} is the actual number of electrons in the system. For an exactly half-filled band system $\alpha=0$, i.e. the Fermi energy is exactly in the middle of the gap.

The result of a general calculation of $\chi T/C$ as a function of the reduced temperature kT/E_c is given in fig.2. It is worth noting that the maximum susceptibility of such a band system is reached for $\chi T/C = 0.50$, or $\chi = \frac{1}{2} C/T$, which is half the Curie susceptibility. In passing it is worth mentioning that DPPH shows this "half" Curie behavior over the temperature range from 2 to 40°K (7). Figure 3 shows a fit of the measured magnetic susceptibility of LiTCNQ versus the temperature, based on eqn(4), Ec/k = 1185 K, g = 0.66 and an excess electrons (or holes) of 3.4 %. The agreement leaves little to be desired. A similar plot is given for CuTCNQ in fig.4 with the values Ec/k =2020°K , g = 0.20 and δn =0.05%.

We can conclude that one-electron theory quite suffices for these compounds.

3.2. The Phase Transition

The dimerization of the TCNQ chains finds its cause in the fact that a split half-filled band-system can at 0°K accomodate all electrons in the bottom ("valence") band. The electronic energy is thus lowered and only a little lattice dimerization (repulsion) energy has to be paid. The Peierls state occurs at 0°K where the sum of those two energies is minimal.

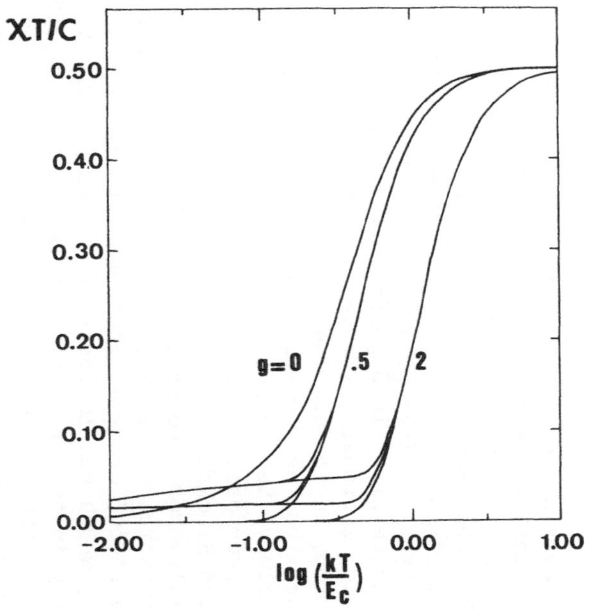

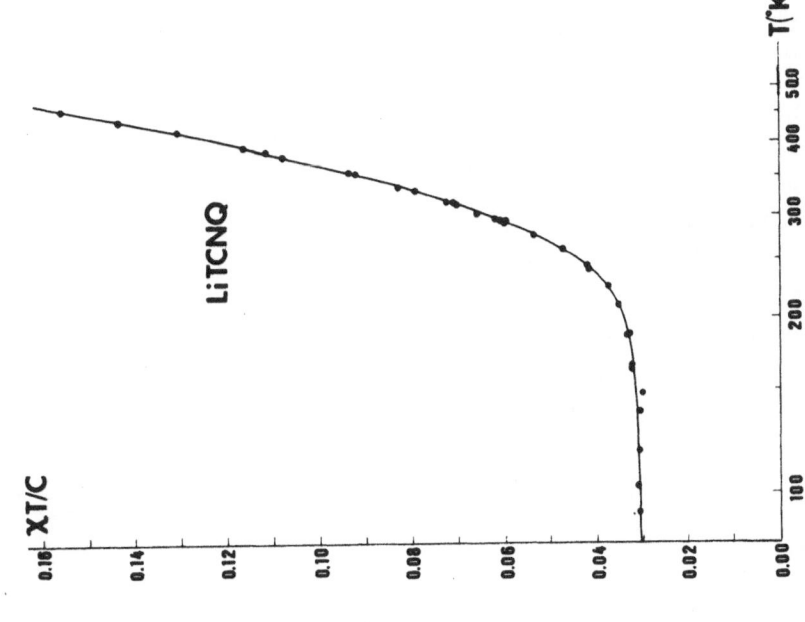

fig. 3. Experimental (dots) and theoretical (drawn line) values of χT/C versus T for Li.TCNQ.

fig. 2. Computer calculated curves of χT/C versus log(kT/Ec) for g = 0.05 and 2 from left to right and n = 0.05, 0.02, and δ0 from top to bottom.

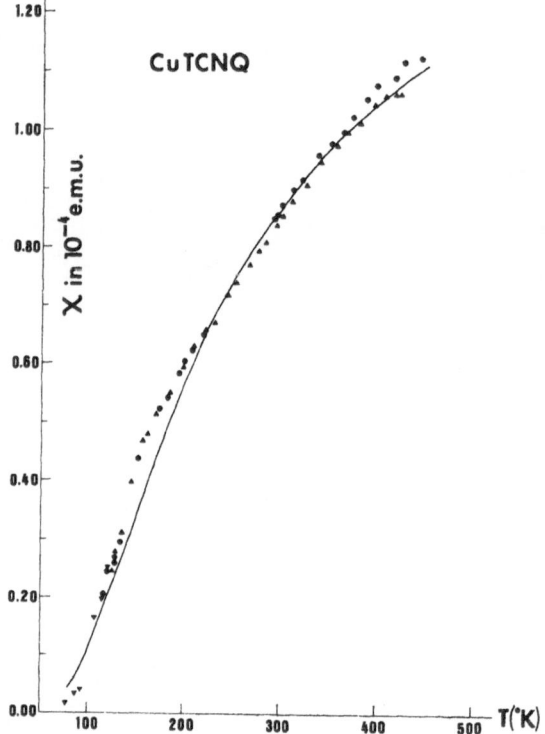

fig.4. Experimental (dots) and theoretical (drawn line) values of
       $\chi T/C$ versus T for Cu.TCNQ.

As the temperature increases the average energy of the electrons
goes up through excitations and the entropy of the electrons (both
spin- and partition entropy) increases, which means it is by no means
certain that the free energy F = U - TS will remain a minimum at the
degree of dimerization at 0 K. This dimerization is defined by the
order parameter $\xi = (r_1-r_2)/(r_1+r_2)$, where r_1,r_2 are the distances
between lattice sites in a point lattice. As ξ decreases the electronic energy increases, the lattice energy decreases and the electronic entropy increases. The behavior of the free energy is thus
undetermined, but as T → ∞ it will have its minimum at $\xi = 0$. There
will thus always be a phase transition, in some cases, however, far
above the decomposition temperature of the TCNQ salt.

Adler and Brooks (8) were the first to describe this transition
theoretically. For a point lattice and a linear dependence of the
transfer integral on distance they concluded that the transition
would be a second order semiconductor to metal transition. For a
TCNQ-chain, which has in addition to its Peierls alternation (r_1,r_2)
an asymmetry in the projections μ_1,μ_2 of the two different TCNQ$^-$-
ions on one another, see fig.5, and with an exponential dependence
of the transfer integral on distance the transitions are second or
first order semiconductor-semiconductor transitions (9). All that

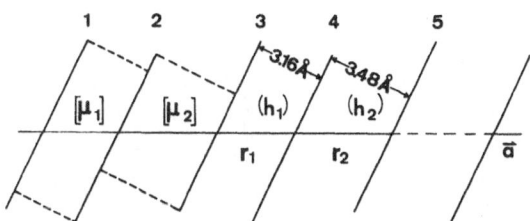

fig.5. Principal characteristics of the stacking of the TCNQ⁻ ions in Rb⁺TCNQ⁻.

happens is that the gap is slowly and/or suddenly reduced.

We will sketch in outline how the free energy and thus the magnetic susceptibility can be calculated in this case.

The electronic partition function has been given in (3). For the total free energy we have to add the repulsive energy term

$$\Delta E_1(\xi) = \tfrac{1}{2} V K_\xi \, \xi^2 \qquad (5)$$

in which V is the volume and K_ξ the "alternation compressibility". The influence of lattice vibrations was found to be negligeable, so we don't have to worry about a lattice entropy term.

To calculate the free energy F as a function of ξ we need an explicit dependence of the transfer integral on ξ . Now we have in these systems $t_{1,2}$ = const.exp(-c r_1,r_2)

$$\text{or } t_{1,2} = t_o \, \exp(\pm \, b\xi\,) \qquad (6)$$

where b depends on the type of orbitals considered. With the aid of

$$E_c = 2(t_1 t_2)^{\tfrac{1}{2}} = 2|t_o| \qquad (7a)$$

and $g = \sinh\left\{\tfrac{1}{2} \ln(t_1/t_2)\right\} = \sinh(b|\xi|)$ (7b)

we transform the electronic energy of eqn (1) into

$$E(k) = \pm \, E_c \left\{ g^2 + \cos^2(\tfrac{1}{2}kr) \right\}^{\tfrac{1}{2}} \qquad (8)$$

which of course through (7b) explicitly depends on ξ . Constructing the partition function $Z = Z_{el}.Z_1$ and with $F = -\beta^{-1} \ln Z$ the free energy can now be calculated as a function of ξ .

For three cases the results of such calculations as a function temperature are given in figure 6. They are parametrized in the reduced temperature kT/E_c and the "reduced hardness" Ω/E_c, where $\Omega = B/r_o^n$ is the Born repulsion between two neighboring sites at $\xi = 0$. Hard lattices (A) yield second order transitions, soft lattices (C) first order, separated by a critical point (B). The reduced transition temperature as a function of the reduced hardness is given in figure 7. It will be noted that for the second-order transitions the transition temperature depends exponentially on the "reduced hardness" Ω/E_c, which is a result also obtained from analyses of electron-phonon coupling exercises. However, in these exercises one usually modulates only the on-site energy in a linear fashion and even when the variation of the transfer integral is

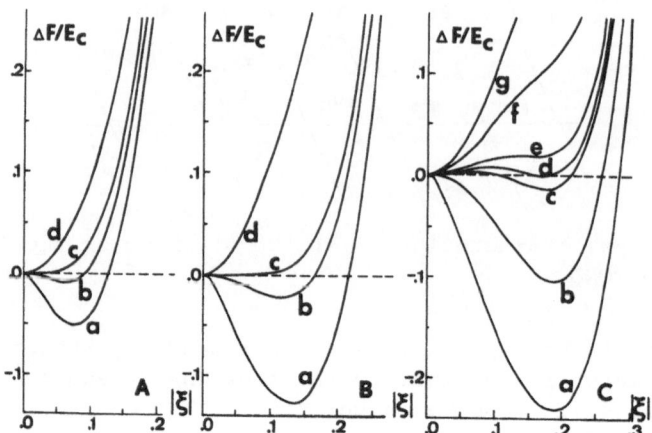

fig.6. Free energies as a function of temperature

    A. Second order transition
    B. Critical transition
    C. First order transition

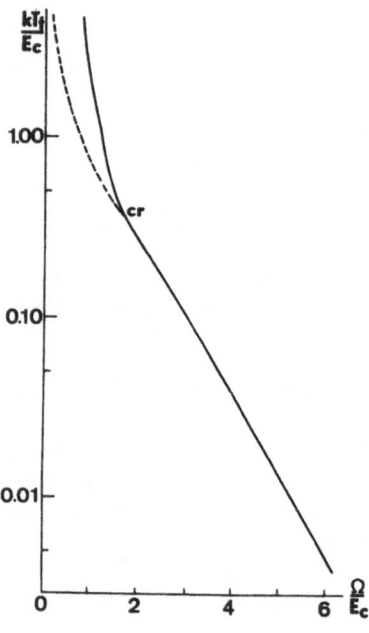

fig.7. First and second order transition lines joined by a critical point.

included, one usually retains only the linear term, which precludes obtaining a first-order phase transition. In our analysis, we do obtain first-order phase transitions but no simple exponential relation is obtained.

Calculation for actual compounds are quite complicated since the skewed arrangement (μ_1 and μ_2, cf.fig.4) as well as the lattice expansion have to be taken into account. Moreover, the only well-known crystal structure of the alkali-TCNQ's is RbTCNQ. We therefore fitted everything at the low temperature RbTCNQ structure and then let the computer find the minimum free energy. For RbTCNQ the result is compared with experiment in figure 8. Similar calculations were performed at higher values of E_c, imitating the closer TCNQ$^-$ approach expected for the smaller alkali ions. The results of these calculations are given in figure 9 and should be compared to the experiments collected in figure 10. The theory can indeed well imitate the real life behavior. Particularly the trend observed in going from larger to smaller alkali-ions is gratifying. There are, however, many enantiomorphs of the simple alkali-TCNQ salts, as f.i. illustrated by Rb.TCNQ of which by now at least three phases are known and therefore care must be exercized in claiming success. One may not be comparing isomorphic structures!

We can conclude that one-electron theory can adequately account for the phase transitions in the alkali TCNQ-salts.

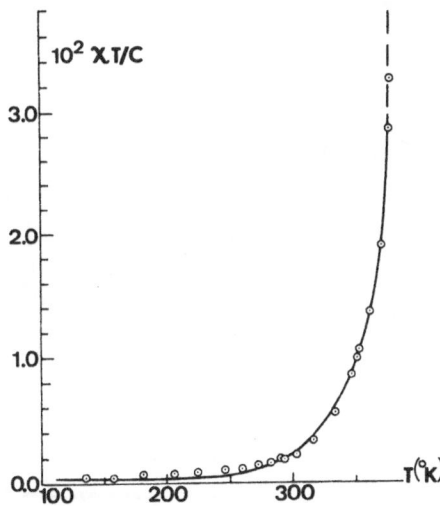

fig.8. Magnetic susceptibility of Rb$^+$TCNQ$^-$ into the phase transition. The drawn line gives the result of a calculation with E_c/k = 2725 K, δn = 0.0004 and g(T) varying between 0.66 and 0.1, where the first value is determined by the known geometry at low temperatures.

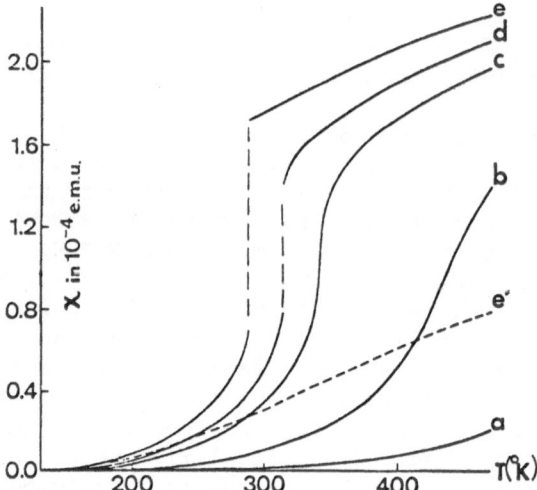

fig.9. Calculated magnetic susceptibilities through the phase
       transition with from a to e decreasing total bandwidth. e'
       shows case e when the parameters are held constant, showing
       clearly the effect of the phase transition.

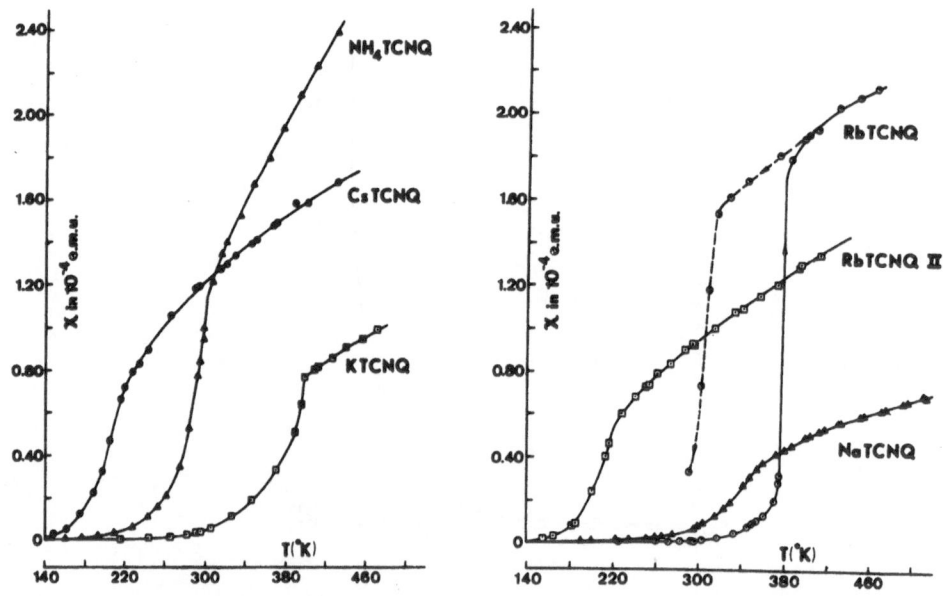

fig.10. Experimental magnetic susceptibilities of various alkali
        TCNQ-salts as a function of temperature. The curves have
        been corrected for small Curie-type impurities.

4. TRIPLET EXCITONS

The two-electron terms we have so far decided to neglect arise from the operator e^2/r_{ij} for the electron-electron repulsion. If this repulsion has any importance, it means that an electron traveling through the solid will repel the other electrons, i.e. create a deficit of negative charge around itself. Of course this deficit of negative charge is an excess of positive charge (or a hole), and thus one can easily see that electron-electron repulsion leads to electron-hole attraction. A bound electron-hole pair is usually called a Wannier exciton, in contrast to the purely localized on-site excitation which has received the name Frenkel-exciton. Wannier excitons are found in semiconductors such as Germanium, Frenkel excitons in Van der Waals solids, such as rare gas crystals or crystals of organic molecules. The Wannier exciton is usually quite "large", its wave function spreading over many sites, the Frenkel exciton is localized on one lattice site. The propagating interaction of the pure Wannier exciton is the one-electron transfer integral t, for the pure Frenkel exciton the two-electron dispersion interaction D.

In the special case of a Peierls unstable one-dimensional chain there is another way in which excitons may come about. Local excitation of an electron and a hole will lead to a de-dimerization, because locally the energy gained by putting two electrons in a bonding orbital is lost. A lattice distortion will then accompany the excitation leading to binding of the electron hole pair. If we consider the original dimer as a single site then we thus create a Frenkel exciton, but due to the fact that we have a finite transfer integral t it will mix in states where the electron and hole are not on the same site (dimer), i.e.Wannier type exciton states. Since the energy of the lattice distortion E_L also prevents the transfer of an electron to another site, it should be added to the effective electron-electron repulsion $U_{eff}=U + E_L$. Below, we shall note, that this effect is by no means negigeable in the alkali TCNQ salts. The existence of excitons implies a finite value of the effective on-site electron-electron repulsion U_{eff}, for $U_{eff} << t$ we will have pure Wannier, and for $t << U_{eff}$ pure Frenkel excitons.

In TCNQ-salts we appear to be in neither limit the results are best described by a linear combination of the two types of exciton, which can then therefore be regarded as Frannier or Wenkel excitons. The extent of mixing of Wannier and Frenkel excitons depends on the value of t/U_{eff}. Conversely, if some idea can be obtained experimentally of the extension (the "Wannier"ness) of the exciton the value of t/U may be estimated. We will discuss this point more quantitatively in the next section.

The linear chains of TCNQ-ions offer good "rails" for excitons to travel on. How often do they switch rail? We will show that a careful study of the line width behavior of the exciton ESR signal reveals a lot about the anisotropy of its motion.

4.1. The Extended Frenkel Exciton

We will develop a description of the Extended Frenkel Exciton
(EFE, a Frenkel exciton with only a little "Wannier"ness) in the
limit of a strongly dimerized linear chain, i.e. with $t_2 \ll t_1$,
where t_1 is the transfer integral within, and t_2 the transfer
integral between the dimers.

We first solve to any desired degree of accuracy the isolated
dimer and find wavefunctions for its ground and excited states (10).
We then realize that in the chain there will be states where an
electron is taken from one dimer and placed on an other. We will
call the energy to perform this act I, and we should realize that I
is both the "on-site" electron-electron repulsion (Hubbard U) for
dimers, and the exciton binding energy of an electron and hole, when
they are both localized on one "site"(dimer). This binding will be
counteracted by the one-electron transfer-integral t_2 between the
dimers which will try to spread the electron and the hole to the
nearest neighbours. Roughly, it will succeed to the extent t_2/U_{eff},
i.e., a fraction t_2/U_{eff} of the density of hole and electron will
reside on the neighbouring dimers.

In ESR we measure the dipolar splitting which is proportional to
$\langle \frac{1}{r^3} \rangle$ of the electron and the hole. As t_2/U_{eff} increases therefore
we expect the dipolar splitting to decrease. More careful analysis
of this problem (11), yields the result given in figure 11, where
the ratio Λ/Λ_o, where Λ_o is the dipolar splitting expected for
the pure Frenkel exciton and Λ the dipolar splitting of the Frannier exciton is plotted as a function of t_2/U_{eff}. If Λ_o were accurately known, the value t_2/U_{eff} could be read off this graph,
after Λ had been determined experimentally.

The best procedure would be to determine Λ_o experimentally in a
really isolated dimer, but such has proved to be impossible. Next
best is to take the fairly decent wavefunctions for $TCNQ^-$ (12,13)
and calculate Λ_o for the known Rb TCNQ structure. With the thus
obtained value of Λ/Λ_o =0,68 we find a value for t_2/U_{eff} of about 1.
We conclude that the intra dimer transfer integral is competitive
with U_{eff}.

4.2. The dimensionality of the excitons

Linewidths of ESR spectra can often give information on the dynamics of the processes studied. This also proves to be the case for
the triplet excitons in TCNQ-salts. We will compare the linewidth
behavior of the excitons in $Rb^+ TCNQ^-$ where the $TCNQ^-$ ion are stacked
as given in figure 12 and tri-methyl-benzimidazolium (TMB^+)-$TCNQ^-$,
where the stacking is as given in figure 13, i.e. where the pairs of

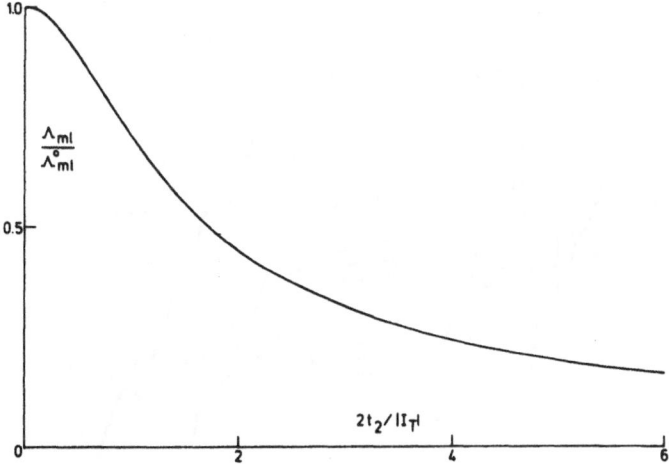

fig.11. The relative dipolar splitting as a function of t_2/I

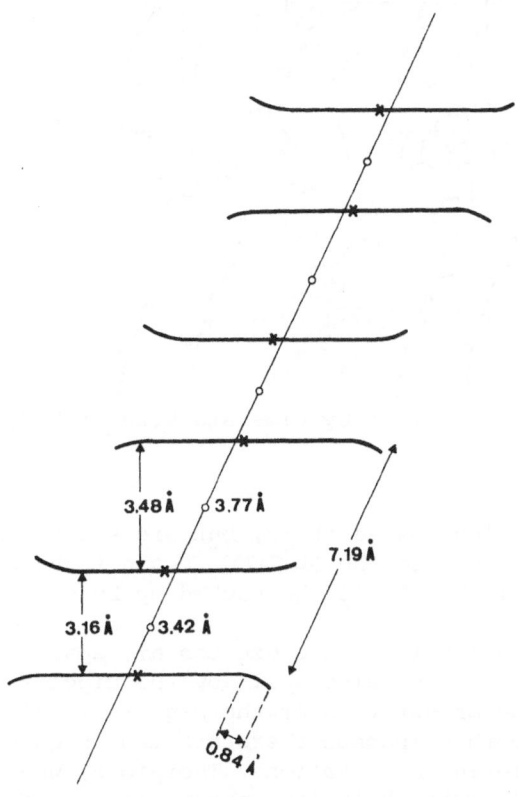

fig.12. Stacking in the Rb TCNQ chain.

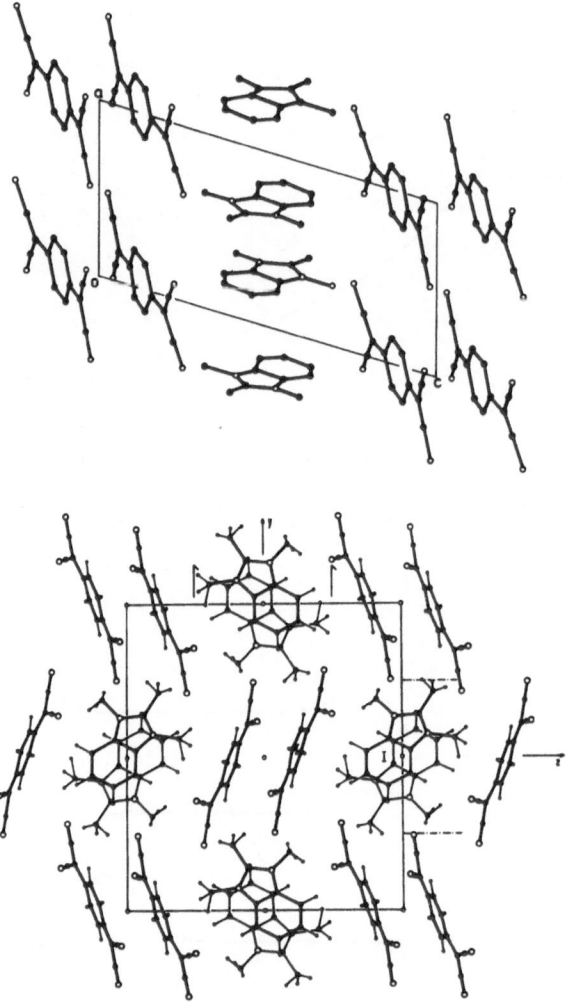

fig.13. Unusual pairwise side by side stacking of TCNQ⁻ ions in
        TMB⁺TCNQ⁻ .

TCNQ⁻ do not form a lengthwise chain, but are stacked pair-wise side
by side. Off-hand we expect the Rb⁺TCNQ⁻ chain to be much more one-
dimensional, a result which will be backed up by our linewidth
considerations.

   The triplet state levels of the exciton are apart from the splitting by the magnetic field, also split by the dipolar interaction,
which depends on the orientation of the magnetic field with respect
to the dimers. In both compounds there are two inequivalent chains
of dimers with different orientations. Therefore, when an exciton
"hops" to an inequivalent chain its energy levels and therefore its
Larmor frequency will in general change. Only when the magnetic

field is in a symmetry plane, i.e. when it has the same orientation
with respect to the dimers of the inequivalent chains this does not
occur. A change of the Lamor frequency leads to dephasing of the
ensemble of spins, therefore gives a contribution to T_2 and therefore to the line width. This contribution disappears when the magnetic field is in a symmetry plane, i.e. when the generally observed
two sets of dipolar split ESR lines coincide. For both Rb^+TCNQ^- and
TMB^+TCNQ^- this is beautifully observed as illustrated in figure 14.
The linenarrowing is a measure for the hopping rate, and as is shown
in figure 15 a temperature dependent measurement yields straight
lines for the logarithm of the transverse hopping rate υ_t versus
$1/T$. This immediately leads to the conclusion, that activation
energies 0.08 eV for Rb^+TCNQ^- and 0.16 eV for TMB^+TCNQ^- are involved
in the exciton transfer. In other words the exciton-polaron binding
energies are 0.08 and 0.16 eV respectively. Since the total frequency
is known we also find the pre-exponential factors, which are 1.1×10^8
and 2.0×10^9, sec^{-1} respectively.

A different situation obtains for the exciton motion along the
chain. At finite densities of excitation collisions will occur. In
the "collision complex" the identity of the triplet sublevels of the
two excitons will be lost and therefore at low collision rates they
will lead to line broadening. At high rates the sublevels will be
intermixed so frequently that for the "slow" ESR experiment they
will have lost all their identy, present themselves as equally
weighted linear combinations and therefore do not give rise to a
dipolar splitting any more. At higher exciton excitation densities
we therefore expect the dipolar lines to merge into a single line,
the width of which is finally determined by the collision rate.
(This quantity is equal to $\tau_{coll}^{-1} = \varrho \upsilon_1$, where ϱ is the excitation
density and υ_1 the longitudinal exciton transfer rate). All these
effects are observed in Rb^+TCNQ^- and TMB^+TCNQ^-, when the magnetic
field is in a symmetry plane, the interchain hopping effect on the
linewidth being eliminated in this way. Figure 16 gives a plot of
the logarithm of the widths of the dipolar split lines as a function
of $1/T$ in this orientation. The activation energies Ea and exponential factors υ_o derived are given in table 4.1. Since the activation
energy of ϱ (E)$_o$ can be obtained separately from intensity measurements, the activation energy for longitudinal hopping E_L^1 (which also
contains the temperature dependence of the collision cross section)
can be obtained by subtraction and compared with the energy E_L required for transverse hopping as determined earlier.

The reasonable agreement obtained shows that the longitudinal
hopping is governed by the same mechanism, and also diffusional,
which was a priori assumed in the analysis.

Of course, not every collision has to be effective. A realistic
expression for the linewidth is given by
$(\Delta\upsilon - \Delta\upsilon_o) = 2/3 \sqrt{3} \, (\upsilon_1 \varrho A)$, where $\Delta\upsilon_o$ is the low temperature
limiting linewidth and A the cross-section. Looking back at figure
13 we note that apart from the symmetry plane effect the Rb^+TCNQ^-

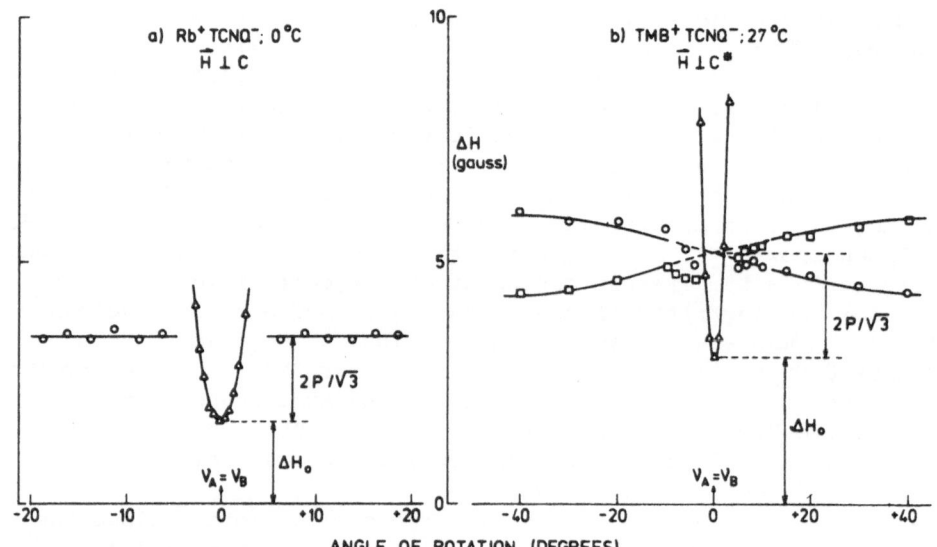

fig.14. Line narrowing at equivalent magnetic field position in
Rb$^+$TCNQ$^-$ and TMB$^+$TCNQ$^-$. Note the isotropy, resp. anisotropy
of the linewidths at inequivalent positions.

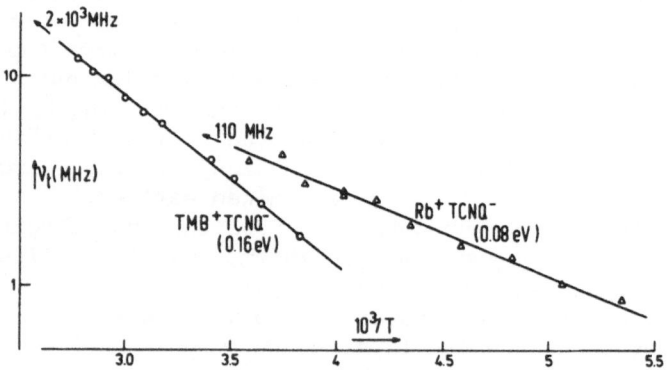

fig.15. Temperature dependence of the linenarrowing in Rb$^+$TCNQ$^-$
and TMB$^+$TCNQ$^-$.

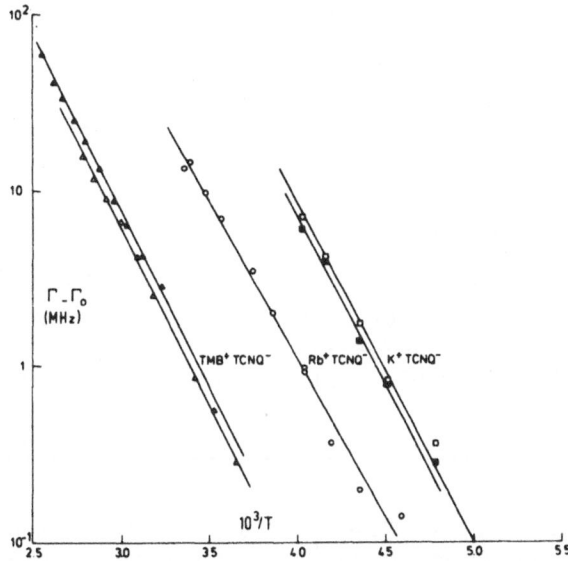

fig.16. Temperature dependence of the linewidths of the dipolar
        split lines.

Table 4.1. Activation energies and pre-exponential factors for the
           line broadening.

	v_o (sec^{-1})	Ea (eV)	E (eV)	E_L^1 (eV)	E_L (eV)
Rb$^+$TCNQ$^-$	2×10^{13}	0.36	0.26	0.10	0.08
TMB$^+$TCNQ$^-$	8×10^{12}	0.40	0.20	0.20	0.16

linewidth is independent of orientation. This means that ϱv_1 is so
high that no remnant of the hyperfine interaction is observed. This
yields a lower limit for v_1 of 10^{11} sec^{-1} and together with
v_1 A $\gtrsim 10^9$ sec^{-1} a value of A $\gtrsim 0.01$ which puts the collisions in the
weak coupling limit (14). On the other hand TMB$^+$TCNQ$^-$ does show an
orientationally dependent hyperfine linewidth, yielding an upper
value of v_1 of $v_1 \gtrsim 10^7$ sec^{-1} and with v_1 A $\gtrsim 2 \times 10^6$ sec^{-1} we find
that A must be close to its maximum value of 2/9, the strong
coupling limit (14). Considerations of the change of the dipolar
splitting as a function of temperature yield essentially the same
results. Even at the lowest temperatures, where triplets can be observed, the dipolar splittings are slightly temperature dependent.
This cannot be due to collisions, but probably derives from the lattice expansion and from the de-dimerization inherent in a Peierls-
unstable solid. Both these effects lead to a decrease in $<\frac{1}{r^3}>$

and therefore to a reduced dipolar splitting. Taking this effect into account removes inconsistencies observed in the past (6), a plot of log $(d_o^2 - d^2)$ versus $1/T$ at higher temperatures approaching the curve calculated from the line broadening (11).

We can now compare the transverse and the longitudinal hopping rates in the two compounds. They are collected in Table 4.2.

Table 4.2   Exciton motion

	transverse			longitudinal		
	υ_o (sec^{-1})	E_L (eV)	υ_1 (200°K)	υ_o (sec^{-1})	E_L (eV)	υ_1 (200°K)
Rb$^+$TCNQ$^-$	1.1×10^8	0.08	1.1×10^6	$>10^{14}$	0.10	$> 10^{11}$
TMB$^+$TCNQ$^-$	2×10^9	0.15	3.3×10^6	10^{12}	0.20	10^7

It is clear that the Rb$^+$TCNQ$^-$ at 200°K is one-dimensional at least in the ratio 10^5:1, while in TMB$^+$TCNQ$^-$ this ratio is at most 3:1. It should be pointed out that this conclusion depends in no way on the macroscopic properties of the crystals, as long as they are single.

5. HOW CAN U$_{eff}$ BE SO SMALL?

As pointed out earlier, the TCNQ$^-$ salts have been treated both in the high and in the low U limit, which should be interpreted as $U_{eff} \gg 4t$ as $U_{eff} \ll 4t$, where t is the transfer integral. Now, all authors appear to agree that t is some number between 0,1 and 0.2 eV and therefore the question is whether U is larger or smaller than 0.4 to 0.8 eV. Since some conductivity is always observed U$_{eff}$ cannot in faith be larger than 1 eV, and, whatever point of view one may take, it is worthwhile considering the question why U$_{eff}$ should be so small, since the "bare" U-value (I-A) is 3.60 or 4.60 eV as calculated by various MO-methods (12,13). Usually, U$_o$ is calculated by $U_o = I-A-P_+-P_-$, where I is the ionization energy, A the electron affinity, and P$_+$ and P$_-$ the lattice polarization energies of the positive and negative species created in the electron transfer. It is the purpose of this section to show that this calculation overlooks the important contribution due to the valence electrons themselves, which of course can also be effective in reducing U. This effect is related to the usual screening effects, but since it does not screen between charges, it is probably better called "solvation", in analogy to solvent molecules reducing the energy of ions in solvents. It is at this point worth recalling that NaCl would not dissolve in water if the solvation energies (in this case the "hydratation" energies) would not be so large. Similar to the water molecules surrounding the ions, the valence electrons will try to reduce the energy of a Hubbard-type electron excitation, by polarizing around the positive and negative charge created. The complication, however, is that this polarization energy itself depends on U. If U is infinitely large polarization of the valence electrons is impossible, but as U decreases, it rapidly increases. If we consider the symmetrical case of one Hubbard excitation, creating one separated electron and hole pair, then we can write very roughly in second order perturbation theory $U_{eff} = U_o - \frac{4Q^2}{U_o - 4t}$, for the energy of this excitation, where U_o is the "dressed" U as given above, Q is the electron-electron repulsion matrix element giving rise to the polarisation and $U_o - 4t$ is the energy required to move an electron from one site to the neighbouring one so as to give a contribution to the polarization. The factor 4 arises since the effect takes place at both sides of electron and hole in a linear chain.

It will be clear that at high values of U_o this effect can be neglected, but as U_o becomes anywhere near 4t (and in TCNQ-salts this means below 1 eV!) than this self-solvation should be dominant. Although perturbation theory then clearly becomes inadequate, the equation shows that on these considerations alone U_{eff} may become 4t or less, which is a result consistently found in the experimental work. Much more careful calculations are required to establish this point further, but very roughly it would seem that values of U_{eff} between 4t and say 8t are excluded, the self-solvation effect taking over in this range and reducing U_{eff} dramatically.

Acknowledgements

The work discussed in the above sections has largely been carried out by Drs.Vegter and Hibma, while working for their Ph.D.degree at the University of Groningen. The author gratefully acknowledges their contributions and the interesting discussions he has had with Dr.G.A.Sawatzky and Mr.P.I.Kuindersma.

References

1. F.H.Herbstein, Perspectives in Structural Chemistry, Vol.4, p.166; edited by J.D.Dunitz and J.A.Ibers (Wiley, 1971).
2. A.Girlando, R.Bozio and C.Pecile, Chem.Phys.Letters 25, 409 (1974).
3. J.G.Vegter and J.Kommandeur, Phys.Rev.Abstracts, 5, No 7 (1974).
4. S.Hiroma, H.Kuroda and H.Akamatu, Bull.Chem.Soc.Japan 44,9 (1971).
5. P.I.Kuindersma, private communication
6. P.L.Nordio, Z.G.Soos and H.M.McConnell, Ann.Rev.Phys.Chem.17,237 (1966)
7. P.A.Fedders and J.Kommandeur, J.Chem.Phys.52,2014 (1970).
8. D.Adler and H.Brooks, Phys.Rev.155, 826 (1967).
9. J.G.Vegter and J.Kommandeur,Phys.Rev.B 8, 2887 (1973).
10. Y.Suezaki, Phys.Letters 38A, 293 (1972).
11. T.Hibma, Thesis, University of Groningen, 1974.

12.H.Th.Jonkman and J.Kommandeur, Chem Phys.Lett. 15, 496 (1972).
13.H.Th.Jonkman, G.A.v.d.Velde and W.C.Nieuwpoort,Chem.Phys.Lett.
                                                     25, 62 (1974).
14.V.Yu.Zitserman, Mol.Phys.20, 1005 (1971).

THE ELECTRONIC PROPERTIES OF TTF-TCNQ[*]

A.J. Heeger and A.F. Garito

Department of Physics and Laboratory for Research
on the Structure of Matter, University of Pennsylvania,
Philadelphia, Pennsylvania 19174, U.S.A.

INTRODUCTION

The class of solids known as organic charge transfer salts has received considerable attention in recent years. This interest arises from the novelty of the systems as well as from the fact that the flat planar molecules involved lead to anisotropic structures and therefore to pseudo one-dimensional electronic properties. The existence of real physical systems[1-3] which have the properties of one-dimensional metals is particularly important because of the exciting possibilities associated with the one-dimensional electron gas.[4-10] The pseudo one-dimensionality of the electronic properties of tetrathiofulvalinium tetracyanoquinodimethan (TTF-TCNQ) has been established experimentally by a variety of measurements.[3,11-15] As a highly conducting organic metal which remains metallic to relatively low temperatures,[11,15,16] this system is at the focus of work directed toward understanding the nature of the metallic state in pseudo one-dimensional conductors. More generally, the existence of a class of materials made up of organic molecules as fundamental units allows the possibility of utilizing the flexibility of organic chemistry to design and synthesize molecules with particular characteristics with the intent of achieving a desired bulk solid state property.

The organic charge transfer salts are composed of a donor molecule, D, and an acceptor molecule, A. In contrast to conventional

[*] Research supported by the U.S. National Science Foundation through the Laboratory for Research on the Structure of Matter and Grant No. GH-39303, and by the Advanced Research Projects Agency through Grant No. DAHC 15-72C-0174.

molecular crystals (that is, crystals made up of neutral organic molecules bonded by van der Waals forces), these salts have unpaired electrons on the acceptor or donor or both as a result of the basic charge
transfer reaction[17]

$$D\ A\quad \rightarrow\quad D^+\ A^-\ .\qquad\qquad\qquad\qquad\qquad (1)$$

In the simplest cases, the energy required for charge transfer is
given by:

$$\Delta E_{ct}\quad =\quad (I_D - A_A)\ -\ (E_M + E_{ex} + E_{pol})\ ,\qquad\qquad (2)$$

where the neutral DA system is taken as the zero of energy. I_D is
the donor ionization potential and A_A is the electron affinity of
the acceptor.

   When $\Delta E_{ct} \ll 0$, the resulting system will be nearly fully ionic
with unpaired electrons outside a filled core. Generally in such
molecules, the affinity level corresponds to the lowest unfilled π-
level of the neutral acceptor molecule so that one is dealing primarily with the $2p_z$ wavefunctions of C and N. The directionality of
the p_z wavefunctions, together with the chain-like crystal structures, leads to the pseudo one-dimensional electronic properties of
these compounds.

   The chemical and structural features required for metallic behavior can be summarized by the following conditions:[17,18]

   --the existence of unpaired electrons within the molecular units
      (charge transfer);

   --a uniform crystal structure consisting of linear, parallel
      stacked columns of flat planar molecules with significant in-
      termolecular overlap leading to bandwidths such that, in the
      absence of electron-electron repulsions, the associated elec-
      tronic band structure would be metallic;

   --molecular units exhibiting intramolecular electron correlation
      to allow charge density to reside on diametrically opposed
      parts of the molecule;

   --highly polarizable molecular units having minimal size to dimi-
      nish costly electron-electron repulsive interactions.

The first condition is straightforward. Generally, the unpaired electron would reside in a single non-degenerate state corresponding to
the lowest available energy level of the π-electronic system of the
flat, planar neutral molecule. In the TCNQ salts, commonly TCNQ⁻
open-shell anions are stacked face-to-face in linear chains separated
by similar chains of cations.[19,20] In contrast to the direct charge
transfer from a donor cation to the TCNQ acceptor in the TCNQ salts,

charge transfer in Pt salts, for example, $K_2Pt(CN)_4Br_{0.3} \cdot 3H_2O$,
results from doping the crystals with a strong electron acceptor such
as Br.[21]

The band structure associated with such linear chain systems
should be metallic; however, slight non-uniformity between molecular
units commonly results in insulating behavior. For example, a weak
structural distortion along the stacked columns can lead to an energy
gap at the Fermi surface and a resulting semiconducting state. This
is the case for the 1:1 alkali TCNQ salts. X-ray studies of the
crystal structure show linear parallel stacked columns of dimerized
TCNQ anions separated by similar alkali ion chains.[19,20] As expected,
their electronic properties are semiconducting. Low temperature x-ray
studies of the Pt-chain salts show that the semiconducting state results from a distorted structure.[21] Thus, in general, a crystal
structure of uniformly spaced molecular units is a necessary condition for metallic behavior.

The effects of repulsive electron-electron interactions have been
at the focus of most of the experimental and theoretical work on the
TCNQ charge transfer salts. It is clear that in order to achieve metallic behavior, molecular design features are required which suppress
the repulsive interactions responsible for localizing the unpaired
electrons onto molecular sites, so that double occupancy fluctuations
characteristic of a metal (i.e., $2M^- \rightarrow M^= + M^0$) are allowed. This
can be achieved by preliminary design of intramolecular correlation
and high molecular polarizability.[17,18]

The TCNQ molecule is a good example of intramolecular correlation
effects. The energy required to transfer a single π electron in the
solid from one $TCNQ^-$ site to an adjacent $TCNQ^-$ site depends on the effective Coulomb repulsion between the two odd electrons on the resulting $TCNQ^=$ dianion. The Coulomb repulsion between two outer electrons
on a single C atom is about 10 eV. However, on a $TCNQ^-$ site, the
characteristic separation distance is greatly increased. In the event
of an ionic fluctuation with two excess electrons on a single TCNQ
molecule, there is a clear tendency for the two electrons to localize
at opposite ends of the molecule and correlate to stay apart in order
to reduce their mutual Coulomb repulsion, U_0. This correlated structure may be schematically described by a generalized Heitler-London-
type wavefunction.

$$\psi \quad = \quad (1/\sqrt{2})[\chi_i(1)\,\chi_j(2) + \chi_i(2)\,\chi_j(1)] \qquad (3)$$

where χ_i denotes the wavefunction in the region of one dicyano me-
thylene group (i) and χ_j denotes the wavefunction at the opposite
group (j). As a result, when electron (1) is on group (i), electron
(2) is on group (j), and vice versa. The characteristic distance now
between electrons is 5.5 - 7 Å , so that the estimated Coulomb re-

pulsion, U_O, is drastically reduced to 2.0 - 2.5 eV.[22] Allowing
delocalization into the ring raises this energy somewhat. From semiempirical (SCFMO) theory, the value of U_O has been estimated as 3.5 -
3.9 eV. As in any self-consistent-field method, however, SCFMO theory
probably underestimates the effect of electron correlation. Thus we
estimate U_O to be about 2.5 - 3.0 eV. The bare Coulomb repulsion,
U_1, between adjacent TCNQ$^-$ anions has been estimated as 3.0 eV also
using SCFMO charge densities for TCNQ$^-$.[22] Therefore, the difference
$U_O - U_1$ is estimated at 0.5 - 1 eV.

Further reduction can be accomplished by using polarizable molecular units. The source of the molecular polarizability arises from
virtual excitations of nearby π-π^* singlet excitons. We reduced the
model problem to that of interacting excitonic polarons, and detailed
calculations indicate the existence of an indirect attractive interaction via these virtual excitons.[23,24] One result is that the total
effective interaction can be written as

$$U_F = U - 2E_B \qquad\qquad\qquad (4)$$

where $E_B = \Gamma^2/\hbar\omega_O$ is the excitonic polaron binding energy arising
from the coupling Γ of the electronic system to the polarizability
expressed in terms of excitons at energy $\hbar\omega_O$. This indirect interaction may be very large, and under favorable circumstances can reduce the net interaction to the point where the metallic state results.
This conditions was experimentally realized in NMP-TCNQ where both the
NMP cation and TCNQ anion are highly polarizable.[2] Even greater polarizability was expected for TTF because of the C-S bond, and TTF-
TCNQ is indeed metallic with Coulomb interactions apparently reduced
to playing a minor role.[3,25]

THE TTF-TCNQ SYSTEM

As part of a continuing effort to stabilize the metallic phase,
a variety of donor cations have been considered. There are two basic
requirements:[17]

1) large cation polarizability, and
2) relatively small size.

The former was discussed above; the latter follows from the fact that
the interaction between the conducting electron and the induced polarization varies as r^{-4}. Based on these arguments, it would appear
that molecular systems like TTF would be prime candidates for stabili-
zing the metallic state, being both relatively small and possessing a
large molecular polarizability as a result of the polarizable C-S
bond. That the resultant TTF cations are also open-shell systems and
thus potentially metallic provides an additional attractive feature.

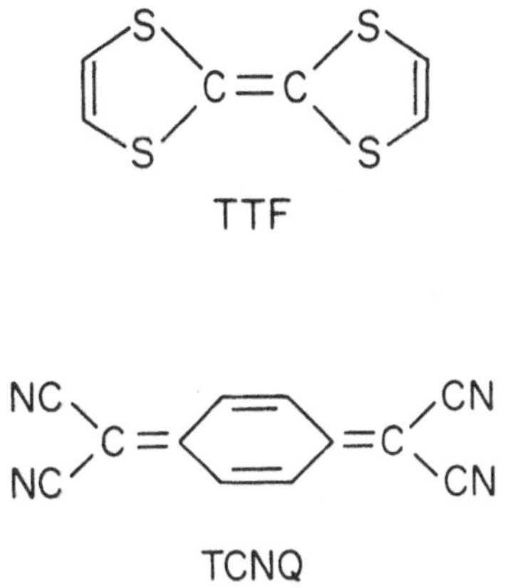

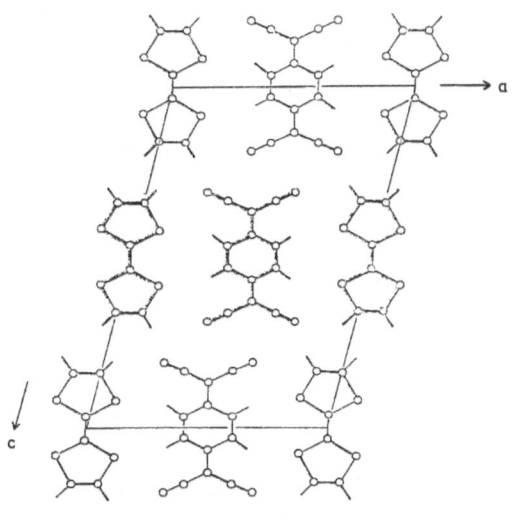

Fig. 1a. Molecular structures
of TTF and TCNQ.

Fig. 1b. Parallel chain crystal
structure of TTF-TCNQ (see Ref.
26).

The molecular structures of TTF and TCNQ are shown in Fig. 1a;
a view of the parallel chain crystal structure of the salt is given
in Fig. 1b.[26] There is ample evidence from analyses of x-ray measurements of molecular bond lengths and of photoemission results[27] suggesting that the charge transfer between TTF and TCNQ is incomplete
($\frac{1}{2}$ - 1) which would lead to the resultant bands not being half filled.

At room temperature, TTF-TCNQ is among the best-known organic
conductors with a conductivity of nearly 10^3 $\Omega^{-1}cm^{-1}$ along its principal conducting (crystallographic b) axis.[11,14-16] Measurements of
the conductivity in the transverse a and c^* directions are shown
in Fig. 2 [11] and indicate an anisotropy $\sigma^b/\sigma^a \sim 500$ to 10^3 at room
temperature, increasing to at least 10^4 near 60 K. These data provide a quantitative measure of the one-dimensional nature of the electronic system.

The thermoelectric power is similarly anisotropic as shown in
Fig. 3.[28] Along the conducting b-axis, the TEP is negative and
(above 140 K) linear with temperature, extrapolating to zero at T =
0 K. The negative values are consistent with incomplete charge transfer in a tight-binding band with width of order 0.5 eV. Below 140 K,
deviations from linearity are observed and the TEP crosses zero at the
metal-insulator transition. The transverse thermopower (a-axis) is
positive and non-linear over the entire temperature range.

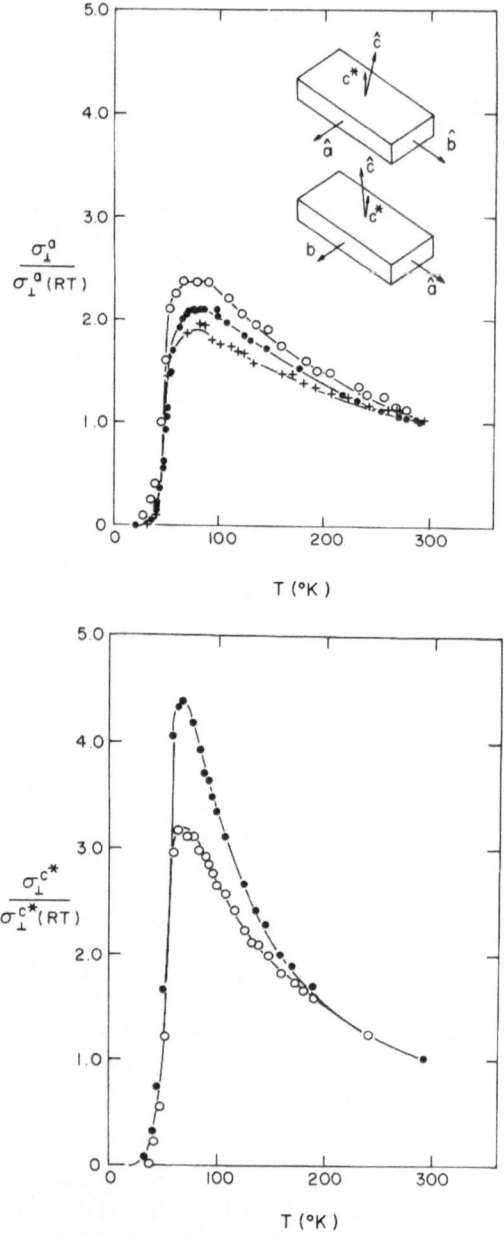

Fig. 2.   Temperature dependence of transverse conductivities nor-
malized to room temperature values: $\sigma_\perp^a \simeq 0.5 - 1\ \Omega^{-1}cm^{-1}$ and $\sigma_\perp^{c*}$
$\simeq 4 - 6\ \Omega^{-1}cm^{-1}$.   (a) $\sigma_\perp^a$ obtained by standard four-probe [•-•-•],
microwave (10 GHz) [o-o-o], and Montgomery [x-x-x] methods. The two
crystal habits of TTF-TCNQ are shown.   (b) $\sigma_\perp^{c*}$ obtained by four-
probe [o-o-o] and Montgomery [•-•-•] methods.

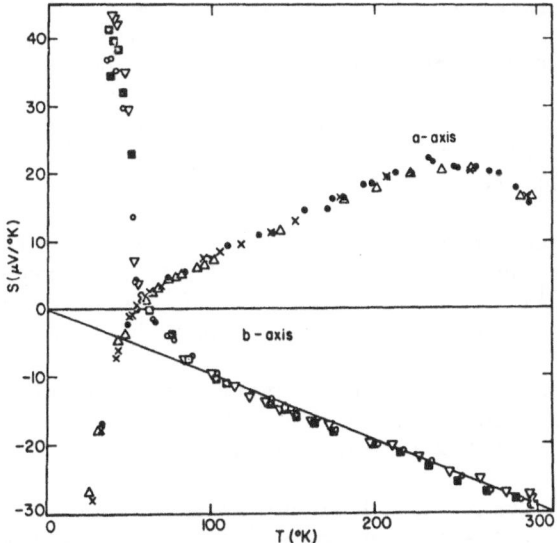

Fig. 3.   Absolute thermoelectric power along <u>a</u> and <u>b</u> axes for single x-tals of TTF-TCNQ (see Ref. 28).

Thus above 60 K, TTF-TCNQ has properties indicative of metallic transport along the <u>b</u>-axis, whereas analysis of the transverse data indicates diffusive transport perpendicular to the chain axes.[14]

The one-dimensional metallic properties have been confirmed by optical studies.[12] In the visible, the transmittance was accurately proportional to $\sin^2 \theta$ where θ is the polarization angle relative to the <u>b</u>-axis as shown in Fig. 4a. The measured anisotropy in the

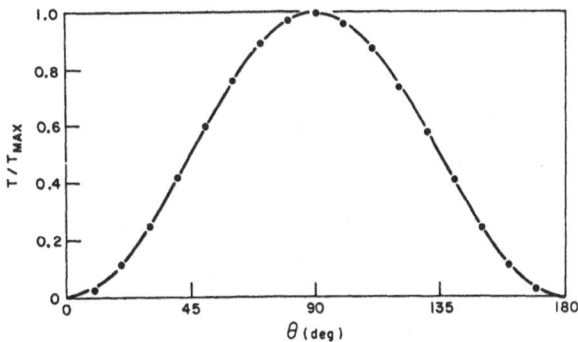

Fig. 4a.   Normalized transmittance of TTF-TCNQ single crystal as a function of polarization angle relative to conducting <u>b</u>-axis. Data points fall accurately on the $\sin^2 \theta$ curve and indicate an anisotropy of greater than 10^3.

transmittance is greater than about 1000. The polarized reflectivity[12],[13] in the near infrared is similarly anisotropic as shown in Fig. 4b. For electric vector polarized along the b-axis, a plasma edge is seen near 1.3 μm with R// rising to metallic-like reflectivity in the infrared. The transverse reflectivity, $R_\perp$, remains of order 10% throughout the infrared, characteristic of an insulating solid with a small dielectric constant.

The overall anisotropy is maintained in the low temperature insulating phase as evidenced by the remarkable anisotropy in the dielectric constant.[14] At 4.2 K, $(\epsilon_1^b/\epsilon_1^a) \geq 500$, indicating that nearly all the oscillator strength is along b .

The data therefore present an overall picture of TTF-TCNQ as a highly anisotropic conductor which may be approached from the idealized point of view of the one-dimensional metal.

In order to clarify the dominant interactions and obtain a measure of the overall scale of energies involved, a variety of experiments have been performed. Estimates of the bandwidth were obtained from studies of the plasma frequency,[12],[13] thermoelectric power,[28]

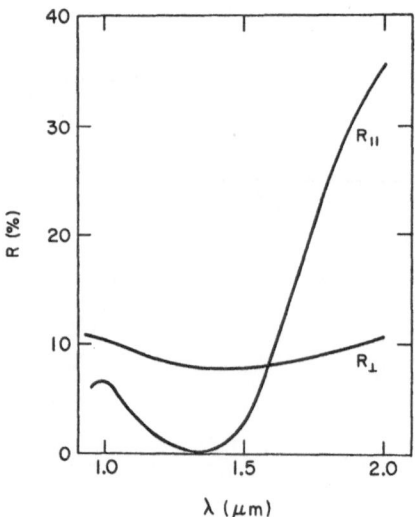

Fig. 4b.   Polarized single crystal reflectance spectra of TTF-TCNQ in the vicinity of the plasma edge.

and magnetic susceptibility.[25] In each case, a value of order 0.1
eV was found for the intrachain (b-axis) transfer integral indicating a bandwidth of order 0.5 eV. Information on the strength of
the electron-electron interaction is available from magnetic susceptibility[25] and nuclear relaxation rate[30] studies. The non-magnetic
ground state and non-enhanced susceptibility in the metallic state
demonstrated that Coulomb correlation effects play a minor role.
On the other hand, that the electron-phonon interaction is strong
may be inferred directly from the mean free path as given by[2]

$$\Lambda \quad = \quad v_F \tau \quad = \quad \sigma \pi \hbar / \, 2 \, N e^2 a \tag{5}$$

where N is the carrier density, a the lattice constant in the
conducting direction, and σ the measured conductivity. With a =
3.8 Å, $N = 4.7 \times 10^{21}$ cm^{-3}, and $\sigma = 10^3 \ \Omega^{-1}$cm^{-1}, Λ is estimated
to be of order one to two lattice constants. Thus the electrons are
very strongly coupled to the lattice or molecular modes, and the system is near the limit of applicability of band theory.

The strong electron-lattice coupling has been confirmed by infrared reflectivity studies of the scattering rate, τ_{ph}^{-1}. Using
Hopfield's expression[31] for the dimensionless coupling constant $\lambda = N(0) \, V_{ph}$

$$\lambda \quad = \quad (\hbar / 2\pi \tau_{ph}) \, (kT)^{-1} \tag{6}$$

(valid for $T > \theta_D$), a strong coupling value of $\lambda = 1.3$ was obtained.[12] Analysis of the temperature dependence of the scattering
rate indicates that $\tau(T)$ is consistent with single-phonon scattering, with a contribution from temperature-independent defect scattering which varies from sample to sample and which increases as the
number of defects increases.[12] The defect scattering was used to
obtain an estimate of the single particle residual resistivity. Comparison with the typical dc data shows that the measured dc resistivity regularly falls below the residual resistivity by an order
of magnitude, implying that the dc conductivity is carried in a
collective manner.

Strong coupling of the conduction electrons to the intramolecular lattice modes has also been suggested.[32] The transferred electron density on the TCNQ anion is concentrated on the terminal di-
cyanomethylene groups with maximum amplitude on the $C \equiv N$ as evidenced by direct NQR studies.[33] As a result, fluctuations in charge
density affect the various bond lengths and strongly couple to the
intramolecular vibrational modes. This coupling has been suggested[34]
as responsible for the structure in the infrared reflectivity near
6.5 μm.[12] The resonant scattering of the conduction electrons from
the $C \equiv N$ and/or $C = C$ optical modes can give a maximum in the scattering rate $[1/\tau(\omega)]$ and a corresponding minimum in reflectivity.

In summary, TTF-TCNQ is a narrow band, one-dimensional metal with relatively strong electron-phonon coupling. Such a system is expected to be unstable toward a soft-mode structural transition driven by the divergent response of the electron gas at $q = 2k_F$, the Peierls instability.[4,5] Given the experimental strong coupling of the electrons to lattice acoustic modes and intramolecular vibrational modes, the Peierls distortion with $q = 2k_F$ could result either in a crystalline distortion characterized by molecular motion (e.g., dimerization in the case of $2k_F = \pi/a$ or longer wavelength analogues for $2k_F < \pi/a$) or by small bond length distortions. The fact that despite considerable effort,[35] no distortion has been seen with x-ray scattering or diffraction suggests that the coupling to the intramolecular bond lengths may be the dominant feature.

Generally, one anticipates a decrease in conductivity due to increased scattering for a system approaching such a transition. Thus, the strong temperature dependence observed in the metallic state has been difficult to understand on the basis of single-particle scattering. The fact that the conductivity above the transition increases rapidly with decreasing temperature was interpreted as evidence for the presence of electron-electron pairing correlations mediated by the soft phonons near $2k_F$.[15] This interpretation led Bardeen[6] to point out that the correlations might indeed result from the Peierls effect; however, the resultant dynamically distorted state could give enhanced conductivity arising from the coupled electron-phonon wave mechanism first described by Fröhlich.[5] The Peierls distortion need not be static with fixed phase from chain to chain, but can be dynamic giving rise to giant density waves in the solid. Fröhlich[5] was the first to point out that coupling of the electrons to these giant density waves in one-dimensional systems can lead to superconducting effects without BCS pairing. In effect, the electrons "surf-ride" on the giant density waves in the underlying lattice, and electron scattering is inhibited by the strong coupling of the electron lattice system. Because of the absence of long-range order in one-dimensional systems and the existence of dissipative processes for the giant density wave mechanism, one can expect large enhanced conductivities but not true persistent currents.

It is useful to look at the Fröhlich state from an energy band point of view. Following Allender, Bray, and Bardeen,[36] we show in Fig. 5 an idealized $\epsilon(k)$ vs k for a one-dimensional tight-binding metal with zero current (5a) and with a finite current (5b). The fact that the energy gap is tied to the Fermi surface in the moving frame of reference suppresses single-particle scattering across $2k_F$ and gives rise to enhanced conductivity. In the actual system, as a result of the one-dimensional fluctuations which are present above the three-dimensional interchain ordering temperature, the energy gap becomes a pseudo-gap; but at least schematically the same ideas apply.

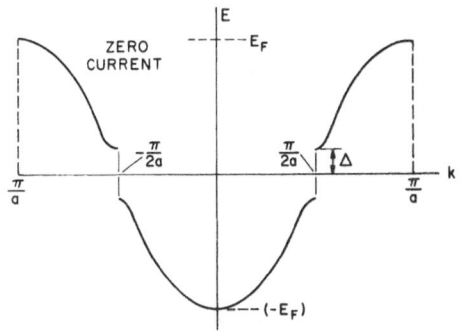

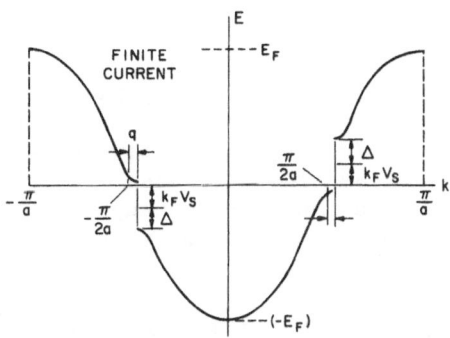

Fig. 5a. $\epsilon(k)$ vs k for the
Peierls-Fröhlich state with zero
current flow (see Ref. 36).

Fig. 5b. $\epsilon(k)$ vs k for the
Peierls-Fröhlich state with finite
current; the distribution has a
center of mass momentum q and
velocity v_S .

Figure 5 suggests the existence of two kinds of electronic excitations in such a Peierls-Fröhlich "superconductor."

1)   A collective mode of the coupled electron lattice system.
     Since the Peierls-Fröhlich state can be viewed as a coherent
     superposition of polarons, it is not surprising that the
     collective mode has a large effective mass, M^*,[5] dominated
     by the electron-phonon coupling. In addition, since the
     many-body collective behavior suppresses scattering, one
     anticipates an extremely long collective mode scattering
     time, τ_c .

2)   Single-particle excitations across the pseudo-gap. These
     are analogous to the simple excitations in a semiconductor
     and correspond to shake-off excitations in which an electron
     is excited leaving the underlying lattice fixed. The single-
     particle excitations should have the simple band mass, m^*,
     and a relatively short single-particle scattering time, τ_{sp} .

Lee, Rice, and Anderson[37] pointed out the extreme sensitivity of the
Fröhlich giant density wave mechanism to pinning. The physical point
is that charged impurities, defects, and intrinsic interchain coupling
all conspire to fix the phase of the collective mode and thereby pin
it and inhibit motion. In this case, the pinning effects provide a
restoring force to the collective mode motion and shift the collective mode oscillator strength from zero frequency to the pinning frequency, ω_T .

Lee, Rice, and Anderson[37] derived the low frequency dielectric
function resulting from the above described excitations,

$$\epsilon - 1 = \frac{2}{3}(\omega_p^2/\omega_G^2) + \Omega_p^2/(\omega_T^2 - \omega^2 - i\omega/\tau_c) \tag{7}$$

where ω_p is the single-particle plasma frequency ($\omega_p^2 = 4\pi Ne^2/m^*$ with m^* the band mass), $\hbar\omega_G$ the single-particle energy gap, $\Omega_p^2 = 4\pi Ne^2/M^*$ the collective mode plasma frequency, M^* the collective mode effective mass [$M^* = 1 + \lambda(\omega_G/\omega(2k_F))^2$ where $\omega(2k_F)$ is the bare phonon frequency], and ω_T is the pinning frequency. We have phenomenologically inserted the imaginary part proportional to the collective mode scattering time, τ_c.

Based on the above analysis, one can construct a schematic diagram of $\sigma_1(\omega)$ vs ω showing the characteristic features of the collective mode and single-particle excitations as shown in Fig. 6. In the conducting state (Fig. 6a) the collective mode is centered at zero frequency with a width determined by the collective mode lifetime, $\tau_c(T)$. The pseudo-gap leads to a broad minimum in $\sigma_1(\omega)$ with the single-particle (semiconductor interband) transitions showing up at higher frequencies. When the frequency exceeds the single-particle scattering rate, the usual Drude roll-off [$\sigma_1(\omega) = (Ne^2\tau/m^*)/\omega^2\tau_{sp}^2$] is expected. Pinning of the collective mode will shift the oscillator strength into the infrared as shown in Fig. 6b.

The fundamental signature of the Fröhlich state is therefore relatively high electrical conductivity, $\sigma_{Fröhlich} = Ne^2\tau_c/M^*$, in the presence of a pseudo-gap in the electronic excitation spectrum. The high conductivity is expected to be extremely sensitive to pinning by defects and impurities. At sufficiently low temperature, interchain Coulomb coupling of the charge density waves will cause a three-dimensional ordering transition at which point the phase of the charge density wave is pinned leading to a high dielectric constant insulating state. Thus, a complete study of $\sigma_1(\omega)$ and the corresponding

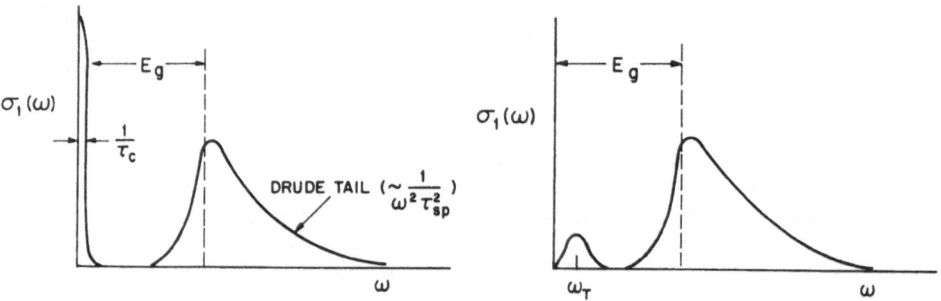

Fig. 6.   Schematic diagram of $\sigma_1(\omega)$ for the Peierls-Fröhlich state.   (a) Without pinning; the collective mode is centered at zero frequency.   (b) Pinned; the collective mode shifts to the infrared.

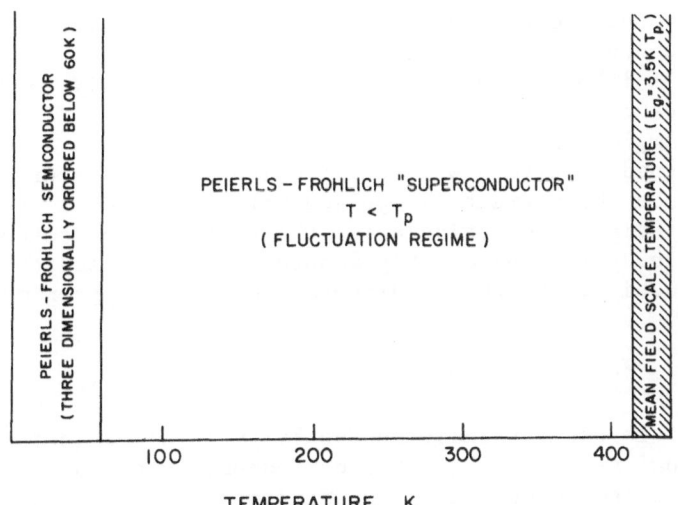

Fig. 7.   Phase diagram of TTF-TCNQ.

dielectric function as a function of temperature is required.

Figure 7 shows a schematic phase diagram of TTF-TCNQ based on the studies of $\sigma_1(\omega)$ described in the following section. The mean field scale temperature (determined from the pseudo-gap) is of order 500 K.[38,39] Above 60 K, $\sigma_1(\omega)$ for TTF-TCNQ has the general features expected of a Peierls-Fröhlich superconductor. The metal-insulator transition at 60 K (accompanied by a well-defined heat capacity anomaly[40] characteristic of a second-order phase transition) results from a three-dimensional phase transition below which the Fröhlich mode is pinned and shifted to the far infrared.

THE FREQUENCY DEPENDENT CONDUCTIVITY OF TTF-TCNQ

DC Conductivity

Standard four-probe electrical conductivity measurements along the crystallographic b axis on all samples showed that, with decreasing temperature, the conductivity rapidly increases to a maximum value which varies typically from crystal to crystal.[11,15] The temperature at which the maximum occurs, T_M, is in the range 58-60 K. Below the maximum, the conductivity suddenly decreases characteristic of a metal-insulator transition. The variation of the conductivity maximum from crystal to crystal we argued was due to the extreme sensitivity of one-dimensional metals such as TTF-TCNQ to crystalline defects, twinning, and impurities.

DC conductivity measurements on TTF-TCNQ have been further

examined using the techniques of Montgomery and co-workers[41] on highly anisotropic conductors. Montgomery's results can be applied to
the question of inhomogeneous currents wherein a nearly zero current
density can result at the surface where the voltage leads contact the
anisotropic sample. In this method, one purposely uses a lead configuration designed to give a well-defined inhomogeneous current distribution. When the current is applied with the leads aligned parallel to the principal conducting axis, the voltage drop between the
opposite two leads is considerably reduced by the anisotropy. The
apparent conductivity in this configuration can be approximated by[11]

$$\sigma_{/\!/}^{app} \; = \; \sigma_{/\!/} \; (G/A^{\frac{1}{2}}) \quad \exp\left[\pi(\ell_{\perp}/\ell_{/\!/})A^{\frac{1}{2}}\right] \tag{8}$$

where $\sigma_{/\!/}(\sigma_{\perp})$ is the intrinsic conductivity parallel (perpendicular)
to the principal axis, $\ell_{/\!/}(\ell_{\perp})$ the corresponding distance between
current and voltage leads, A the anisotropy $\sigma_{/\!/}/\sigma_{\perp}$, and G a
simple geometric factor. The apparent conductivity would then be determined by the magnitude and temperature dependence of the aniso-
tropy $\sigma_{/\!/}/\sigma_{\perp}$. Consequently, detailed measurements of the temperature dependence of the electrical anisotropy have been carried out.[11]

The b-a anisotropy $(\sigma_{/\!/}^{b}/\sigma_{\perp}^{a})$ is shown in Fig. 8 as a function
of temperature.[11] The magnitude of the anisotropy implies pseudo-

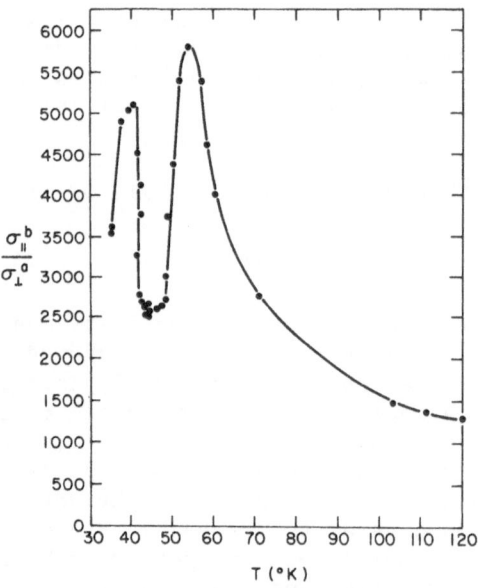

Fig. 8.   Anisotropy $\sigma_{/\!/}^{b}/\sigma_{\perp}^{a}$ using the Montgomery technique showing a double maximum; $\sigma_{/\!/}^{b}/\sigma_{\perp}^{a}(RT) \simeq 500$.

one-dimensional behavior as described above. The experimental observation of double maxima in the anisotropies has been used as a
convenient internal check on the validity of $\sigma_{//}^{b}(T)$ four-probe
measurements. We have generated false giant conductivity maxima
using a completely misaligned lead configuration similar to that
used by Schafer et al.,[42] which is nothing more than a slightly
modified Montgomery configuration. The fact that the anisotropy
dominates the measurement is clearly signified by the double maximum in the false conductivity. The sensitivity of the double anisotropy maxima as a self-consistent check on the validity of $\sigma_{//}^{b}$
data has been shown experimentally.[11] The data show that even a
slight enhancement due to inhomogeneous currents is accompanied by
the clear signature of the second peak in the data near 40 K.

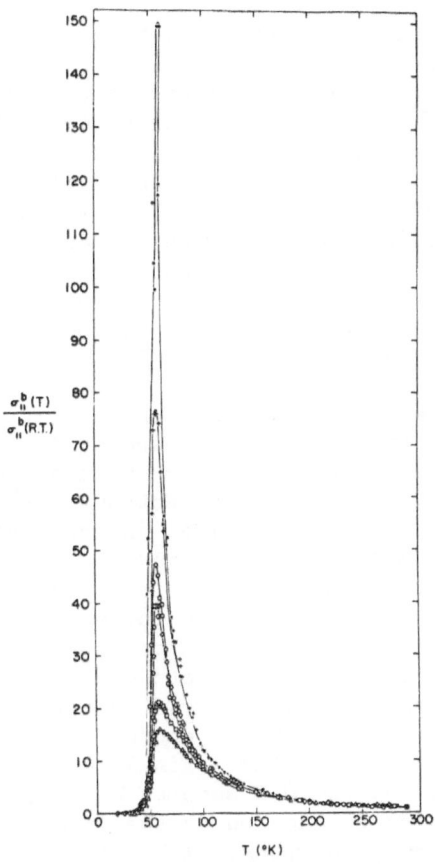

Fig. 9. Normalized four-probe data showing the b-axis conductivity
as a function of temperature. There is no trace of a second low
temperature maximum.

Normalized four-probe dc conductivity data $\sigma_{//}{}^b(T)/\sigma_{//}{}^b(300)$ are given in Fig. 9. We find these results intrinsic to TTF-TCNQ undistorted by anisotropy-related effects. Included are data which reach peak normalized values of 17, 21, 40, 48, 75, and 150. The normalized conductivities are nearly identical over the temperature range from 300 down to 140 K, where, for crystals which show larger values of $\sigma_{//}{}^b(T_M)/\sigma_{//}{}^b(300)$, the conductivity continues to increase with decreasing temperature while the lower values successively nest underneath. Each crystal cleanly undergoes the metal-insulator transition near 58 K, and $\sigma_{//}{}^b/\sigma_{//}{}^b(300)$ sharply decreases below 58 K. There is no trace of a second low-temperature maximum arising from anisotropy effects.

The absolute value of the intrinsic room-temperature conductivity is required, since it sets the scale for all the data. Measurements from a large number of crystals yield an average value of 550 $\Omega^{-1}cm^{-1}$, with values ranging to above 10^3. The measured crystals contain microcrystalline imperfections, from imperfect growth or sample handling and mounting, which leads to smaller effective cross-sectional areas than inferred from the actual dimensions. Thus, the intrinsic value of $\sigma_{//}{}^b(300)$ is greater than the unrestricted average. Direct evidence comes from the fact that for crystals with room temperature values differing by a factor of 5, the normalized temperature dependences (150 < T < 300 K) are identical. The maximum room-temperature value obtained with contactless microwave measurements is 1000 $\Omega^{-1}cm^{-1}$, where the true microwave value is necessarily higher since edge and corner effects enhance the internal field over that for a simple ellipsoid. We conclude that the intrinsic value of $\sigma_{//}{}^b(300)$ is approximately 10^3 $\Omega^{-1}cm^{-1}$.

The value of the maximum normalized conductivity, $\sigma_{//}{}^b(T_M)/\sigma_{//}{}^b(300)$ varies from crystal to crystal and is associated with the extreme sensitivity of one-dimensional metals to crystalline defects, twinning, and impurities.[11,15] The intrinsic anisotropy $\sigma_{//}{}^b/\sigma_\perp{}^a \gg 10^4$ near 58 K makes TTF-TCNQ extremely sensitive, and crystal perfection at the level of parts per million is required, for any defects will either remove a given chain altogether, force carriers to tunnel through the defect, or force transverse current flow between chains. We note, further, that the strain field around a defect can be expected to be effective in scattering both electrons and $2k_F$ phonons in a narrow band anisotropic solid. An extreme example of crystal imperfection limiting the maximum conductivity is presented in Fig. 10,[11] where $\sigma_{//}{}^b/\sigma_{//}{}^b(300)$ for a single crystal temperature cycled several times from 300 to 50 K becomes essentially equivalent to that of a polycrystalline compaction in the final cycle. This kind of degradation is commonly observed, with the highest conductivity specimens being particularly sensitive.

The problem of crystal perfection has not yet been solved, and intrinsic conductivity maxima in the range $\sigma_{//}{}^b(T_M)/\sigma_{//}{}^b(300)$ > 20-150

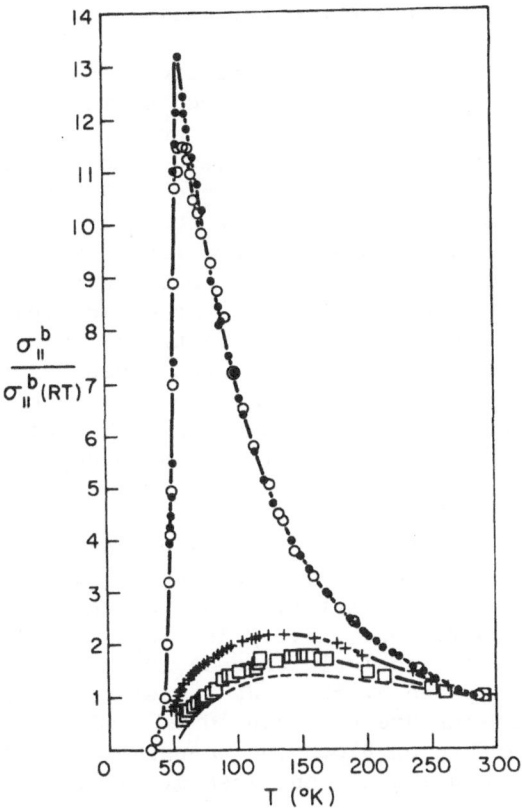

Fig. 10.   Cycling experiment showing degradation of the conductivity maximum: [•-•-•] first cycle (σ_{RT} = 685 $\Omega^{-1}cm^{-1}$); [o-o-o]
fourth cycle (σ_{RT} = 685 $\Omega^{-1}cm^{-1}$); [+-+-+] sixth cycle (σ_{RT} = 325
$\Omega^{-1}cm^{-1}$); [□-□-□] eighth cycle (σ_{RT} = 165 $\Omega^{-1}cm^{-1}$); [-----] polycrystalline data.

have been observed in five out of approximately 150 crystals measured.
Of this total, we estimate that only a small fraction (~10%) are of
suitably high quality as judged by close visual inspection. Maxima
with $\sigma_{//}{}^{b}(T_M)/\sigma_{//}{}^{b}(300)$ = 35 [43] and $\sigma_{//}{}^{b}(T_M)/\sigma_{//}{}^{b}(300)$ = 60 [44] have
been observed in other laboratories. Microscopic crystal perfection
remains as one of the most important unsolved materials problems in
this area of research.

Our conclusion is that dc conductivity at room temperature is
approximately 10^3 $\Omega^{-1}cm^{-1}$ and that there is strong evidence of peak
conductivities (near 60 K) exceeding 10^5 $\Omega^{-1}cm^{-1}$. Peak values exceeding 10^4 $\Omega^{-1}cm^{-1}$ have been observed in many laboratories. The
sensitivity of the dc conductivity in a one-dimensional system to
defects and impurities is not surprising, and should in fact be anticipated as a feature of the problem.

Microwave Conductivity

An experimental study of the microwave properties of pure crystals of the organic salt TTF-TCNQ has been carried out[14] using the cavity perturbation technique of Buravov and Shchegolev.[45] Included are electron spin resonance and a complete study of the dielectric function, $\epsilon_1 - i\epsilon_2$, from room temperature to 4.2 K as measured along the principal conducting b-axis and the transverse a-axis using the highest purity material synthesized thus far. The spin resonance line is asymmetric, characteristic of a metal with skin depth less than the sample dimension, when the rf magnetic field is perpendicular to the b-axis, and symmetric (Lorentzian) when the rf magnetic field is parallel to the b-axis. These results are understood in terms of the Dyson-Bloembergen theory of resonance lineshape as applied to the pseudo one-dimensional metal. The dielectric measurements (E//b) are consistent with the skin depth limiting behavior.

In the purest samples, cavity Q measurements show negligible loss over a relatively wide temperature range (50 < T < 100 K), indicative of very high conductivity. Based on these data, one can infer that the microwave conductivity is approximately 10^3 $\Omega^{-1}cm^{-1}$ at room temperature and regularly exceeds 10^4 $\Omega^{-1}cm^{-1}$ below 100 K, with higher values implied by the wide temperature range over which the conductivity remains greater than 10^4 $\Omega^{-1}cm^{-1}$. This is to be contrasted with the work of Bloch and co-workers, who reported modest conductivities and significant loss at all temperatures.[46]

We have recently extended the cavity perturbation technique of Buravov and Shchegolev[45] into the surface impedance regime.[47] This allows, for the first time, direct analysis of the cavity Q data in the high conductivity regime where the skin depth is less than the sample dimensions.

In the surface impedance regime, the loss in the cavity due to inserting the metallic sample is given by

$$L = A / \sigma\delta \tag{9}$$

where δ is the classical skin depth $\delta = c/\sqrt{2\pi\sigma\omega}$. The constant A is a geometrical factor determined by the shape and filling factor of the sample. For an ellipsoid of revolution (semi-axes a and b)

$$A = (3/2^9) \, \omega^2 E_o^2 \, (a^5/b) \, [\ln(2a/b) - 1] \tag{10}$$

where E_o is the unperturbed microwave electric field at the antinode of the TE_{101} cavity. Equations (9) and (10) have been checked quantitatively using lengths of copper alloy (Evanohm) of known conductivity with magnitude similar to that of TTF-TCNQ.

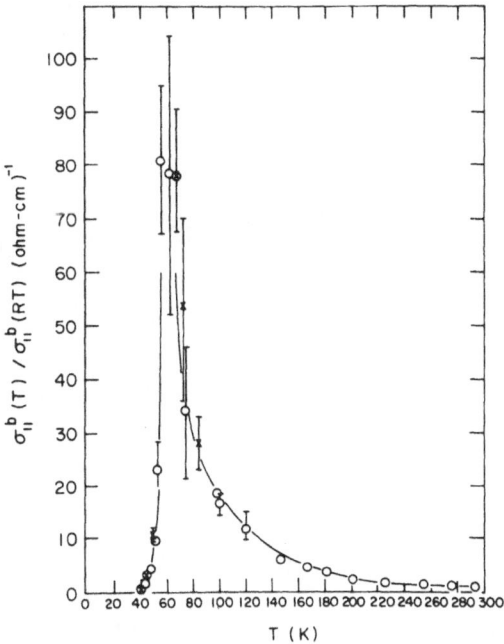

Fig. 11. Microwave conductivity of a single crystal as obtained from experiments in the surface impedance regime.

    Using the surface impedance technique, we have verified that the room temperature microwave conductivity is approximately 10^3 $\Omega^{-1}cm^{-1}$ and have observed conductivities with peak values much greater than 10^4 $\Omega^{-1}cm^{-1}$. An example is shown in Fig. 11 where the peak value exceeds 50 times the room temperature value. Normalized microwave (10^{10} Hz) conductivities in this range can be obtained routinely on pure samples with high quality surfaces. The data show many of the same features observed earlier in the four-probe dc conductivity studies. With decreasing temperature, the conductivity rapidly increases to a maximum value which varies from crystal to crystal. Below the maximum, the conductivity suddenly decreases due to the metal-insulator transition. The normalized microwave and dc conductivities are nearly identical over the temperature range 300 K down to 140 K, independent of the ultimate maximum value at 58 K. Additional microwave measurements confirm the same extreme sensitivity of one-dimensional metals to crystalline defects and imperfections, as was found in the earlier dc studies. For example, experimental studies have shown that chemical impurities (e.g., from contact with air for time periods on the scale of hours) cause a serious degradation of the microwave conductivity.

    We emphasize that the microwave measurement is a contactless technique in which the conductivity is determined from the I^2R

losses in the cavity. Thus, the observation of microwave peak conductivities exceeding $5 \times 10^4 \ \Omega^{-1}\text{cm}^{-1}$ for TTF-TCNQ is particularly significant.

Optical Conductivity: Visible and Near Infrared

The optical conductivity has been studied by utilizing several single crystal and thin film techniques.[12,13,48,49] The near infrared single crystal reflectivity data were presented earlier (Fig. 4b). The plasma edge is clearly evident in $R_\parallel$ with the Drude-like rise at longer wavelengths.

The near infrared (>4000 cm^{-1}) reflectance data were fit to the standard reflectivity expression,

$$R = \frac{1 + |\epsilon| - [2(|\epsilon| + \epsilon_1)]^{\frac{1}{2}}}{1 + |\epsilon| + [2(|\epsilon| + \epsilon_1)]^{\frac{1}{2}}} \qquad (11)$$

using the semiclassical Drude dielectric function for a metal.

$$\epsilon(\omega) = \epsilon_{core} - \omega_p^2/(\omega^2 + i\omega\tau^{-1}) \equiv \epsilon_1 + i\epsilon_2 , \qquad (12)$$

where

$$\omega_p^2 = 4\pi Ne^2/m^* , \qquad (13)$$

ϵ_{core} is the residual dielectric constant at high frequency arising from core polarizability, τ is the electronic relaxation time, ω_p is the plasma frequency, N is the electron density, and m^* is the optical effective mass. A non-linear least-squares computer routine was used with results shown in Fig. 12.[12] The best fit at room temperature was obtained with $\epsilon_{core} = 2.40$, $\omega_p = 1.79 \times 10^{15}$ sec^{-1} ($\hbar\omega_p = 1.2$ eV), and $\tau = 3.1 \times 10^{-15}$ sec. The value of ϵ_{core} is consistent with expected core polarizability contributions for molecular crystals.

Polarized single crystal reflectance spectra were measured at several discrete temperatures from 300 K to 4.2 K (see Fig. 12).[12] The results were analyzed in the same way as for the room temperature data to find the optimum values of parameters ϵ_{core}, ω_p and τ . The plasma frequency does not change throughout this temperature range. In particular, there is no change in ω_p below the transition temperature.[12,13] From this fact it can be concluded that the magnitude of the energy gap in the low temperature phase is much less than the photon energies in the optical experiments, i.e., $\omega_g \ll \omega_p$.

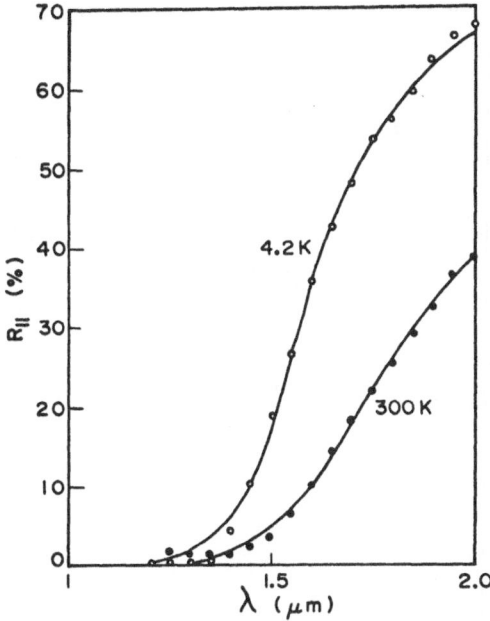

Fig. 12.   Single crystal reflectance of TTF-TCNQ for light polarized
parallel to the conducting axis at room temperature and at 4.2 K.
Solid lines are least squares fits of data points to eq. (12).

This is consistent with all other results regarding the energy gap
in TTF-TCNQ. In addition, the lack of change in ω_p rules out many
possible mechanisms for the metal-insulator transition. For example,
it has been suggested that either double charge transfer or back
transfer to the neutral molecules would lead to filled bands and an
insulating ground state.[50] Both of these models imply that the conduction electron density vanishes at low temperatures. The lack of
any temperature dependence to the plasma frequency demonstrates con-
clusively that no such change in the electron density occurs.

The values of the single-particle scattering time, τ_{sp} , are
temperature dependent and increase as temperature decreases. However, our results indicate that the measured value of τ for a particular sample invariably decreased after repeated temperature cycling, indicating an extreme sensitivity to strains and defects induced by differential thermal contractions relative to the copper
mounting plate. This defect sensitivity is expected for pseudo one-
dimensional systems with large anisotropy since individual chain
breaks would have a drastic effect, and is observed in other experiments as described above. Consequently, to obtain a meaningful temperature dependence required acquisition of a full set of data on
each temperature run. It may be noted in Fig. 12 that the reflec-

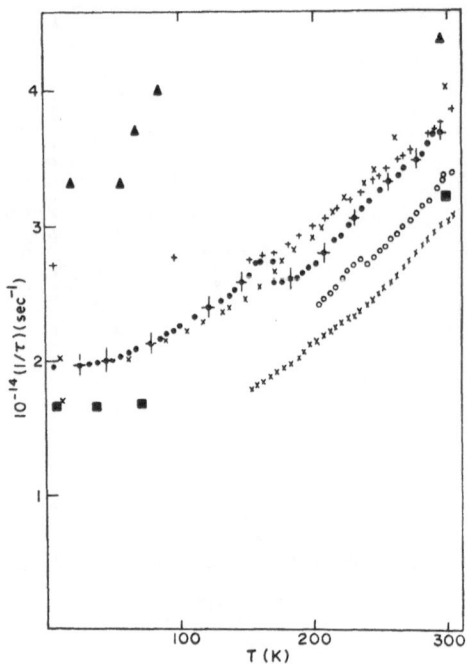

Fig. 13.   Reciprocal of scattering time as a function of temperature
for several samples. Breaks in some curves are caused by strain-in-
duced defects in the samples. ■■■ , least squares fits to single
crystal data; ▲▲▲ , data from Ref. 13; ✕✕✕ , two runs on a
given sample showing residual resistivity effect. Other symbols represent different samples (see Ref. 12).

tance at 2 μm is very sensitive to the value of τ_{sp} (recalling that
ω_p and ϵ_{core} do not vary appreciably). Thus, by measuring the reflectance at this wavelength as the temperature is varied, the temperature dependence of τ_{sp} may be monitored continuously.[12] The
results of this procedure for several samples are shown in Fig. 13,
where τ_{sp}^{-1} is presented as a function of temperature. Also shown
are the discrete temperature results for the first run on a single
sample. The agreement between the two measurement techniques is seen
to be good. It may be noted that at high temperatures (T $\gtrsim$ 100 K)
the relaxation time varies inversely with temperature, while at low
temperatures τ_{sp} saturates and approaches a constant value. The
linear temperature dependence of τ_{sp}^{-1} above 100 K was attributed
to scattering from thermal acoustic phonons and used to obtain a measure of the electron(acoustic phonon)-coupling constant, $\lambda_{ph} \simeq 1.3$.

When the phonon emission contribution is subtracted from the τ_{sp}^{-1} curve, the defect contribution is seen to be small compared to the phonon scattering contribution at high temperatures in the best samples, but does contribute significantly in poorer quality samples, especially after multiple temperature cycling. From Fig. 13 we see that $\tau_{def}^{-1} \gtrsim 1 \times 10^{14}$ sec^{-1}, implying an optical residual resistivity such that $\sigma_{def} \lesssim 3000 \ \Omega^{-1} cm^{-1}$. This value represents the maximum conductivity due to single-particle scattering processes. The peak value of the dc conductivity, however, is several times higher than this value, even for crystals showing typical behavior with $\sigma_{/\!/}^b (T_M) / \sigma_{/\!/}^b (300) \simeq$ 15-20. This result strongly implies that there is excess dc conductivity associated with a many-body collective effect in all TTF-TCNQ crystals, and that the low frequency transport is dominated by a collective mode.

Also shown on Fig. 13 are the scattering rates obtained from infrared reflectivity studies of samples prepared by the IBM (San Jose) group.[13] Of particular interest is the result that $\tau \simeq 2.3 \times 10^{-15}$ sec, roughly a factor of 1.5 shorter than the room temperature value obtained from our best samples. Comparing these data with those reported here, and noting the sensitivity to temperature cycling with the associated growth of a temperature-independent contribution, leads to the conclusion that the IBM samples are highly defected either by impurities or chain perturbations, such that the residual resistivity is given by

$$\rho_o \quad = \quad (\omega_p^2 \tau_o /4\pi)^{-1} \quad = \quad 1.5 \times 10^{-3} \ \Omega \ cm \quad . \tag{14}$$

Since this value is comparable to the typical room temperature values obtained at dc, one expects to observe essentially no temperature dependence in the normalized conductivity obtained from such samples. On the contrary, conductivity studies on such crystals demonstrate $\sigma_{/\!/}^b (T_M) / \sigma_{/\!/}^b (300) \simeq 16$, and higher values have been observed. The point is clear. It appears that the measured dc resistivity regularly falls below the measured residual resistivity (from single-particle scattering off defects and/or impurities) by more than an order of magnitude, implying that the dc conductivity is carried in a collective manner.

The different magnitudes and temperature dependences found for the optical and dc conductivity are to be expected in the Fröhlich state as indicated in the schematic diagram of Fig. 6. The crucial question is the existence of the pseudo-gap at lower frequencies.

Optical Conductivity: Infrared

To investigate the important frequency range between microwave and the near infrared, we have carried out studies of the optical properties of thin films and single crystals in the spectral range 20 cm^{-1} to 4000 cm^{-1}. [48,49]

Figure 14 shows the frequency dependence of $\sigma_1(\omega)$ as obtained from transmission and reflection measurements on thin films made by vacuum sublimation.[47] Since we are treating the polycrystalline films as an effective medium, the experiment measures the average value of the anisotropic conductivity, $\langle \sigma_1 \rangle = \frac{1}{3}(\sigma_1{}^b + \sigma_1{}^a + \sigma_1{}^{c*})$. At all frequencies below the plasma frequency where measurements have been made, the conductivity along the b-axis is more than two orders of magnitude larger than in the transverse directions. Thus, we have plotted $3\langle \sigma_1 \rangle$ to make comparison with values obtained from single crystal measurements. The solid bars indicate the conductivity

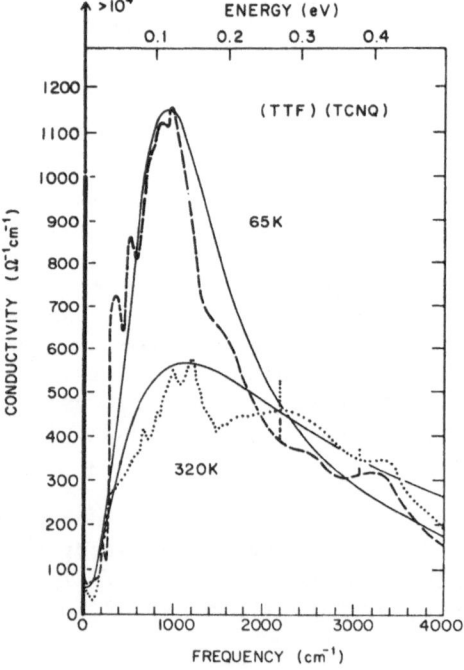

Fig. 14.   Conductivity versus frequency at 65 K (dashed line) and 320 K (dotted line). Heavy bars at zero frequency indicate the dc and microwave values, $\sim 10^3$ at 320 K and $> 10^4$ at 65 K.

obtained from dc and microwave measurements at the two temperatures.[11,14,15]

The general form of the conductivity is that of an electronic
system with an energy gap, $\omega_g \simeq 10^3$ cm^{-1} (0.14 eV). However, as
described above in detail, it has been demonstrated with both dc
and 10 GHz experiments that in the low frequency limit the intrinsic
principal axis conductivity, $\sigma_{//}{}^b$, is approximately 10^3 Ω^{-1}cm^{-1} at
300 K and by 65 K exceeds 10^4 Ω^{-1}cm^{-1}. Thus, the frequency-dependent conductivity shown in Fig. 14 indicates the existence of quasi-
particle excitations above an energy gap with a collective mode centered at zero frequency. $\sigma_1(\omega)$ is thus qualitatively similar to
that obtained from infrared studies of superconductors, and is in
general agreement with the schematic diagram shown in Fig. 6 for the
Fröhlich superconductor.

Because of the large energy gap, any description of $\sigma_1(\omega)$ must
include not only the direct transitions across the gap (proportional
to the product of the initial and final densities of states), but also
the indirect transitions involving multiple phonon processes which
give rise to the familiar Drude absorption in metals. We approximate
the frequency-dependent conductivity with a Drude-Lorentz model appropriate to a one-dimensional system with a large density of states
peak at the band edge,

$$\sigma(\omega) \quad = \quad \frac{\omega}{4\pi} \left\{ \omega_p{}^2 \tau \; / \; [\omega - i(\omega_g{}^2 - \omega^2)\tau] - i(\epsilon_{core} - 1) \right\} + \sigma_M \quad ,$$

$$(15)$$

where σ_M represents the small residual conductivity in the pseudo-
gap region. The solid lines on Fig. 14 represent fits of eq. (15)
to the data with ω_p = 9560 cm^{-1} and ϵ_{core} = 2.4 as obtained from
single crystal reflectivity measurements near the plasma edge. On
cooling, ω_g decreases from 1150 cm^{-1} (320 K) to 950 cm^{-1} (65 K),
and the single-particle scattering time (τ_{sp}) at high frequencies
increases from 1.8×10^{-15} sec (320 K) to 3.8×10^{-15} sec (65 K).[48]
The temperature dependence of τ, the magnitude of ω_p, and the
Drude form of $\sigma_1(\omega)$ are all consistent with the single crystal reflectivity measurements described above.[12]

Kramers-Kronig analysis of single crystal data leads to similar
results as shown in Fig. 15. The general form of $\sigma_1(\omega)$ is in good
agreement with the film data. The molecular features are sharper
and the single-particle scattering time (τ_{sp}) is longer, as expected.

The dominant molecular feature is the strong minimum in $\sigma_1(\omega)$
near 1600 cm^{-1} which has been attributed by Gutfreund et al. to
resonant scattering of the electrons by $2k_F$ optical phonons arising

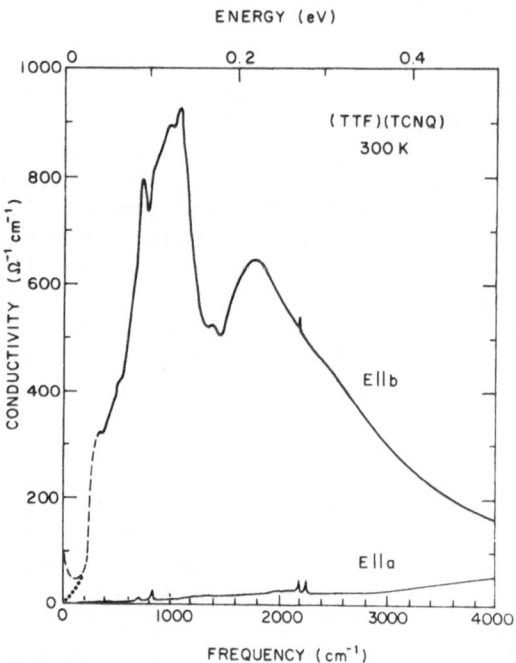

Fig. 15.   Frequency-dependent conductivity as obtained from Kramers-Kronig transform of single crystal data.

from the $C \equiv N$ molecular vibrations.[34] C^{13} labeling experiments are underway to verify this assignment.

The far infrared results at 65 K and 4.2 K are shown in Fig. 16.[48] The overall magnitude of $\sigma_1(\omega)$ at 4.2 K is reduced in the spectral range below 200 cm^{-1}, but the principal feature in the low temperature data is the broad maximum centered at 80 cm^{-1}. In addition, several sharp temperature-independent molecular lines are clearly seen as peaks in $\sigma_1(\omega)$.

According to Lee, Rice, and Anderson,[37] the giant density wave collective mode will be pinned below the three-dimensional ordering temperature leading to a metal-insulator transition, and the collective mode contribution to σ_1 will shoft away from zero frequency to a characteristic pinning frequency, ω_T .   Transport[11-16] and thermodynamic[25,40] studies do show a metal-insulator transition near 58 K below which the conductivity falls dramatically. We identify the broad maximum centered at 80 cm^{-1} in the low temperature data (Fig. 14) as the pinned collective mode. The total oscillator strength in this maximum corresponds to an effective mass of about 300 times

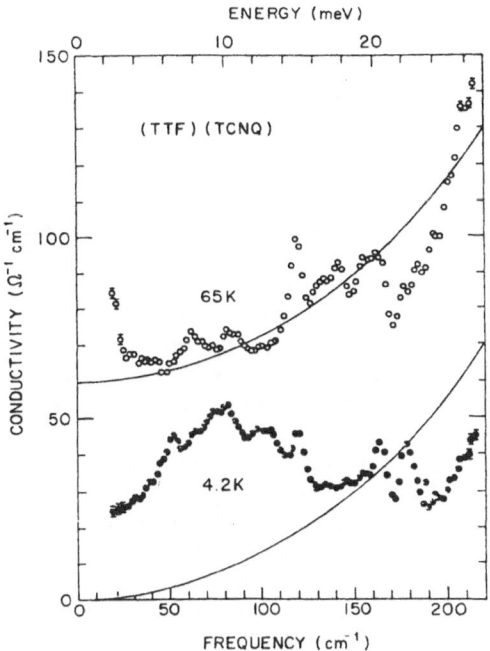

Fig. 16.   Conductivity versus frequency.   Open circles are at 65 K
and closed circles are at 4.2 K.   Solid lines are from eq. (15) with
$\sigma_M = 0$ at 4.2 K and $\tau = 7 \times 10^{-15}$ sec.

the band mass, and agrees in order of magnitude with our estimate of
the collective mode contribution at 65 K based on the magnitudes of
$\sigma_1(0)$, $\sigma_1(10\,GHz)$, and the far infrared data.   The low temperature
microwave results showing the conductivity exceeding the dc values
by three orders of magnitude are understood as resulting from the
tail of the 80 cm^{-1} broad maximum.   Thus, the experiments yield direct evidence for the pinning of the collective mode at low temperatures.

Additional evidence of the pinned Fröhlich mode is obtained from
microwave studies of TTF-TCNQ at low temperatures.   Values of the
conductivity and dielectric constant along different crystallographic
axes were determined using the cavity perturbation method of Buravov
and Shchegolev on crystals having normal or transverse crystallographic habits.   The low temperature (T < 58 K) results show the microwave conductivity along the principal conducting b-axis exceeds the
dc values by three orders of magnitude and the real part of the dielectric constant is unusually large $[\epsilon_1{}^b(4.2K) = (3.2 \pm 0.7) \times 10^3]$.
In contrast, the dielectric constant in the transverse a-direction
was found to be $\epsilon_1{}^a(4.2K) = 6 \pm 2$.   The large value for $\epsilon_1{}^b$ and

the dielectric anisotropy have recently been confirmed with dielectric resonator techniques.[51]

The unusually large b-axis dielectric constant, $\epsilon_1^b(4.2K) \simeq 4 \times 10^3$, is consistent with the existence of the pinned mode in the infrared. The low temperature dielectric constant (ϵ_1^b) was found to increase in magnitude with increasing sample purity,[14] a result also noted recently for the electronically one-dimensional salt $K_2Pt(CN)_4Br_{0.3} \cdot 3H_2O$.[52] The effect of trace amounts of non-specific impurities in TTF-TCNQ is to decrease the low temperature value of ϵ_1^b from approximately 3×10^3 to $(0.5-1) \times 10^3$ while simultaneously diminishing the excess microwave conductivity characteristic of known pure samples. Both of these effects are consistent with increased pinning at low temperatures due to impurities.

As shown in eq. (7), the low frequency dielectric function of the pinned Fröhlich state contains two contributions, a single-particle contribution (proportional to ω_p^2/ω_G^2) and a collective mode contribution. The infrared optical studies indicate that more than 99% of the single-particle oscillator strength lies at frequencies above 150 cm^{-1}, contributing less than 100 to $\epsilon_1^b(4.2K)$. Thus, the major contribution to $\epsilon_1^b(4.2K)$ is associated with the pinned mode. Estimates of ϵ_1^b from the oscillator strength based on the far infrared and microwave conductivities yield a pinned mode contribution of $(0.5-1) \times 10^3$. The width and central frequency of the pinned mode are expected to be sensitive to defect and surface pinning mechanisms which would be largest in thin films. Direct optical absorption measurements on single crystals of TTF-TCNQ in the far infrared range are currently being carried out to study in further detail the frequency dependence of the pinned mode and the contributions to $\epsilon_1^b(\omega)$.

The studies of $\sigma_1(\omega)$ (dc, microwave, infrared) thus provide strong evidence of a collective mode below a pseudo-gap with single-particle excitations at higher frequencies.

Our present understanding of the data is based on one-dimensional fluctuations of a Peierls-Fröhlich dynamically distorted state at temperatures well below the Peierls transition. Using the conventional mean field result, $\hbar\omega_g = 3.5 k_B T_p$, with $\omega_g = 1050 \pm 100$ cm^{-1}, we obtain $T_p = 430$ K, in reasonable agreement with previous estimates of T_p from other measurements.[11,25] One-dimensional fluctuations at temperatures below T_p smear the density of states such that the peak at the gap edge does not renormalize monotonically.[53] The metal-insulator transition at 60 K results from three-dimensional ordering with an associated pinning of the Fröhlich mode. The schematic phase diagram is given above in Fig. 7.

The pseudo-gap in $\sigma_1(\omega)$ is well defined even at room temperature where fluctuation effects would suggest a relatively high density

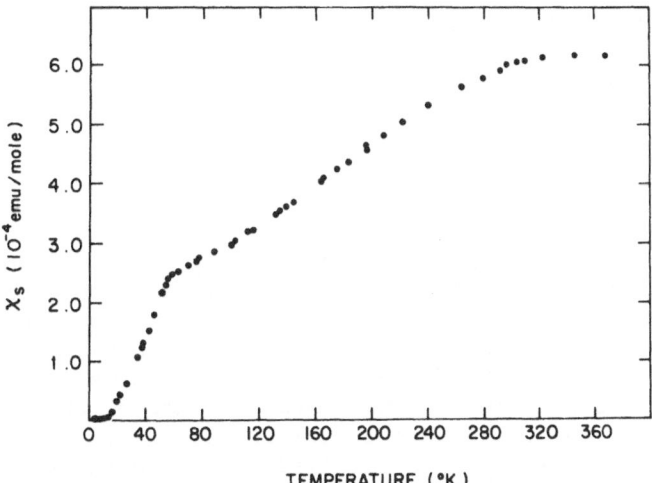

Fig. 17.   Temperature dependence of the spin susceptibility of TTF-
TCNQ.

of states in the gap region. That there are states in the gap can
be inferred from the magnetic susceptibility as shown in Fig. 17.[25]
The susceptibility shows the qualitative features predicted by Lee,
Rice, and Anderson,[53] and would imply that at room temperature the
density of states near E_F is within 20% of the unperturbed value.
In contrast, $\sigma_1(\omega)$ in the gap region below 200 cm^{-1} is more than
an order of magnitude below the gapless Drude curve. We conclude
that the pseudo-gap in $\sigma_1(\omega)$ is in essence a mobility gap caused
by the strong dynamic fluctuations in local potential due to the
inherent fluctuations in the one-dimensional Peierls-Fröhlich state
below the mean field temperature. Below the three-dimensional ordering temperature (60 K), the ordered state would be expected to
lead to a true semiconducting gap, and indeed the residual background
conductivity below 200 cm^{-1} decreases significantly.

The collective mode lifetime, τ_c , has not been measured directly. However, an estimate can be obtained from the measured dc conductivity and the effective mass, M^*,

$$\sigma_{collective} \quad = \quad Ne^2\tau_c \ / \ M^* \quad . \tag{17}$$

Taking $M^* \simeq 300 \ m^*$, as inferred from the low temperature pinned
mode oscillator strength, and assuming that near 60 K the collective
mode dominates the dc conductivity ($\sigma_{dc} \sim 10^4$-$10^5 \ \Omega^{-1}$cm^{-1}) leads to
a value of $\tau_c^{-1} \lesssim 10^{11}$ sec. Thus, the collective mode lifetime is

nearly four orders of magnitude greater than the single-particle
scattering time. This comparison represents one of the clearest
indications of the coherence in the "metallic" state of TTF-TCNQ.

CONCLUSION

We have attempted to summarize the experimental data obtained
from TTF-TCNQ in the context of the developing understanding of the
one-dimensional Peierls-Fröhlich state. Many features remain to be
studied in detail. The existence of two separate and potentially
conducting chains (TTF and TCNQ) is well known. However, the detailed role played by each of these individual chains in the various
electronic properties is not known. Measurements of the local susceptibility using NMR techniques on selectively deuterated samples
should contribute toward a more detailed understanding.[54] Similarly,
the labeling by g-values[55] and detailed intramolecular vibrations
can be used.

Although an estimate of the magnitude of the charge transfer is
available as described above,[27] a more precise value is important
given the expected sensitivity of the Fröhlich conductivity to com-
mensurability pinning.[56] However, surely the most important gap in
our present experimental knowledge is in the detailed structural aspects of TTF-TCNQ. The crystal structure is known. However, the
subtle distortions anticipated for the Peierls-Fröhlich state have
not yet been observed. It has been suggested by Morrel Cohen and
co-workers[57] that an a-axis ferroelectric distortion would cause an
energy gap at the Fermi surface of the three-dimensional band structure and thus cause the 60 K metal-insulator transition. Such a distortion, internal to the unit cell, would cause no superlattice lines
and would therefore be difficult to detect with diffraction techniques. However, the small a-axis low temperature dielectric constant,
$\epsilon_1^a(4.2K) = 6 \pm 2$, [14,51] appears to be inconsistent with such a model. In addition, infrared studies on films and single crystals show
no evidence of strong optical phonons polarized in the a-direction[48],
[49] The a-axis dielectric constant is simply due to molecular core
polarizability and several characteristic intramolecular modes of the
TTF and TCNQ molecular units. A more plausible explanation is that
the density wave primarily involves intramolecular bond lengths as
discussed briefly above. However, resolution of this important question must await additional experimental studies.

We have emphasized the many-body aspects of this strongly coupled electron-phonon system, but the band structure and one-electron
properties are also of fundamental importance.[58,59] The diffusive
nature of the transverse transport properties[14] implies that one-
dimensional models are a good first approximation. However, the
clarification of the role of three-dimensional coupling and the detailed three-dimensional band structure requires further study.[28]

The controversy surrounding the study of TTF-TCNQ and related systems centered on the dc electrical conductivity. However, it is now generally agreed that the intrinsic conductivity exceeds 10^4 $\Omega^{-1}cm^{-1}$ and there is strong evidence that the peak value exceeds 10^5 $\Omega^{-1}cm^{-1}$ both at dc and microwave frequencies. The sensitivity of the one-dimensional conductor to impurities and defects has been established, so that one can only assume that these values are lower bounds. However, it is most important to view the dc and microwave conductivities in the context of our overall experimental knowledge of $\sigma_1(\omega)$ as presented in the previous section. <u>TTF-TCNQ is not a simple metal; it has the optical spectrum of a semiconductor, but conducts at zero frequency due to a collective mode</u>.

Continued research on this area promises to be exciting and will inevitably involve even more detailed studies of presently known systems as well as the development of fascinating new materials. Studies have already begun on the first metallic polymer, $(SN)_x$[60,61] The collective many-body enhancement of the conductivity discovered in TTF-TCNQ may be anticipated in a new class of solid state materials, one-dimensional metals.

ACKNOWLEDGMENT

The work of many people at the University of Pennsylvania is summarized in this brief review. We particularly wish to acknowledge and to thank Dr. A.A. Bright, Marshall J. Cohen, Prof. Michael Cohen, L.B. Coleman, W.J. Gunning, C.S. Jacobsen, Dr. S.K. Khanna, Dr. E.F. Rybaczewski, J.C. Scott, L.S. Smith, Dr. D.B. Tanner, T.-S. Wei, and Dr. F.G. Yamagishi.

REFERENCES

1.   I.F. Shchegolev, Physica Status Solidi <u>12</u>(a), 9 (1972).

2.   A.J. Epstein, S. Etemad, A.F. Garito, and A.J. Heeger, Phys. Rev. B <u>5</u>, 952 (1972).

3.   A.F. Garito and A.J. Heeger, Nobel Symp. <u>24</u>, 129 (1973).

4.   R.E. Peierls, <u>Quantum Theory of Solids</u> (Clarendon Press, Oxford, 1955), p. 108.

5.   H. Fröhlich, Proc. Roy. Soc. <u>A223</u>, 296 (1954).

6.   J. Bardeen, Solid State Commun. <u>13</u>, 357 (1973).

7.   Yu. A. Bychkov, L.P. Gorkov, and I.E. Dzyaloshinskii, Zh. Eksp. Teor. Fiz. <u>50</u>, 738 (1966) [Sov. Phys. JETP <u>23</u>, 489 (1966)].

8.   A.M. Afanas'ev and Yu. Kagan, Zh. Eksp. Teor. Fiz. 43, 1456 (1963) [Sov. Phys. JETP 16, 1030 (1963)].

9.   M. Weger, Rev. Mod. Phys. 36, 175 (1964).

10. W. Little, Phys. Rev. 134, A1416 (1964).

11. Marshall J. Cohen, L.B. Coleman, A.F. Garito, and A.J. Heeger, Phys. Rev. B 10, 1298 (1974).

12. A.A. Bright, A.F. Garito, and A.J. Heeger, Solid State Commun. 13, 943 (1973); ibid., Phys. Rev. B 10, 1328 (1974).

13. P.M. Grant, R.L. Greene, G.C. Wrighton, and G. Castro, Phys. Rev. Lett. 31, 1311 (1973).

14. S.K. Khanna, E. Ehrenfreund, A.F. Garito, and A.J. Heeger, Phys. Rev. B 10, 2205 (1974).

15. L.B. Coleman, M.J. Cohen, D.J. Sandman, F.G. Yamagishi, A.F. Garito, and A.J. Heeger, Solid State Commun. 12, 1125 (1973).

16. J. Ferraris, D.O. Cowan, V. Walatka Jr., and J.H. Perlstein, J. Am. Chem. Soc. 95, 948 (1973).

17. A.F. Garito and A.J. Heeger, Accts. Chem. Res. 7, 232 (1974).

18. A.J. Heeger and A.F. Garito, AIP Conf. Proc. 10, 1476 (1973).

19. F.H. Herbstein in Perspectives in Structural Chemistry, ed. J.D. Dunitz and J.A. Ibers (Wiley, New York, 1971), Vol. IV, pp. 166-395.

20. R.P. Shibaeva and L.O. Atovmyan, Zhurnal Strukturnoi Khimii 13, 546 (1972).

21. H.R. Zeller, Advances in Solid State Physics (Pergamon Press, New York, 1973).

22. J.G. Vegter, J. Kommandeur, and P.A. Fedders, Phys. Rev. B 7, 2929 (1973); H.T. Jonkman and J. Kommandeur, Chem. Phys. Lett. 15, 496 (1972); J.G. Vegter and J. Kommandeur, AIP Conf. Proc. 10, 1525 (1973).

23. P.M. Chaikin, A.F. Garito, and A.J. Heeger, Phys. Rev. B 5, 4966 (1972).

24. P.M. Chaikin, A.F. Garito, and A.J. Heeger, J. Chem. Phys. 58, 2336 (1973).

25. J.C. Scott, A.F. Garito, and A.J. Heeger, Phys. Rev. B (October, 1974).

26. T.E. Phillips, T.J. Kistenmacher, J.P. Ferraris, and D.O. Cowan, Chem. Comm. 14, 471 (1973).

27. P. Nielsen, A.J. Epstein, and D.J. Sandman, Solid State Commun. 15, 53 (1974).

28. J.F. Kwak, P.M. Chaikin, A.A. Russel, A.F. Garito, and A.J. Heeger, Solid State Commun. (in press).

29. P.M. Chaikin, J.F. Kwak, T.E. Jones, A.F. Garito, and A.J. Heeger, Phys. Rev. Lett. 31, 601 (1973).

30. E.F. Rybaczewski, A.F. Garito, and A.J. Heeger, Bull. Am. Phys. Soc. 18, 450 (1973); also, E.F. Rybaczewski, Ph.D. Thesis, University of Pennsylvania, 1974.

31. J.J. Hopfield, Comments on Solid State Physics 3, 48 (1970).

32. H. Gutfreund, B. Horowitz, and M. Weger, J. Phys. C 7, 383 (1974).

33. J. Murgich and S. Pissanetzky, Chem. Phys. Lett. 18, 420 (1973).

34. H. Gutfreund, B. Horowitz, and M. Weger, Solid State Commun. 15, 849 (1974).

35. Several groups have carried out studies of the crystal structure at and below room temperature. As of this writing, no extra lines (indicative of a change in unit cell) have been observed. This work is continuing. We note, however, that one must be careful not to draw permature conclusions since the low z atoms involved are weak scatterers and, especially if $2k_F$ is incommensurate with the lattice, the corresponding superlattice lines might be extremely weak.

36. D. Allender, J.W. Bray, and J. Bardeen, Phys. Rev. B 9, 119 (1974).

37. P.A. Lee, T.M. Rice, and P.W. Anderson, Solid State Commun. 14, 703 (1974).

38. C.G. Kuper, Proc. Roy. Soc. A227, 214 (1955).

39. M.J. Rice and S. Strässler, Solid State Commun. 13, 125 (1973).

40. R.A. Craven, M.B. Salomon, G. DePasquali, R.M. Herman, G. Stucky, and A. Schultz, Phys. Rev. Lett. 32, 769 (1974).

41. H.C. Montgomery, J. Appl. Phys. 42, 2971 (1971).

42. D.E. Schafer, F. Wudl, G.A. Thomas, J.P. Ferraris, and D.O.
     Cowan, Solid State Commun. 14, 347 (1974).

43. R.D. Groff, A. Suna, and R.E. Merrifield, Phys. Rev. Lett. 33,
     418 (1974).

44. D. Jerome, W. Müller, M. Weger, and B.A. Scott, J. de Phys.
     Lettres 35, L77 (1974).

45. L.J. Buravov and I.F. Shchegolev, Prib. Tek. Eksp. 2, 171 (1971).

46. A.N. Bloch, J.P. Ferraris, D.O. Cowan, and T.O. Poehler, Solid
     State Commun. 13, 753 (1973).

47. S.K. Khanna, Michael Cohen, W.J. Gunning, A.F. Garito, and A.J.
     Heeger (to be published).

48. D.B. Tanner, C.S. Jacobsen, A.F. Garito, and A.J. Heeger, Phys.
     Rev. Lett. 32, 1301 (1974); also, ibid., Lake Arrowhead Con-
     ference on 1D Conductors, May, 1974.

49. D.B. Tanner, C.S. Jacobsen, A.F. Garito, and A.J. Heeger, Phys.
     Rev. Lett. (submitted).

50. J.H. Perlstein, J.P. Ferraris, V.V. Walatka, D.O. Cowan, and
     G.A. Candela, AIP Conf. Proc. 10, 1494 (1973).

51. S.K. Khanna, A.F. Garito, A.J. Heeger, and R.C. Jaklevic, Solid
     State Commun. (in press).

52. R.C. Jaklevic and R.B. Saillant, Solid State Commun. (in press).

53. P.A. Lee, T.M. Rice, and P.W. Anderson, Phys. Rev. Lett. 31,
     462 (1973).

54. E.F. Rybaczewski, E. Ehrenfreund, A.F. Garito, and A.J. Heeger
     (to be published).

55. Y. Tomkiewicz, B.A. Scott, L.J. Tao, and R.S. Title, Phys. Rev.
     Lett. 32, 1363 (1974).

56. J.R. Schrieffer, Nobel Symp. 24, 142 (1973).

57. Morrel Cohen, J.A. Hertz, P.M. Horn, and V.K.S. Shante, Proc.
     Intl. Symp. on Atomic, Molecular and Solid-State Theory and
     Quantum Biology, Sanibel, Florida, 1974.

58. V. Bernstein, P.M. Chaikin, and P. Pincus (to be published).

59. A.J. Berlinsky, J.F. Carolan, and Larry Weiler, Solid State
     Commun. 15, 795 (1974).

60. V.V. Walatka, M.M. Labes, and J.H. Perlstein, Phys. Rev. Lett.
     31, 1139 (1973).

61. A.A. Bright, Marshall J. Cohen, A.F. Garito, A.J. Heeger, C.M.
     Mikulski, P.J. Russo, and A.G. MacDiarmid, Phys. Rev. Lett.
     (submitted).

THERMAL AND MAGNETIC BEHAVIOR OF ONE-DIMENSIONAL MAGNETS

Bernard C. Gerstein

Ames Laboratory-USAEC and Department of Chemistry

Iowa State University, Ames, Iowa 50010

This portion of the discussion will deal with static thermal and magnetic behavior of one dimensionally interacting magnetic solids. Magnetic heat capacities and susceptibilities of such systems will be treated. In keeping with the spirit of the ASI as a school", the material contained in this portion will be quite elementary, with the idea of introducing the reader to the concept of how non-interacting three dimensional systems behave, how this behavior differs from that expected for lower dimensional systems, and to the techniques used to calculate thermodynamic properties of lower dimensional systems. We will stress the manner in which thermodynamic behavior changes as 1-D and higher dimensional interactions become important. Classic calculations to illustrate this point will be considered in detail. The primary purpose of this discussion will therefore be to enable the reader to attack the literature in the field with understanding. This paper will not be a review of the static properties of lower dimensional systems. For this type of review see de Jong and Miedema,[25] Ackerman, Cole, and Holt,[26] and Ginsberg.[27]

Here we will specifically discuss, I) the physical basis of the "exchange" interaction, II) superexchange, III) initial susceptibility and magnetic heat capacities of systems with zero exchange (Curie Law behavior and Schottky heat capacities), and IV) the transfer matrix technique for Ising Model calculations, with, and without a molecular field.

I.   THE HEISENBERG - DIRAC - VAN VLECK INTERACTION

The basic idea is that we use quantum mechanics to obtain
the energy levels of a representative system of the ensemble at
which we are looking, and then use statistical mechanics to
calculate the macroscopic properties of the ensemble.[1]   The
hinge on which the entire treatment swings, therefore, is the
quantum mechanical interaction.   The operator used to describe
quantum mechanical interactions between two systems with spin
$\vec{S_1}$ and $\vec{S_2}$ is [2]

$$\mathcal{K} = -2J_{12}\,\vec{S_1} \cdot \vec{S_2} \quad . \tag{1}$$

This is the Heisenberg-Dirac-Van Vleck, or H. D. V. V. interaction.   The best known simplification of this interaction is that
of the Lenz-Ising model, in which only z components are
allowed to exist. $\mathcal{K}_{Ising} = 2J_{12}S_{z1}S_{z2}$. Other variations, to be
discussed later, allow the x, y, and z components in (1) to enter
with different coefficients.

The problem of determining the behavior of an infinite
system with (1) included as an interaction is in general insoluble.
The charm of systems with dimensions lower than three is that
an approach to solving this problem exists, and in some cases,
solutions may be obtained in closed form.   Table I lists some
Hamiltonians which have been used in 1-D systems as a function
of spin dimensionality.   The starred (*) models have been solved
in closed form.   Behavior of infinite systems for the other
models must be inferred from calculations on a finite number
of spins.

The physical basis of the H. D. V. V. interaction is fundamental to an understanding not only of direct magnetic exchange,
but of indirect, or superexchange as well, and may be viewed as
follows.   If there exist two orthonormal orbital wave functions,
ϕ and χ, available to a system, the spin-orbital products possible
for a two electron system can be arranged into spin triplet and
singlet states.   We designate the three triplet spin functions by
$|1\rangle$ and the singlet spin function by $|0\rangle$.   The triplet spin-
orbital will be

$$\frac{1}{\sqrt{2}}\left[\, \phi(1)\,\chi(2) \,-\, \chi(1)\,\phi(2)\,\right]|1\rangle$$

$$\equiv \frac{1}{\sqrt{2}}\,(\phi\chi - \chi\phi)\,|1\rangle \quad \equiv \Psi_A\,|1\rangle \quad \equiv \Psi_{triplet} \quad .$$

Table I.   1-D Models

$-\mathcal{H}$ (not incl. Zeeman)	S	Name, Ref.
$\sum_i 2J \underset{i}{S}_{zi} S_{zi+1} (n.n.)*$	$^1/2, 1, {}^3/2, 2, {}^5/2, 3$	Lenz-Ising (n.n Ising)
n.n. Ising + $g\mu_B \sum_{i=1}^{N} H_L^M S_i^{z*}$	1/2	n.n. Ising plus Mol. Field[12]
$\sum_i 2J \sum S_i^z S_{i+1}^z$ $+ \gamma(S_i^x S_{i+1}^x + S_i^y S_{i+1}^y)$	1/2	Bonner-Fisher[17] Anisotropic Finite Chains, n.n.
"	1	Weng & Griffiths;[18] Finite Chains, n.n.
$2|J| \sum_{i=1}^{N/2} S_{zi} S_{zi+1}$ $+ A S_{zi+1} S_{zi+2})$	1/2	Duffy-Barr;[19] o–o–ooo–o
$-2J \sum_{i=2}^{2N} (S_{i-1}^x S_i^x + S_{i-1}^y S_i^y)$		X-Y Model[21]

Random Ising; Higher Spin*, [20]

Ising-Arbitrary Range of Interaction*, [22]

Heisenberg Model, *, [23] S = ∞

Ising; Genl Spin*, [24]

The singlet spin-orbital is

$$\frac{1}{\sqrt{2}}[\phi\chi + \chi\phi]|0\rangle \equiv \Psi_S|0\rangle \equiv \Psi_{singlet} \quad .$$

The S and A subscripts on the orbital functions indicate that these two particle functions are symmetric and antisymmetric, respectively, with respect to interchange of like particles. The energies of a system with these spin-orbital states are given by the solutions of

$$|H_{ij} - S_{ij}W| = 0$$

with

$$H_{ij} = \langle \Psi_i |\mathcal{K}| \Psi_j \rangle$$

and

$$S_{ij} = \langle \Psi_i | \Psi_j \rangle \quad .$$

If $\mathcal{K}$ only contains kinetic and electrostatic potential energy terms, i.e.,

$$\mathcal{K} = \hat{T}_1 + \hat{T}_2 + \hat{V}_1 + \hat{V}_2 + \hat{V}_{12} \quad^\dagger$$

where T_i and V_j are kinetic and potential energy operators for the i and j^{th} electrons, then the three triplet spin-orbitals are degenerate. $\Psi_{triplet}$ and $\Psi_{singlet}$ are orthogonal and diagonalize $\mathcal{K}$, so the two solutions of the secular determinant are

$$W_{triplet} = H_{11} = Q - J$$
$$W_{singlet} = H_{22} = Q + J \qquad\qquad (2)$$

where Q is the "coulomb integral"

$$Q = (\phi\chi |\mathcal{K}| \phi\chi)$$

and J is the "exchange integral"

$$J = (\phi\chi |\mathcal{K}| \chi\phi) \quad .$$

If the single particle energies of the orbitals ϕ and χ are degenerate, then the triplet state will be the ground state. An example of such a case is the ground state of molecular oxygen, where $\phi = \pi_x^*$ and $\chi = \pi_y^*$. If the single particle energies of ϕ

$\dagger$ We ignore the internuclear potential which, while important to a consideration of the total energy, is immaterial to the present argument.

and χ are far apart, e.g., greater than one e.v., then the ground state will be a singlet. An example of such a case is the hydrogen molecule, where ϕ is the $1s\sigma$ molecular orbital and χ is $1s\sigma^*$. In this case the difference between the single particle energies of ϕ and χ is so great that it pays for both electrons to occupy ϕ in the ground state, and the only possible spin function is the singlet, $|0\rangle$. In any event, we have, from (2),

$$\Delta E_{triplet-singlet} = -2J \quad . \tag{3}$$

If $\hat{S}_1$ and $\hat{S}_2$ are the spin angular momentum operators for electrons 1 and 2 respectively, we have

$$\hat{S}^2 = (\hat{S}_1 + \hat{S}_2)^2 = \hat{S}_1^2 + \hat{S}_2^2 + 2\hat{S}_1 \cdot \hat{S}_2 \quad .$$

Then, since the expectation value of the square of the total spin, $\langle \hat{S}^2 \rangle = S(S+1)$, (with $\hbar = 1$) we have

$$\langle \hat{S}_1 \cdot \hat{S}_2 \rangle = \frac{1}{2}(S(S+1) - \langle \hat{S}_1^2 \rangle - \langle \hat{S}_2^2 \rangle)$$

$$= \frac{1}{2}(S(S+1) - \frac{3}{2}) \quad .$$

Therefore for $S = 1$, the triplet state, we have

$$\langle \hat{S}_1 \cdot \hat{S}_2 \rangle = \frac{1}{4} \quad ,$$

and for the $S = 0$, singlet

$$\hat{S}_1 \cdot \hat{S}_2 = -\frac{3}{4} \quad . \tag{4}$$

Making the identification

$$\mathcal{H} = -2J \, \hat{S}_1 \cdot \hat{S}_2$$

we see $\langle \mathcal{H} \rangle = -\frac{3}{2}J$ for the singlet, and $\langle \mathcal{H} \rangle = \frac{1}{2}J$ for the triplet. These provide the correct singlet-triplet separation, and the quantum mechanical basis for the "magnetic exchange" interaction.[3]

II. SUPEREXCHANGE

An important point to be seen from the above discussion is that ϕ and χ are functions of electron spacial coordinates (x, y, z). Therefore if electron (1) has coordinates $x\,y\,z$ and at the same time electron (2) has the same spacial coordinates, $\Psi_{singlet}$ has a very large value, and $\Psi_{triplet}$ has zero value.

Therefore the singlet wavefunction has its largest values when the electrons occupy the same or nearly the same spacial positions, whereas the triplet wavefunction is zero or nearly zero under these conditions. <u>Because of coulombic repulsions</u>, therefore, the electronic energy of $\Psi_{triplet}$ is lower than that of $\Psi_{singlet}$. This idea forms the basis for an understanding of the Pauli exclusion principle, and of the rules governing spin alignment when two centers interact via a third, intervening center, i.e., via magnetic "superexchange". These rules, based ideas of Anderson,[4] Slater,[5] Goodenough,[6] and Kanamori,[7] and discussed by Rundle[8] may be illustrated by the cases of $CsNiCl_3$ and $(CH_3)_4N NiCl_3$[9a,b] $(TMANiCl_4)$ in which there exist chains of face shared $NiCl_6^-$ octahedra. The Ni-Ni distances in these chains is too great for direct exchange to be effective. The magnetic information must therefore be transmitted via the bridging chlorines. Here, Ni(II) is a d^8 system which is octahedrally coordinated. Only the unfilled e_g manifold will therefore contribute to magnetic exchange. There are two mechanisms for assigning spin on one nickel relative to its neighbors via interaction thru an intermediate chlorine. These are interactions of the e_g orbitals on the nickels with either the s, or with the p orbitals on the bridging chlorine. When two e_g orbitals on neighboring nickels interact with two orthogonal p (e.g., a p_x and a p_y) orbitals on a bridging chlorine, the sigma two center bond between each Ni and the chlorine will consist of a filled bonding (e_g^b) and a half-filled antibonding (e_g^a) combination. Then an unpaired electron in the predominantly nickel-like e_g^a state on one nickel polarized, say in a ↑ orientation will polarize the electrons in the orthogonal e_g^a orbital between the neighboring Ni and the bridging Cl in a ↑ orientation (since coulombic repulsion is reduced when electrons in orthogonal orbitals form spin states with maximum degeneracy).

On the other hand, the two e_g orbitals on neighboring nickels can interact with a neighboring Cl via an empty Cl s orbital to form the three-center bonding, non-bonding, and anti-bonding combinations, which we call e_g^b, e_g^n, and e_g^a. Both the e_g^b and the e_g^n levels will be filled. By the Pauli principle if an electron localized near one nickel is polarized up, that in the same wavefunction near the other will be polarized down, and the resulting spin alignment will be antiferromagnetic. The total interaction can therefore by ferromagnetic or antiferromagnetic depending upon the relative importance of chlorine s and p orbitals to the bonding. It is seen to be antiferromagnetic in $CsNiCl_3$, with Ni-Cl-Ni bridging angle 74° 41', and ferromagnetic in TMANiCl$_3$ with bridging angle 78° 49'. The 90° d-p interaction is clearly more important in the latter case.

In order to account for the fact that the system under consideration is in fact infinite, the localized bonding picture must be replaced by a band picture. One idea of the beginnings of a translation between the two pictures has been published by the author.[10]

III. A MAGNETIC ENSEMBLE WITHOUT EXCHANGE

A. Curie's Law

Once the quantum mechanical problem of determining system energies utilizing an interaction such as (1) has been solved, one uses standard statistical mechanics to obtain thermal properties. The pertinent formulae are:

$$Z = \sum_i e^{-\epsilon_i/kT} \quad , \tag{5}$$

$$A = -kT \ln Z \quad , \tag{6}$$

$$U = A - T \frac{\partial A}{\partial T}\bigg)_V \quad , \tag{7}$$

$$C_H = \frac{\partial U}{\partial T}\bigg)_H = 2kT \frac{\partial \ln Z}{\partial T} + kT^2 \frac{\partial^2 \ln Z}{\partial T^2} \quad , \tag{8}$$

$$X_H = \frac{M}{H} = \frac{-kT}{H} \frac{\partial A}{\partial H}\bigg)_{T,H} = \frac{kT}{H} \frac{\partial \ln Z}{\partial H} \quad . \tag{9}$$

An illustrative example of the use of these relations is the calculation of the magnetic susceptibility and heat capacity of an ensemble of N spin systems in a solid, interacting with a magnetic field by the Zeeman term,

$$\mathcal{K}_z = -\mu_B \vec{H} \cdot \Sigma \vec{S} \quad . \tag{10}$$

μ_B is the Bohr Magneton, 0.9×10^{-20} ergs-gauss^{-1}. The "exchange" term, (1), is absent for the present calculation. To arbitrary orders in perturbation theory, the quantum mechanical solution to (10) is a power series in H, i.e., the ith system energy is given by

$$\epsilon_i = \epsilon_i^o + \frac{b_i H}{2} + c_i H^2 + \cdots \quad , \tag{11}$$

with $b_i \sim \mu_B$, $c_i \sim \mu_B^2$, etc. The partition function for a solid of N weakly interacting systems with system partition function

$\sigma = \sum_i e^{-\epsilon_i \beta}$ ($\beta = (kT)^{-1}$) is given by $Z = \sigma^N$. So $\chi = \dfrac{NkT}{H} \dfrac{\partial \ln \sigma}{\partial T}$.

Each term in σ will be of the form $e^{-[\epsilon_i^o + \frac{b_i H}{2} + c_i]/kT}$,

and with $b_i \sim \mu_B \cong 10^{-20}$ ergs-gauss^{-1}, we have for $T > 1$ K and $H \sim 1$ gauss, $b_i/kT \le 10^{-4} << 1$, and $c_i/kT <<< 1$. Therefore to better than 0.1% under the stated conditions

$$e^{-(\epsilon_i^o + \frac{b_i H}{2} + c_i H^2)\beta}$$

$$= e^{-\epsilon_i^o/kT}(1 - \beta\frac{b_i H}{2} + \beta^2 \frac{b_i H^2}{2} - \beta^2 c_i H^2)$$

to terms in H^2. Then

$$\chi = \frac{NkT}{H}\frac{\partial \ln \sigma}{\partial H} = \frac{N\beta^{-1}}{H\sigma}\frac{\partial \sigma}{\partial H} \qquad (12)$$

$$= \frac{N\beta^{-1}}{H\sum e^{-\epsilon_i^o/kT}} \sum_i (-b_i\beta + 3b_i H^2 - 2\beta c_i H)\, e^{-\epsilon_i^o/kT} .$$

To have a zero susceptibility, and therefore zero magnetic moment in zero applied field, the sum $\sum_i -b_i\beta$ must vanish.

$$\chi = \frac{N\sum(b_i^2\beta - 2c_i)\, e^{-\epsilon_i^o/kT}}{\sum e^{-\epsilon_i^o/kT}} . \qquad (13)$$

The same expansion has been used for both the denominator and numerator, keeping only terms of the same order in both. For a system with spin S and with $(\epsilon_j^o - \epsilon_o^o)/kT << 1$, $j \neq 0$, the first term in the sum of (13) becomes

$$\frac{N\sum\limits_{-s_z}^{s_z} \frac{b_{s_z}^2}{kT}}{2S+1} = \frac{N\mu_B^2}{3kT} S(S+1) \equiv \frac{C}{T} , \qquad (14)$$

Curie's Law. The second term is the perturbation term due to paramagnetism of higher states, or the "temperature independent paramagnetic" term, $N\alpha$.

B.   The Curie-Weiss Law

The above treatment assumed the system has no spontaneous moment. To investigate the origin of magnetic ordering, we postulate, (Weiss), that in a magnetic body there is a "Weiss field" proportional to the magnetization, i.e., $H_w = \lambda_M$, tending to align the spins. Further, we say that above the temperature of alignment, the effective field, H_E, is the external field H_0, plus the Weiss field, H_W; $H_E = H_0 + H_W$. We then assume that Curie's Law, (14), holds if H is replaced by H_E,

$$\frac{C}{T} = \frac{M}{H_E} = \frac{M}{H_0 + \lambda M} \qquad (15)$$

Then

$$X = \frac{M}{H_0} = \frac{C}{T - C\lambda} \qquad (16)$$

At $T_c = C\lambda$, therefore, M will be finite at zero applied field. The expression

$$\chi = \frac{C}{T - T_c} \qquad (17)$$

is the Curie-Weiss Law, which describes the susceptibility of many materials above their ordering temperature. Clearly

$$\lambda^{-1} = C/T_c = Ng^2\mu_B^2 S(S+1)/3kT_c \qquad (18)$$

The origin of the Weiss field is seen to be identified with the term

$$\mathcal{H}_{ij} = -2J_{ij}\vec{S}_i \cdot \vec{S}_j$$

since for n nearest neighbors with spin S,

$$E_{exchange} \cong -2Jn\langle S^2 \rangle$$

$$\equiv -g\mu_B\langle S \rangle H_E = -g\langle S \rangle \mu_B (\lambda\mu_B\langle S \rangle \Omega^{-1})$$

The term in parenthesis is equal to M, and Ω is the atomic volume. Therefore

$$J = \lambda g^2\mu_B^2/2n\Omega \qquad ,$$

and with $N = \Omega^{-1}$, we have by (17)

$$J = \frac{3kT_c}{2nS(S+1)} \qquad (19)$$

A positive T_c, as inferred from behavior where (17) applies, is taken to indicate a positive J (ferromagnetic exchange). A negative T_c is taken to indicate antiferromagnetic exchange. In summary, then the relation

$$\chi = \frac{C}{T - T_c} + N\alpha \qquad (20)$$

describes the behavior of many materials where $(\epsilon_i^o - \epsilon_o^o)/kT \ll 1$. $N\alpha$ is obtained from a plot of $\underline{\chi}$ vs. T^{-1} extrapolated to $T = \infty$. A plot of $(\chi - N\alpha)$ vs. T yields T_c, and in the range where the relation holds, a plot of $(\chi - N\alpha)(T - T_c)$ yields a constant value, C.

C. Heat Capacity

The contribution to the zero field heat capacity of such a system of weakly interacting spins in obtained from (8). For $Z_o = (\Sigma\, e^{-\epsilon_i^o \beta})^N$, differentiation yields

$$C_H = \frac{R}{Z_o^2} \left\{ Z_o \sum_{i=1} \left(\frac{\epsilon_i^o}{kT}\right)^2 \exp(-\epsilon_i^o/kT) \right.$$

$$\left. - \left[\sum_{i=1}\left(\frac{\epsilon_i^o}{kT}\right)\exp(-\epsilon_i^o/kT)\right]^2 \right\} \quad .$$

(21)

In the low temperature limit, C_H increases exponentially with temperature, and in the high temperature limit decreases as T^{-2}. Between these limits is a maximum. For an ensemble of systems, each having one energy level Δ above a non degenerate ground state, the maximum is at $(C_H/R) = 0.44$, and $(RT/\Delta) = .44$. The height of the peak measures the degeneracy of the levels, and the temperature of the peak measures the splitting. An example of such a "Schotty" type heat capacity for a 13 level system is shown in Fig. 1.[28] The positions of the nine lowest levels are indicated as vertical bars on the temperature axis.

IV. BEHAVIOR OF ONE-DIMENSIONAL SYSTEMS; THE TRANSFERMATRIX TECHNIQUE

A. 1-D Ising Model, S=1/2

In the early 1960's, thermal and magnetic measurements on 3d transition metal salts such as $KCuCl_3$[11] and $CrCl_2$[12] yielded results different from that expected for behavior predicted by (20) or (21). The behavior also differed from that of systems with long range order in 3-D, namely, sharp peaks in the heat capacity and three-dimensional coherent magnetic scattering of slow neutrons. Structures of these compounds characteristically consisted of ligand bridged chains of 3-d ions, and it became clear that such solids might offer laboratories in which to test models of magnetic exchange.

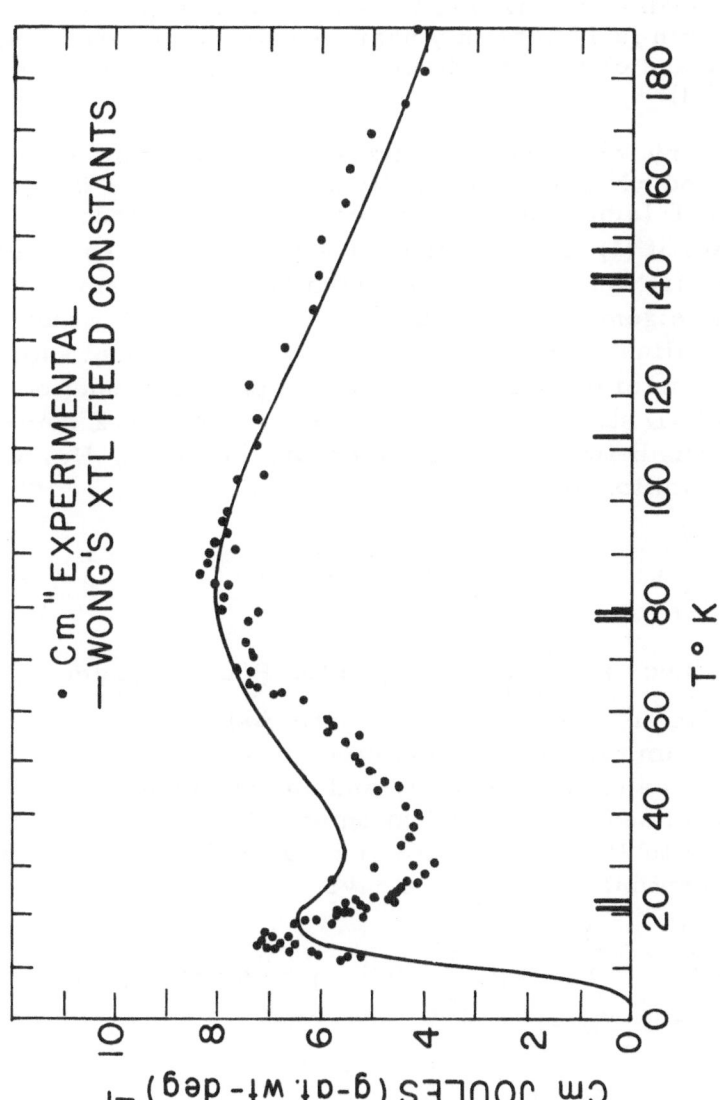

Fig. 1. Schottky Heat Capacity of a Thirteen Level System, $Tm(C_2H_5SO_4)_3$. Positions of first Nine Levels Above Ground Indicated as Vertical Lines on the T Axis.

The appealing feature of "real" systems in which one-dimensional interactions dominate is that thermodynamic functions of such a system may be obtained in closed form, given the interaction (1) with some severe, but perhaps physically reasonable, restrictions. The restrictions are upon a) the dimensionality of the spin, b) the range of the interaction, c) the anisotropy of the assumed interaction, and d) the magnitude of the external field. The simplest possible case is that for a one-dimensional system in which a) the spin is 1/2, b) only nearest neighbor interactions are considered, and c) only the Z component is considered (the Lenz-Ising model).

There are many techniques for attacking the problem at hand. A powerful method which was originally used by Onsager for the solution of the 2-D Ising model, which only takes into account the Z component of (1) is that of the transfer matrix. The basic idea is that the partition function for a spin $\frac{n}{2}$ Ising system is found from the largest eigenvalue of a $(2n+1) \times (2n+1)$ "transfer matrix". The technique is illustrated using the spin 1/2 Ising model with nearest neighbor interactions. To obtain the partition function, $Z_N(T, H)$, for a 1-D chain of spin 1/2 systems with only Z component interactions between nearest neighbors in a field, H_Z, we start with the Hamiltonian

$$\mathcal{K} = -\beta^{-1} \sum_{i=1}^{N} \left[j S_i S_{i+1} + \frac{\eta}{2} (S_i + S_{i+1}) \right] \quad . \tag{22}$$

Here, $j = \frac{2J}{4kT}$, and $\eta = \frac{g \mu_B H_Z}{2kT}$. μ_B is the Bohr magneton, g is the electron g factor, and H is the magnetic field. The factors of 4 and 2 in the denominators arise because the spin is 1/2. $\beta = (kT)^{-1}$. In these units, $\langle S_i \rangle = \pm 1$, and an ensemble energy, ϵ_k, will be determined by a specification of $\langle S_i \rangle$ for each S_i, i = 1, 2 ... N. With H in the Z direction, we will be calculating the parallel susceptibility, $\chi_{\parallel}$. We have

$$Z_N = \sum_k e^{-\beta \epsilon_k} = \sum_{S_1=-1}^{1} \sum_{S_N=-1}^{1} \exp \left\{ \sum_{i=1}^{N} [j S_i S_{i+1} + \frac{\eta}{2} (S_i + S_{i+1})] \right\} . \tag{23}$$

We define

$$U(S_i, S_{i+1}) = -\beta^{-1} \left\langle j S_i S_{i+1} + \frac{\eta}{2} (S_i + S_{i+1}) \right\rangle$$

and

$$F(S_i, S_{i+1}) = \exp \left\{ -\beta U(S_i, S_{i+1}) \right\} \quad .$$

Therefore,

$$Z_N = \sum_{S_i=-1}^{1} \cdots \sum_{S_N=-1}^{1} F(S_1, S_2) F(S_2, S_3) \cdots F(S_N, S_1) .$$

We then define the matrix

$$T = \begin{pmatrix} T_{++} & T_{+-} \\ T_{-+} & T_{--} \end{pmatrix} = \begin{pmatrix} e^{j+\eta} & e^{-j} \\ e^{-j} & e^{j-\eta} \end{pmatrix}$$

with matrix elements

$$T_{S_i, S_{i+1}} \equiv T_{\pm, \pm} = F(S_i = \pm 1, \; S_{i+1} = \pm 1) \quad .$$

By the rule for matrix multiplication,

$$(AB)_{ii} = \sum_k A_{ik} B_{ki} \quad ,$$

so

$$(ABC)_{ii} = \sum_\ell \sum_k A_{in} B_{n\ell} C_{\ell i} \qquad \text{etc.}$$

Clearly then,

$$Z = \sum_{S_1=-1}^{1} \cdots \sum_{S_N=-1}^{1} F(S_1, S_2) F(S_2, S_3) \cdots F(S_N, S_1)$$

$$\equiv \sum_{S_1=-1}^{1} \cdots \sum_{S_N=-1}^{1} T_{S_1, S_2} T_{S_2, S_3} \cdots T_{S_N, S_1}$$

$$= \sum_{S_1=-1}^{1} (T)^N_{S_i, S_i} = \text{Trace}(T^N) \quad .$$

The trace of a matrix is the sum of the diagonal elements, and also the sum of its eigenvalues. The eigenvalues of T^N are $\lambda^N_\pm$, where $\lambda_\pm$ are the eigenvalues of T, obtained by a solution of

$$\begin{vmatrix} e^{j+\eta} - \lambda & e^{-j} \\ e^{-j} & e^{j-\eta} - \lambda \end{vmatrix} = 0$$

which are

$$\lambda_\pm = e^j \cosh \eta \pm (e^{2j} \sinh^2 \eta + e^{-2j})^{1/2} \quad . \tag{24a}$$

The partition function is therefore

$$Z_N(T, H) = \lambda_+^N + \lambda_-^N \quad . \tag{24b}$$

We note $\lambda_+ > \lambda_-$, so $\lambda_+^N \gg \lambda_-^N$, and (24b) becomes

$$\underset{N \to \infty}{\text{Lim}} \quad Z_N(T, H) = \lambda_+^N [1 + \left(\frac{\lambda_-}{\lambda_+^N} \right)] \cong \lambda_+^N \quad . \tag{25}$$

The susceptibility is obtained from (9) as follows:

$$\chi = \frac{\beta^{-1}}{H} \left. \frac{\partial \ln Z}{\partial H} \right)_T = \frac{N g^2 \mu_B^2 \beta}{4 \eta \sigma} \frac{\partial \sigma}{\partial \eta}$$

with $\sigma = e^j \cosh \eta + (e^{2j} \sinh^2 \eta + e^{-2j})^{1/2}$. As in the derivation of (13), we expand in powers of η, to terms in η^2. With $\cosh \eta = (1 + \eta^2/2)$, and $\sinh^2 \eta = \eta^2$,

$$\frac{1}{\eta \sigma} \frac{d\sigma}{d\eta} = \frac{e^j + e^{2j}(e^{2j} \eta^2 + e^{-2j})^{-1/2}}{e^j(1 + \eta^2/2) + (e^{2j} \eta^2 + e^{-2j})^{1/2}} \quad .$$

For the zero field case, we have

$$\frac{1}{\eta \sigma} \frac{d\sigma}{d\eta} = \frac{e^j + e^{3j}}{e^j + e^{-j}} = \frac{1 + e^{2j}}{1 + e^{-2j}} = e^{2j}$$

with the result[12]

$$\chi_{\parallel} = \frac{C}{T} e^{J/kT} \equiv \frac{N g^2 \mu_B^2}{4kT} \exp(J/kT) \quad . \tag{26}$$

With $J < 0$ (antiferromagnetic case) we see that χ is a maximum at $J = kT$. A convenient method of determining if a given physical system matches this behavior is to plot $(|J|\chi \exp{}'Ng^2\beta^2)$ vs. (kT'/J), as shown for $(CH_3)_2NH_2CuCl_3$[9b] in Fig. 2. On such a plot, one obtains one theoretical curve which is a function of (kT'/J). Adjusting the value of (J/k) with which the experimental points are multiplied, therefore, determines the best fit of the experiment to the theory.

Equation (26) becomes infinite at $T = 0$ if $J > 0$ (ferromagnetic coupling) and is zero at $T = 0$ for $J < 0$ (antiferromagnetic coupling). There is no singularity in χ for $T > 0$, and therefore no long range order at finite T. If $J = 0$, (26) becomes Curie's Law, (14), with $S = 1/2$. The zero field heat capacity of the 1-D spin $1/2$ Ising model is obtained from (7), (23), and (25). In the absence of a field,

$$Z = \sigma^N = (e^j + e^{-j})^N \equiv (e^{J/2kT} + e^{-J/2kT})^N,$$

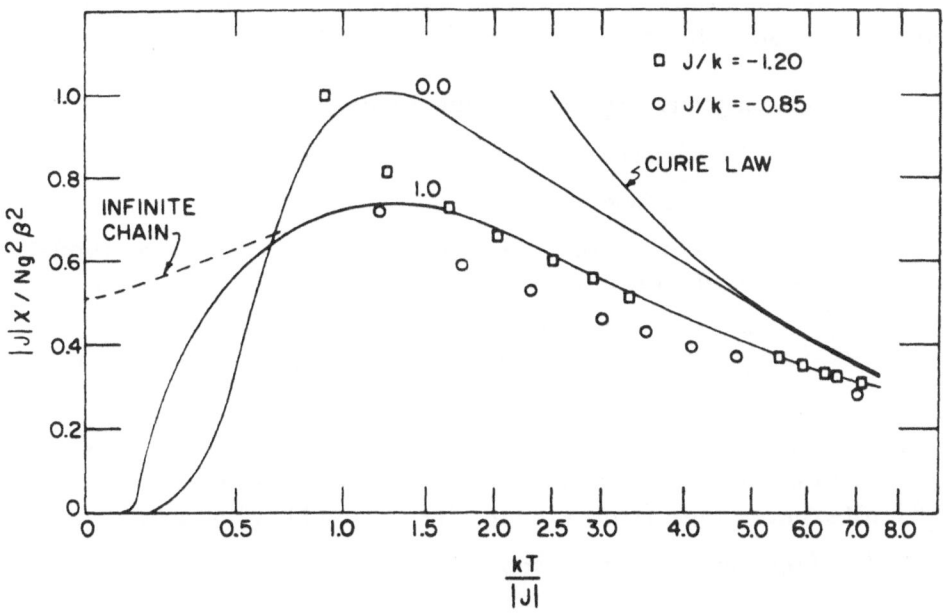

Fig. 2. Reduced Susceptibility as a Function of Reduced Temperature. Model Fitting in $(CH_3)_2NH_2CuCl_3$.

and

$$C_H/R = 2(j \text{ sech } j)^2 \cdot \qquad (27)$$

with $j = J/2kT$, we see that the limiting behavior at $T = 0$ and $T = \infty$ is identical to a system of non-interacting spins. The heat capacity is at a maximum at $kT/J = 0.218$, where $C_H/R = 1$. For perpendicular susceptibility of the spin 1/2 Ising model with nearest neighbor interactions, the Hamiltonian is taken to be

$$\mathcal{K} = -kT\left[\sum_{i=1}^{N} j\, S_{zi}\, S_{zi+1} - \frac{\eta}{2} \Sigma(S_{xi} + S_{xi+1})\right] \cdot \qquad (28)$$

The perpendicular susceptibility is found to be[14]

$$\chi_{\perp} = \frac{Ng\mu_B^2}{2|J|}\,(\tanh\frac{|J|}{kT} + \frac{|J|}{kT}\,\text{sech}^2\frac{J}{kT})$$

which yields the same result for both the ferro and antiferromagnetic cases. The susceptibility of $T = 0$ is found to be finite, and a maximum in χ is found at $(kT/2J) = 0.41677$. The ratio $(\chi_{\perp}, \max/\chi_{\perp}(0))$ is 1.1997. Such behavior is not found for 1-D organic donor acceptor complexes, such as TMPD-TCNQ, discussed by Zoltan Soos elsewhere in this volume.

B. Antiferromagnetic Spin 1/2 Ising Model with Molecular Field;
Only Nearest Neighbor Interactions.[12]

Parallel Susceptibility and Heat Capacity.

The addition of a molecular field to the spin 1/2 Ising model leads to a 3-D ordering temperature, T_c. The Hamiltonian is

$$\mathcal{K} = (-\sum_{i=1}^{N} 2J\, S_i S_{i+1} + g\mu_B H\, S_i + g\mu_B H_i^M S_i) , \qquad (29)$$

with $S_i \equiv S_{zi}$. If the ordered state is antiferromagnetic, there are two interpenetrating sublattices, and the molecular field, H_i^M, alternates sign from lattice site to neighboring site.

This fact may be expressed as

$$H_i^M = \gamma[(-1)^i\langle\mu_0\rangle + \langle\delta\mu\rangle] \cdot$$

γ is a molecular field constant, $\langle \mu_0 \rangle$ is the mean magnitude of the magnetization per atom in the absence of an external field, and $\langle \delta\mu \rangle$ is the field induced magnetization. $\langle \mu_0 \rangle$ on alternating members of the chain is equal and opposite, so the net magnetization of the chain vanishes. $\langle \delta\mu \rangle$ will, however, be equal in magnitude and sign for all members of the chain. Because of the alternation of the molecular field, $\mathcal{K}$ is written as

$$\mathcal{K} = \sum_{j=1}^{N/2} -2J(S_{2j-1}S_{2j} + S_{2j}S_{2j+1}) + \frac{1}{2} g\mu_B (H + \gamma \langle \delta\mu \rangle)$$

$$(S_{2j-1} + 2S_{2j} + S_{2j+1}) + \frac{1}{2} g\mu_B \delta \langle \mu_0 \rangle (-S_{2j-1} + S_{2j} - S_{2j+1}) . \quad (30)$$

The transfer matrix has matrix elements

$$P_{SS'} = \exp[4KSS' - B(S+S') + L(S-S')] \quad (31)$$

with $K = J/2kT$, $B = g\mu_B (H + \gamma \delta\mu)/2kT$, and $L = g\mu_B \gamma \mu_0 /2kT$. The partition function is $\lambda_M^{N/2}$, where λ_M is the largest eigenvalue of $P\tilde{P}$. $\tilde{P}$ is the transpose of P. In the absence of long range order, $L = 0$, and $P = \tilde{P}$. Then, $Z_N = [\lambda_M(P)]^N$.

With $H = 0$, $B = 0$, and $P\tilde{P}$ is the square of Q, with

$$Q_{SS'} = \exp[-4KSS' + L(M+M')] .$$

Then

$$Z = [\lambda_M(Q)]^N ,$$

and we find

$$\lambda_M(Q) = e^{-k} \cosh L + e^{-2k} \sinh^2 L + e^{2k})^{1/2} . \quad (32)$$

In zero applied field, the mean magnitude of the magnetization averaged over a sublattice is

$$\frac{2kTL}{g\mu_B\gamma} = \langle \mu_0 \rangle = (g\mu_B/2N)(\partial \ln Z/\partial L) . \quad (33)$$

To find T_c, we use (32) and (33) and obtain

$$\frac{4kTL}{\gamma g^2 \mu_B^2} = \frac{\sinh L}{(\sinh^2 L + e^{4k})^{1/2}} . \quad (34)$$

At T_c, the derivatives with respect to L of both sides of (34) are equal, so

$$kT_c = \frac{1}{4}[\gamma g^2 \mu_B^2 \exp(-J/kT_c)] \quad . \tag{35}$$

γ may then be eliminated between (34) and (35) to yield

$$\frac{Te^{-J/2kT_c}L}{T_c} = \frac{\sinh L}{(\sinh^2 L + e^{4k})}$$

which may be solved numerically for $T < T_c$ to obtain L, and therefore $|H^M|$, $\langle\mu_o\rangle$, and via appropriate differentiation, C_H. Below T_c, there is an exponential rise in C_H. Above T_c, $L = 0$ and the magnetic heat capacity is given by (27). The resulting form of C_H is shown in Fig. 3. Agreement with experimental results in the case where the molecular field is not much smaller than the 1-D interaction between spins is poor, although the theory clearly describes the qualitative features of the experiment.

Above T_c the susceptibility is obtained from the largest eigenvalue of P with $L = 0$. We find

$$Z_N = [e^K \cosh B + (e^{2K} \sinh B + e^{-2K})^{1/2}]^N$$

and

$$\chi = \frac{Ng^2 \mu_B e^{2K}}{4k[T \mp T_c e^{2(K+K_c)}]} \quad . \tag{36}$$

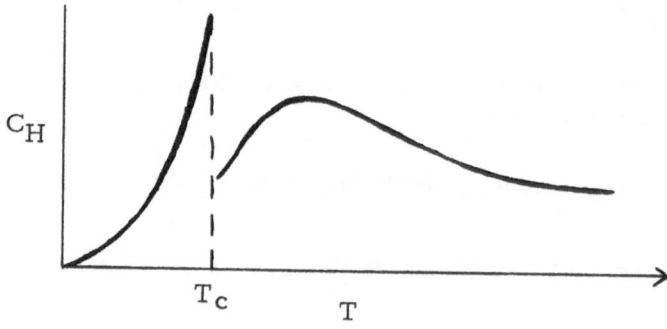

Fig. 3. Heat Capacity of S = 1/2 Ising Model with Molecular Field. 3-D Ordering at $T = T_c$.

C.   Other 1-D Models

Spin 1 Ising Model: Nearest Neighbor Interactions[14, 15]

The Hamiltonian is again given by (22), but now with $j = 2J/kT$, and $\eta = g\mu_B H$. S_i and S_{i+1} may assume the values 1, 0, or -1. The transfer matrix equation to be solved is

$$|T(j, \eta)| = \begin{vmatrix} e^{j+\eta}-\lambda & e^{\eta/2} & e^{-j} \\ e^{\eta/2} & 1-\lambda & e^{-\eta/2} \\ e^{-j} & e^{-\eta/2} & e^{j-\eta}-\lambda \end{vmatrix} = 0 \quad . \qquad (37)$$

We first solve the zero field case. The eigenvalues of $T(j, \eta = 0)$ are

$$\lambda_\pm = \cosh j + \frac{1}{2} \pm (\cosh^2 j - \cosh j + \frac{9}{4})^{1/2} \quad ,$$

$$\lambda_3 = -2 \sinh j \quad .$$

The largest root is λ_1, so $Z_N = \lambda_1^N$. To find the susceptibility, we are concerned with a small change in λ_1 when a field is present. We let $\lambda_1(H) = \lambda_1(H = 0) + \delta$ where $\delta \ll \lambda_1$. We then solve (37) for δ. We find

$$\delta = \frac{1}{2} \frac{H^2}{RT} \frac{(2 \exp j - 1)}{(\lambda_1-\lambda_2)(\lambda_1-\lambda_3)} \quad . \qquad (38)$$

χ is obtained from (9).

$$\chi = \frac{Ng^2\mu_B^2}{kT} \frac{2\exp(j)[\lambda_1+e^{-j}-1]}{(\lambda_1-\lambda_2)(\lambda_1-\lambda_3)} \quad . \qquad (39)$$

An alternate method of obtaining χ, of course, is to use a computer to determine the eigenvalues of (37) as a function of η, fit λ_{max} to a polynomial in η, and carry out the appropriate differentiation. This technique lacks charm, but is probably easiest in the long run for larger values of spin. The appropriate programs for solving the thermodynamic functions of aniso-tropic exchange Hamiltonians have been developed by the group of Professor D. J. Klein, Department of Physcs, at the University of Texas at Austin, Texas. The parallel susceptibilities and heat capacities of the antiferromagnetic spin 3/2, 2, 5/2, and

3 Ising model are given by Van Veen et al. [16] and references
therein contained. The values of C_{max}/R, $kT_{max}|J|$,
$|J|\chi_{max}/Ng^2 u_B^2$, and kT_{max}/J for these models are given in
Table II. [16]

Table II. Values of the Maxima of $\chi_{||}$ and C_H/R for Antiferro-
magnetic Ising 1-D Systems from Spin 1'2 to Spin 3.

S	C_{max}/R	$kT_{max}/|J|$	$|J|\chi_{max}/Ng^2\rho^2$	$kT_{max}/|J|$
1/2	0.439	0.417	0.092	1.00
1	0.941	1.223	0.098	2.552
3/2	1.265	2.321	0.100	4.700
2	1.471	3.717	0.1010	7.46
5/2	1.605	5.412	0.1015	10.82
3	1.697	7.407	0.1018	14.80

REFERENCES

1. See, for example, Chapter 1, Section 1 of "Statistical
    Mechanics, A Set of Lectures", R. P. Feynman, W. A.
    Benjamin, Inc. (1972).

2. J. H. Van Vleck, "Theory of Electric and Magnetic Suscepti-
    bilities", Oxford (1932).

3. For a criticism of the original Dirac derivation (Proc. Roy.
    Soc., A123, 714 (1929)) see J. C. Slater, Revs. Mod. Phys.,
    25, 199 (1953).

4. P. W. Anderson, Phys. Rev., 79, 350 (1950).

5. J. C. Slater, Quart. Prog. Rept., M. I. T., July 15, 1;
    Oct. 15, 1 (1953).

6. J. B. Goodenough, Phys. Rev., 100, 564 (1955).

7. J. Kanamori, J. Phys. Chem. Sol., 10, 87 (1959).

8. R. E. Rundle, Survey of Progress in Chemistry, 1, 81 (1963).

9. a) Joan Smith, B. C. Gerstein, S. H. Liu, and Galen Stucky,
       J. Chem. Phys., 53, 418 (1970);
    b) B. C. Gerstein, F. D. Gehring, and R. D. Willett, J.
       Appl. Phys., 43, 1932 (1972).

10. B. C. Gerstein, J. Chem. Ed., 50, 316 (1973).

11. G. J. Mass, B. C. Gerstein and R. D. Willett, J. Chem.
    Phys., 50, 758 (1964).

12. J. W. Stout and R. C. Chisholm, J. Chem. Phys., 36, 979
    (1962).

13. G. F. Newell and E. W. Montrell, Rev. Mod. Phys., 25,
    353 (1953).

14. T. Obokata and T. Oguchi, J. Phys. Soc. Japan, 25, 322
    (1968).

15. J. Smith, B. C. Gerstein, S. H. Liu and G. Stucky, J.
    Chem. Phys., 53, 418 (1970).

16. J. A. R. Van Veen, H. T. Witteveen and W. J. Vermin,
    Solid State Commun., 13, 1235 (1974).

17. J. C. Bonner and M. E. Fisher, Phys. Rev., 135, 640
    (1964).

18. C. Weng, "Finite Exchange Coupled Magnetic Systems",
    Microfilm Copy, Unpublished Ph. D. Thesis, Carnagie-
    Mellon University (1968).

19. W. Duffy Jr. and K. P. Barr, Phys. Rev., 165, 647 (1968).

20. Fumitaka Matsubara and Koh'ichi Yoshimura, Prog. Theor.
    Phys., 50, 1824 (1973).

21. E. Leib, T. Schultz and D. Mattis, Ann. Phys., 16, 407 (1961); S. Katsura, Phys. Rev., 127, 1508 (1962); 129, 2835 (1963).

22. T. Morita, J. Phys. A: Math. Nucl. Ge., 7, 289 (1974).

23. M. E. Fisher, Am. J. Phys., 32, 343 (1964).

24. M. Suzuki, B. T. Sujiyama and S. Katsura, J. Math. Phys., 8, 124 (1967).

25. L. J. deJong and A. R. Miedema, Adv. in Phys., 23, 1 (1974).

26. J. F. Ackerman, G. M. Cole and S. L. Holt, Inorganic Chimica Acta, 8, 323 (1974).

27. A. P. Ginsberg, Inorganic Chimica Acta, 5, 45 (1971).

28. B. C. Gerstein, L. D. Jennings and F. H. Spedding, J. Chem. Phys., 37, 1496 (1962).

CONSEQUENCES OF EXCHANGE IN LOW-DIMENSIONAL COMPOUNDS: HIGH TEM-

PERATURE SPIN DYNAMICS AS SAMPLED BY MAGNETIC RESONANCE

PETER M. RICHARDS

Sandia Laboratories

Albuquerque, New Mexico 87115, USA

I. INTRODUCTION

The isotropic exchange interaction $2J\vec{S}_i \cdot \vec{S}_j$ between a pair of spins i and j has several consequences as it forms the dominant "glue" of interaction which makes the system of 10^{22} or so spins a strongly coupled one rather than a collection of independent paramagnetic particles. One may roughly divide the effects of J into static and dynamic ones, as can be seen from a simple consideration of two spins only, each with $S = \frac{1}{2}$. For antiferromagnetic coupling (J > 0 in our notation) the antiparallel (singlet) arrangement is lower in energy than the parallel (triplet) alignment by an amount $2J(1/4) - 2J(-3/4) = 2J$. As a consequence the spins will tend to align antiparallel at a sufficiently low temperature $T \sim 2J/k_B$. This effect is the root of static properties such as the magnetic susceptibility which will be discussed by Professor Gerstein in subsequent lectures.

The above example has pictured J as representing a difference in energy between parallel and antiparallel spins. For a discussion of spin dynamics it is convenient to regard J in terms of a spin flipping frequency. Toward this, consider now a situation in which spin i is "down" and spin j is "up" at time t = 0, denoted by the state $|-+\rangle$. This antiparallel alignment is not an eigenstate of the exchange Hamiltonian $2J\vec{S}_i \cdot \vec{S}_j$, however. Rather, it is a linear combination

$$|-+\rangle = \frac{|-+\rangle + |+-\rangle}{2} + \frac{|-+\rangle - |+-\rangle}{2} \qquad (1.1)$$

147

of a triplet state $(|-+\rangle + |-+\rangle)$ with energy $J/2$ and the singlet
state $(|-+\rangle - |+-\rangle)$ with energy $-3J/2$. Thus if the time-dependent
wave function is

$$\psi(t) = a(t)|-+\rangle + b(t)|+-\rangle \tag{1.2}$$

it is a simple exercise in graduate school Quantum Mechanics to
show that

$$|a(t)|^2 = \cos^2(Jt/\hbar) \quad, \tag{1.3}$$

$$|b(t)|^2 = \sin^2(Jt/\hbar) \tag{1.4}$$

That is, the state of the system oscillates between $|+-\rangle$ (i "up",
j "down") and $|-+\rangle$ (i "down", j "up") at the exchange frequency $2J/\hbar$.

The focus of this paper is on the role of J as an exchange
frequency and the consequences of low dimensionality on the dynamics
associated with J. For the two spin problem, the solution is
straightforward, as we have seen. But when one considers the case
at hand of a chain (1 dimension, 1d) or plane (2 dimensions, 2d)
containing an effectively infinite number of spins, there is no
exact solution known and approximations and physical insights have
to be used liberally. The problem contains many features of interest to basic statistical mechanics and fluctuation phenomena and has
occupied workers for quite some time now. Aside from its fundamental
interest to quantum theoreticians, the question of spin dynamics is
of importance to, for example, structural chemists for the following
reason. One can in principle measure J from dynamical properties
and therefore relate it to such facets of the structure as chemical
bonding provided that the theory is good enough to relate an observation such as electron spin resonance (ESR) linewidth to a quantitative value for the exchange interaction. [Although J is most often
determined from the static susceptibility, there can be cases, particularly for very small J's such as involved in weak interchain or
interplane coupling, where ESR can be a more sensitive probe.] On
the other hand, approximate theories of spin dynamics can best be
tested in a simple structure for which there is a single exchange
constant J which can be measured from the susceptibility. Thus the
efforts of theoretical physicists and synthetic chemists can greatly
complement each other, and there is a strong interdisciplinary character to this field.

II.   DYNAMIC CORRELATION FUNCTIONS

The quantity of prime interest in spin dynamics is a spin time
correlation function such as $\langle S_i^z(t)S_j^z(0)\rangle$ where the triangular
brackets indicate a thermal average and time dependence is with

respect to the exchange Hamiltonian $\mathcal{K}_e$,

$$S_i^z(t) = e^{i\mathcal{K}_e t/\hbar} S_i^z(0) e^{-i\mathcal{K}_e t/\hbar} \quad . \tag{2.1}$$

A physical interpretation of the correlation $\langle S_i^z(t)S_j^z(0)\rangle$ is as follows. Suppose we start with a thermal equilibrium distribution of spins, and then at time $t = 0$ force the spin j to be "up" without disturbing the other spins. That is, a disturbance from equilibrium has been introduced at site j at $t = 0$. The correlation function $4\langle S_i^z(t)S_j^z(0)\rangle$ then represents the probability that this disturbance is located at site i at time t.

Probably the most direct way of attaining $\langle S_i^z(t)S_j^z(0)\rangle$ is from inelastic neutron scattering since the cross section for magnetic scattering of a neutron with momentum change $\hbar\vec{k}$ and energy loss $\hbar\omega$ is proportional to $\langle S_k^z S_{-k}^z\rangle_\omega$, the spatial and frequency Fourier transform of $\langle S_i^z(t)S_i^z(0)\rangle$ at wave vector $\vec{k}$ and frequency ω. This has proved to be a powerful tool for studying dynamics of low-dimensional systems[1-3] but it is a technique which is not readily available to most workers. Also, several of the low-dimensional compounds of interest have a large number of non-magnetic scatterers and J's less than 0.1 meV, both of which present serious signal-to-background and resolution problems for neutrons. Thus we will not treat neutron scattering in any detail, but rather examine the effects of the dynamic correlation on NMR and on ESR.

A simple case which illustrates the consequences of exchange on magnetic resonance is that of the NMR of a nucleus of spin I coupled to an electron spin $\vec{S}_i$ through an anisotropic hyperfine interaction $\hbar A I^z S_i^z$ where z coincides with the direction of applied field H. The Hamiltonian is then

$$\mathcal{K} = \gamma_N \hbar H I^z + \hbar A I^z S_i^z + \mathcal{K}_z + \mathcal{K}_e \tag{2.2}$$

where γ_N is the nuclear gyromagnetic ratio, $\mathcal{K}_z$ is the electronic Zeeman Hamiltonian, and $\mathcal{K}_e$ describes exchange coupling of the electron spin system.

The transverse spin operator $I^+ = I^x + iI^y$ obeys the equation of motion

$$\frac{d}{dt} I^+ = \frac{i}{\hbar} [\mathcal{K}, I^+] = i\omega_N I^+ + i\Delta\omega(t) I^+ \tag{2.3}$$

in which $\omega_N = \gamma_N H$ and $\Delta\omega(t) = AS_i^z$. Time dependence of $\Delta\omega(t)$ occurs because of $\mathcal{K}_e$. If we assume that $\Delta\omega(t)$ may be regarded as a classical function of time rather than an operator, then the solution to (2.3) is

$$I^+(t) = I^+(0)e^{i\omega_N t}\exp[i\int_0^t \Delta\omega(t')dt'] \qquad (2.4)$$

Equation (2.4) is not strictly correct because it ignores time ordering — $\Delta\omega(t_1)$ does not commute with $\Delta\omega(t_2)$ for $t_1 \neq t_2$. However, the above formula is adequate within the approximations used later. The NMR is given by a statistical average of Eq. (2.4) over all configurations of the electron spins. That is, we need $\langle \exp[i\int_0^t \Delta\omega(t')dt']\rangle$. Consider first the case of no exchange interaction so that $\Delta\omega(t) = \Delta\omega$ is independent of time and can take on the values $\pm A/2$. We then have, at infinite temperature where all states are equally probable,

$$\langle \exp[i\int_0^t \Delta\omega(t')dt']\rangle = \langle e^{i\Delta\omega t}\rangle = \tfrac{1}{2}[e^{i\frac{1}{2}At} + e^{-i\frac{1}{2}At}] \qquad (2.5)$$

from which it is evident that the spectrum consists of lines at $\omega_N \pm \frac{1}{2}A$.

What happens when the exchange interaction is turned on? $\Delta\omega(t)$ now becomes a function of time which fluctuates roughly at the rate $2J/\hbar$. If $\mathcal{K}_e$ were to consist only of coupling between the spin i and another one j, we could deduce the spectrum of $\Delta\omega(t)$ from (1.3). But in the real situation of interest $\mathcal{K}_e = 2J\sum S_j \cdot S_{j+1}$ for an infinite chain of spins. Instead of a single frequency which characterizes the dynamic correlation as in (1.3) and (1.4), there will now be many frequencies corresponding to the large number of levels of the complete chain. $\Delta\omega(t)$ then becomes more nearly a random function of time which can assume any value between $A/2$ and $-A/2$. If the rate, approximately $2J/\hbar$, at which $\Delta\omega$ changes between $A/2$ and $-A/2$ is greater than the splitting A, then the two lines are expected to merge into a single unresolved line centered at ω_N. The reason is that $\Delta\omega$ does not remain at a single value, say $A/2$, long enough for the nuclear spin to make a complete precession about that effective field, and thus it cannot "know" that there is a distinct transition frequency at the value $\omega_N + \frac{A}{2}$. The manner in which the spectrum collapses into a single line is treated in text books[4] though for somewhat different systems. Here we will assume that there is only a single line and use (2.4) both to estimate its width and to show how the width is related to the dynamic correlation function.

The result in this limit of rapid exchange modulation was derived by Kubo and Tomita:[5]

$$\langle \exp[i\int_0^t \Delta\omega(t')dt']\rangle \approx \exp[-\int_0^t (t-\tau)\langle \Delta\omega(\tau)\Delta\omega(0)\rangle d\tau \qquad (2.6)$$

Kubo[6] later showed that (2.6) is exact for the case in which $\langle \Delta\omega(\tau)\Delta\omega(0)\rangle$ decays rapidly to zero in a time τ_c much less than the observing time t, which is of the order of the NMR relaxation time.

The resonance absorption at a frequency ω relative to the central component at ω_N is given by the Fourier frequency transform of Eq. (2.6). Although (2.6) has been derived here for a special case, it turns out to be applicable in most cases provided we let[5]

$$\langle \Delta\omega(\tau)\Delta\omega(0)\rangle = \langle [\mathcal{K}'(\tau), M_+][M_-, \mathcal{K}']\rangle/\langle M_+M_-\rangle \qquad (2.7)$$

where $M_\pm$ are the raising and lowering operators associated with the magnetization whose resonance is being observed. That is, $M_\pm = \gamma_N\hbar I_\pm$ for NMR and $M_\pm = \gamma_e\hbar S_\pm$ for ESR where γ_e is the electronic gyromagnetic ratio and $I_\pm$ and $S_\pm$ refer to the total spins of the crystal (equivalent spins, i.e. all spins having the same γ_e or γ_N, are assumed). Time dependence of the perturbation $\mathcal{K}'$ should include the Zeeman Hamiltonian $\mathcal{K}_z$ as well as $\mathcal{K}_e$. It is a straightforward exercise to show that (2.7) is consistent with the previous definition $\Delta\omega = AS_i^z$ from the Hamiltonian of (2.2) in which $\mathcal{K}' = \hbar AI^zS_i^z$.

The Kubo-Tomita expression (2.6) has been criticized[7] for use in exchange-narrowed low-dimensional systems for a reason which will become apparent. However, its use has produced good agreement with experiment and the proposed corrections[7] appear to be fairly negligible. Thus Eq. (2.6) will form the basis of subsequent discussion. Note, however, that (2.6) definitely is not correct if $\Delta\omega(t)$ does not fluctuate rapidly in time since it predicts a single line with a Gaussian shape [if $\Delta\omega(t)$ is independent of t, then we have $\langle \Delta\omega(\tau)\Delta\omega(0)\rangle = \langle \Delta\omega^2\rangle = A^2/4$ for spin $\frac{1}{2}$; and $\int_0^t (t-\tau)\langle \Delta\omega^2\rangle d\tau = A^2\tau^2/8$ so that the line shape is the Fourier transform of $e^{-A^2\tau^2/8}$, which is itself a Gaussian] whereas we know that the correct description is two lines at $\omega_N \pm A/2$.

III. FEATURES OF $\langle S_i^z(t)S_i^z\rangle$ AND LOW-DIMENSIONAL EFFECTS

Within the Kubo-Tomita formalism, the NMR problem reduces to finding $\langle S_i^z(t)S_i^z\rangle$, and the general magnetic resonance problem is given in terms of the correlation function shown in Eq. (2.7). We can make the following observations about $\langle S_i^z(t)S_i^z\rangle$. First, if there are N spins in the system, then we expect $4\langle S_i^z(t)S_i^z\rangle \to 1/N$ as $t \to \infty$, so that the correlation decays to zero in an effectively infinite chain. The reason may be seen from the physical interpretation given under Eq. (2.1). At very long times the disturbance, which initially was confined to i, should be spread evenly throughout the system, and thus the probability of its being on any one site is 1/N. Secondly, the short time behavior can be computed exactly from a power series expansion

$$\langle S_i^z(\tau)S_i^z(0)\rangle = \langle S_i^{z^2}\rangle + \tau\,\langle \dot{S}_i^z S_i^z\rangle + \frac{\tau^2}{2}\,\langle \ddot{S}_i^z S_i^z\rangle + \dots \qquad (3.1)$$

where $\dot{S}_i^z$ and $\ddot{S}_i^z$ are the first and second time derivatives at $\tau = 0$, and correlations are evaluated at $\tau = 0$. At infinite temperature the static correlations in (3.1) can be evaluated, although their complexity rapidly increases with successive powers of t, and it has not proved practical to go beyond the t^8 term[8] for a three-dimensional lattice. A simplification at infinite temperature is that all coefficients of odd powers of τ in (3.1) are zero. Between the short times, $\tau \lesssim \hbar/J$, and infinite time is an unfortunately large gap where information is needed most. In the typical exchange-narrowed system we have $\gamma\Delta H \ll J/\hbar$ where ΔH is the observed linewidth. Thus the observing times t in (2.6) are very much longer than the times over which an expansion such as (3.1) is useful. Equivalently, the measurable part of the magnetic resonance spectrum consists of frequencies (relative to the line center) which are much smaller than the exchange frequency which characterizes short-time spin flips.

One needs some way of estimating $\langle S_i^z(t)S_i^z(0)\rangle$ for times longer than $\sim \hbar/J$. A common practice has been to fit the first two terms of (3.1) to a Gaussian,

$$\langle S_i^z(\tau)S_i^z\rangle \approx \langle S_i^{z^2}\rangle e^{-\frac{1}{2}\omega_e^2\tau^2} \qquad (3.2)$$

in which

$$\omega_e^2 = -\langle \ddot{S}_i^z S_i^z\rangle/\langle S_i^{z^2}\rangle \qquad (3.3)$$

by comparison with (3.1). For the more general case of (2.7) we would have

$$\omega_e^2 = -\langle \Delta\ddot{\omega}\Delta\omega\rangle/\langle \Delta\omega^2\rangle = \langle |[\mathcal{H}_e,[\mathcal{H}',M_+]]|^2\rangle/\langle |[\mathcal{H}',M_+]|^2\rangle \qquad (3.4)$$

It is important to note that (3.3) and (3.4) do not in general give the same value for ω_e, even in the same lattice with the same exchange interaction, since the particular form of $\mathcal{H}'$ matters. An ω_e computed from a contact $AI^z S_i^z$ interaction will not be the same as computed from the more general coupling $\sum_{\alpha\beta j} A_j^{\alpha\beta} I^\alpha S_j^\beta$ where the α component of $\vec{I}$ is coupled to the β components of several spins $\vec{S}_j$; and it will also differ from an ESR ω_e where $\mathcal{H}'$ involves electronic dipole-dipole coupling.

Use of (3.2) in (2.6) gives

$$\exp[-\int_0^t (t-\tau)\langle\Delta\omega(\tau)\Delta\omega(0)\rangle\,d\tau] \approx e^{-\gamma\Delta Ht} \tag{3.5}$$

with

$$\gamma\Delta H = \int_0^\infty d\tau\ \langle\Delta\omega(\tau)\Delta\omega(0)\rangle = \left(\frac{\pi}{2}\right)^{\frac{1}{2}}\frac{\langle\Delta\omega^2\rangle}{\omega_e} \tag{3.6}$$

The line is thus Lorentzian (Fourier transform of an exponential) with a half-width in frequency units given by $\gamma\Delta H$. The above assumes that $t \gg 1/\omega_e$ so that the upper limit may be extended to ∞ and $t \gg \tau$ throughout the region where $\langle\Delta\omega(\tau)\Delta\omega(0)\rangle$ is significantly non-zero.

Equation (3.6) has frequently been used to describe exchange narrowed NMR and ESR and has generally met with success in three-dimensional (3d) systems. The reason for the at least qualitative success is that, although the Gaussian form may not be strictly correct, as long as the correlation does decay in a characteristic time of the order of $1/\omega_e$, (3.6) should still be reasonable even though the coefficient $\left(\frac{\pi}{2}\right)^{\frac{1}{2}}$ may be somewhat different. In partic-ular, relative variations such as angular, temperature, and fre-quency dependence of ΔH may be adequately treated by (3.6). Even in 3d, however, care has to be exercised in making numerical esti-mates of ΔH, and the Gaussian form may give ΔH as much as a factor of 2 too small.[9]

The short time expansion, upon which the exchange narrowing formula (3.6) is based, is quite insensitive to dimensionality. ω_e depends only on the number of neighbors coupled by the interac-tion, not on their geometrical arrangement. The effect of the number of dimensions d is particularly strong if, as is believed, the correlation function is characterized by diffusion by high temperatures. Remembering that $\langle S_i^z(\tau)S_i^z\rangle$ represents the probability that the disturbance (or spin deviation) remains on site i at time τ, one sees that $\langle S_i^z(\tau)S_i^z\rangle \propto \tau^{-d/2}$ for $\tau \to \infty$ since this is the dependence predicted by the classical diffusion equation. As has been pointed out several times now in the literature,[10] the integral in (2.6) converges at the upper limit for d = 3 but not for d = 1 or 2. As a consequence, the line shape is predicted to be non-Lorentzian in less than 3 dimensions and the width may have quite anomalous dependences.

Since the $\tau^{-3/2}$ decay in 3d is sufficiently rapid to make $\int^\infty \langle\Delta\omega(\tau)\Delta\omega(0)\rangle\,d\tau$ converge, we see that the long time diffusive tail does not seriously alter the exchange narrowing formula (3.6). On the other hand, for d = 1 we would have

$$\int_0^t (t-\tau)\langle \Delta\omega(\tau)\Delta\omega(0)\rangle \, d\tau \propto \frac{\langle \Delta\omega^2\rangle}{\omega_e^{\frac{1}{2}}} \, t^{3/2} \quad , \tag{3.7}$$

for $t \gg 1/\omega_e$, assuming the correlation to be proportional to $\langle \Delta\omega^2\rangle (\omega_e\tau)^{-1/2}$ for long times, and the characteristic decay time $1/\Gamma$, when (3.7) is unity is

$$\Gamma \sim \frac{\langle \Delta\omega^2\rangle^{2/3}}{\omega_e^{1/3}} \sim \gamma\Delta H \tag{3.8}$$

ΔH thus has a different dependence on the parameters in a low dimensional system. Actually, the difference can be even more drastic than indicated by (3.8). Equation (3.8) assumes that the coefficient of the asymptotic $\tau^{-d/2}$ decay is the same as that of the short-time Gaussian dependence, namely $\langle \Delta\omega^2\rangle$. This is not true, as may be seen by an expansion of the correlation function into wave vector normal modes

$$\langle S_i^z(\tau)S_i^z\rangle = N^{-1} \sum_k \langle S_k^z(\tau)S_{-k}^z\rangle \tag{3.9}$$

For the case of a more general hyperfine interaction $\hbar\sum_j A_j I^z S_j^z$, the pertinent function would be

$$\sum_{j,\ell} A_j A_\ell \langle S_j^z(\tau)S_\ell^z\rangle = N^{-1}\sum_k |A_k|^2\langle S_k^z(\tau)S_{-k}^z(0)\rangle \tag{3.10}$$

where $A_k = \sum_j A_j e^{i\vec{k}\cdot\vec{r}_j}$, $S_k^z = N^{-\frac{1}{2}} \sum_j S_j^z e^{i\vec{k}\cdot\vec{r}_j}$.
The $\tau = 0$ value is

$$\langle \Delta\omega^2\rangle = N^{-1} \sum_k |A_k|^2 \langle S_k^z S_{-k}^z\rangle$$

and thus consists of contributions over the whole Brillouin zone. For long times, however, the $k \to 0$ modes dominate because of their much slower rate of decay. Diffusion means that $\langle S_k^z(\tau)S_{-k}^z(0)\rangle = \langle S_k^z S_{-k}^z\rangle e^{-Dk^2\tau}$ where D is the diffusion coefficient. Upon converting the sum in (3.10) to an integral we then get

$$\sum_k |A_k|^2 \langle S_k^z(\tau) S_{-k}^z(0) \rangle \propto \int k^{d-1} dk |A_k|^2 \langle S_k^z S_{-k}^z \rangle e^{-Dk^2\tau}$$

$$\text{(3.11)}$$

$$\propto |A_0|^2 \langle S_0^z S_0^z \rangle (D\tau)^{-d/2}$$

where the second equality holds at long times for which only wave vectors with $k \approx 0$ contribute. As well as reconfirming the $\tau^{-d/2}$ decay, Eq. (3.11) shows that the coefficient is not $\langle \Delta\omega^2 \rangle$ but rather the $k = 0$ part of $\langle \Delta\omega^2 \rangle$, which can have a different angular and temperature dependence.

IV. PRIME EXPERIMENTAL EXAMPLES

The previous section has pointed to two qualitative differences expected in the exchange-narrowed magnetic resonance of low-dimensional compounds. (i) the line shape should be non-Lorentzian, given by the Fourier transform of $\exp(-t^{3/2})$ for a 1d system and (ii) the angular and temperature dependence may be different from that predicted by $\langle \Delta\omega^2 \rangle$ since only the $k = 0$ part, $\langle \Delta\omega^2 \rangle_{k=0}$, contributes to the long time diffusive decay of the correlation function, whereas all wave vectors k are important for the short time behavior. (For 1d and 2d the long time decay is the most important.) In this section we point to the main triumphs of relating the theory to experiment, and temporarily forego some of the pitfalls.

The $\exp(-t^{3/2})$ line shape has been well verified in[11] spin 5/2 $(CH_3)_4NMnCl_3$ (TMMC) and in[12] spin $\frac{1}{2}$ CuNSal, both of which are highly one-dimensional. The dependence upon $\langle \Delta\omega^2 \rangle_{k=0}$ has been documented[13] in the 2d compound K_2MnF_4. For a 2d system the angular dependence (angle of the field H with respect to the plane) of $\sum_k |F_k|^2$ is quite different from that of $|F_0|^2$, where F_k is the dipolar factor appropriate to ESR, and in K_2MnF_4 one sees an angular dependence more nearly related to $|F_0|^2$ than to $\sum_k |F_k|^2$, which confirms the importance of the long time $k \to 0$ diffusive modes. Furthermore, the temperature dependence[13] of ΔH in K_2MnF_4 also reflects that of $\langle \Delta\omega^2 \rangle_{k\to 0}$ rather than the complete wave vector sum. These examples confirm the basic concepts of resonance in low-dimensional exchange-coupled systems. There are, however, difficulties and subtle effects to be discussed in the next sections which can cloud the picture.

V. THE PROBLEM OF FOUR-SPIN CORRELATIONS

The prime examples in the previous section were for ESR, in which $\mathcal{K}'$ is the dipole-dipole interaction, whereas the treatment in Sects. II and III were explicitly for NMR. In NMR the general hyperfine perturbation of the form $A_j^{\alpha\beta} I^{\alpha} S_j^{\beta}$ leads to a $\langle \Delta\omega(\tau)\Delta\omega(0)\rangle$ as defined in (2.12) which involves two electron spins. For the ESR problem $\mathcal{K}'$ is quadratic in the electron spin operators and (2.7) takes the form

$$\langle \Delta\omega(\tau)\Delta\omega(0)\rangle \propto \sum_{ijk\ell} \sum_{\substack{\alpha\beta \\ \alpha'\beta'}} F_{ij\alpha\beta}^{\alpha\beta} F_{k\ell}^{\alpha'\beta'} \langle S_i^{\alpha}(\tau)S_j^{\beta}(\tau)S_k^{\alpha'}(0)S_{\ell}^{\beta'}(0)\rangle \quad (5.1)$$

A major concern is whether an expression such as (5.1) is proportional to $\tau^{-d/2}$ for long times, even given that the two-spin correlation does have this dependence. Experimentally, the fact that an $\exp(-t^{3/2})$ line shape has been observed in 1d compounds lends credence to the conclusion that the four-spin function Eq. (5.1) is diffusive (i.e. goes like $\tau^{-d/2}$). But theoretically there has been no definitive work of which we are aware. A common approximation has been to factor the four-spin function into a product of two two-spin ones according to, for example,

$$\langle S_i^z(\tau)S_j^x(\tau)S_k^z(0)S_{\ell}^x(0)\rangle = \langle S_i^z(\tau)S_k^z(0)\rangle\langle S_j^x(\tau)S_{\ell}^x(0)\rangle \quad (5.2)$$

If (5.2) is used in (5.1) and a spatial Fourier transform is performed, one finds that

$$\langle \Delta\omega(\tau)\Delta\omega(0)\rangle \propto \sum_k |F_k|^2 \langle S_k^z(\tau)S_k^z\rangle^2 \quad (5.3)$$

where we have further assumed isotropy of the correlations, $\langle S_k^{\alpha}(\tau)S_k^{\alpha'}(0)\rangle = \langle S_k^{\alpha}(\tau)S_k^{\alpha}(0)\rangle\delta_{\alpha\alpha'}$, (the same for α = x, y or z), which is valid for the Heisenberg model. Now, if $\langle S_k^z(\tau)S_k^z\rangle \propto e^{-Dk^2\tau}$, which produces a $\tau^{-d/2}$ behavior for the two-spin function as in Eq. (3.11), then it is evident that (5.3) will also be proportional to $\tau^{-d/2}$. In fact, the only difference between (5.3) and (3.11) would be that the effective diffusion coefficient is 2D rather than D.

Given the approximation (5.2) one can numerically calculate the linewidth in a linear chain system, provided $\mathcal{K}'$ is known. The expression is given later in Sect. VII. TMMC seems to be a nearly ideal crystal in which to make a quantitative comparison. Since the S-state Mn^{++} ion is involved, the dominant broadening mechanism

should be the classical dipole-dipole interaction, with little complications arising from anisotropic exchange or spin-lattice relaxation. Furthermore, the intrachain exchange interaction J is known
from magnetic susceptibility[14] and neutron scattering measurements.[1]
The calculated linewidth[7] exceeds the experimental one[7,11] by a factor
of 1.8. This, together with the observed line shape, may suggest
that the four-spin function (5.1) does indeed decay like $\tau^{-d/2}$, as
predicted by the factorization (5.2), but the coefficient of $\tau^{-d/2}$
is smaller than given by (5.2) and (5.3). However, it is not completely clear that only the dipolar interaction is involved. It
has been proposed[15] that an appreciable single ion anisotropy
DS_i^{z2} is present, and the sign of D is such as to subtract from the
dipolar interaction at k = 0; and thus this interaction could
actually make the line narrower.

VI. INTERCHAIN AND INTERPLANE COUPLING

Although the $\exp(-t^{3/2})$ line shape has been observed in some
1d compounds, there is a goodly number of other seemingly 1d systems for which Lorentzian lines, characteristic of 3d, are observed.
As well as transition-metal-ion salts, the list includes all the
organic charge transfer crystals for which ESR has been studied in
detail. Susceptibility, specific heat and, in some cases, neutron
scattering measurements have shown that the samples are indeed
quasi 1d, with interchain exchange interactions J' often less than
the intrachain interaction J by a factor of 10^{-2} or less.

The answer to the above dilemma is that it apparently takes
only a very small J' to change the observed line shape to Lorentzian.
Eventually the correlation function $\langle \Delta\omega(\tau)\Delta\omega(0) \rangle$ will decay as $\tau^{-3/2}$
characteristic of diffusion in 3d, because of the finite coupling
between chains. Spin-lattice relaxation and/or intrachain dipole
coupling can also force a decay more rapid than $\tau^{-1/2}$. The departure
from a $\tau^{-1/2}$ decay can roughly be described by a cutoff time t_o, as
shown in Fig. 6.1. We assume that t_o is sufficiently long that

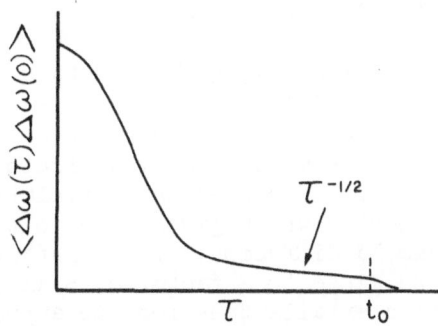

Fig. 6.1. Diffusive time correlation cut off at time t_o.

$\langle \Delta\omega(t_0)\Delta\omega(0)\rangle \ll \langle \Delta\omega^2\rangle$. Then the exact dependence of $\langle \Delta\omega(\tau)\Delta\omega(0)\rangle$ for $\tau > t_0$ is unimportant as long as the integral in (2.7) converges at the upper limit, and we have

$$\int_0^t \langle \Delta\omega(\tau)\Delta\omega(0)\rangle(t-\tau)d\tau = \left\{ \begin{array}{l} (\Gamma_{1d}t)^{3/2} \quad , \quad \frac{1}{\omega_e} \ll t \leq t_0 \\[3mm] (\Gamma_{1d}t_0)^{3/2}\left[\frac{3}{2}\left(\frac{t}{t_0}\right) - \frac{1}{2}\right] , \end{array} \right\}$$

$$t > t_0 \qquad\qquad (6.1)$$

where Γ_{1d} is the decay rate which would be observed in an ideal Heisenberg linear chain.

The crucial parameter is $\Gamma_{1d}t_0$. If $\Gamma_{1d}t_0 \gg 1$, then the observed magnetization decays to a very small value before the cutoff effects come into play, and the resonance is indistinguishable from that of a pure 1d system. But if $\Gamma_{1d}t_0 \ll 1$, the cutoff effects take over before the magnetization has had a chance to decay appreciably, and the line becomes Lorentzian since the decay is exponential for all times of interest. In the intermediate region $\Gamma_{1d}t_0 \sim 1$, the line shape is non-Lorentzian but the departure from Lorentzian is less than for the $\exp(-t^{3/2})$ behavior of the ideal chain.

The fact that a finite cutoff time t_0 would change the decay from $\exp(-t^{3/2})$ to exponential was recognized by Soos[16] several years ago in connection with the ESR of linear chain organic free radials, where Lorentzian lines were seen. More recent efforts [17-19] have focused on obtaining a relation between t_0 and J' and using a more realistic correlation function than one which is simply cut off at t_0. Probably the most important conclusion to have emerged is that

$$t_0^{-1} \propto |J'|^{4/3}/|J|^{1/3} \qquad\qquad (6.2)$$

This dependence bears striking similarity to the dependence in Eq. (3.9) of Γ on $\langle \Delta\omega^2\rangle^{1/2}$ and J ($J \propto \omega_e$), and is what one should expect — at least after some deep thought. In calculating Γ, the perturbation rate $\langle \Delta\omega^2\rangle^{1/2}$ is modulated by diffusion at a rate proportional to J which, in 1d, causes the correlation to decay as $\tau^{-1/2}$. Here the perturbation which gives rise to t_0^{-1} is J', and it is modulated by the same 1d diffusion. The constant of proportionality in (6.2) varies by close to a factor of 5 among the various theories,[19] each of which entails questionable approximations.

The value of $\Gamma_{1d} t_o$ depends on the strength of the dipole coupling and on J', but is independent of J'. Calculations[17,20] for typical intrachain distances show that if $|J'|/k_B \gtrsim 0.03°K$, the line will be nearly indistinguishable from Lorentzian, whereas if $|J'|/k_B \lesssim 0.003°K$ a non-Lorentzian line should be observable.

The extent to which J' and t_o can be measured will now be discussed. The most direct evidence of a quasi 1d compound becoming 3d in character comes from the transition to a magnetically ordered state, generally antiferromagnetic, at a critical temperature T_c. Since there can be no long range order in a purely 1d system, the existence of T_c, as determined by a specific heat and/or susceptibility anomaly, is directly related to J'. The precise relation, however, is somewhat in doubt since there is no exact theory which gives T_c in terms of J' and J. The most widely used expression, based on the random phase approximation and Green functions, is[17,21]

$$k_B T_c = 2.08 \left| JJ' \right|^{\frac{1}{2}} S(S+1) \qquad (6.3)$$

for the case of a tetragonal lattice with a single interchain exchange constant between the 4 nearest neighbors perpendicular to the chain axis. Modifications of (6.3) for different lattice geometries[17,20], more than one interchain exchange constant,[17,22] and other Green function decouplings[17,20,23] have been treated in the literature. J may be determined from the susceptibility or specific heat at temperatures above T_c where there is a broad maximum associated with short range ordering. Exact or nearly exact theories are available here, so the values of J obtained by this method are expected to be quite accurate. This topic will be covered by Prof. Gerstein. The extensive review by de Jongh and Miedema[24] is also recommended. We are aware of one instance,[25] in $CsMnCl_3 \cdot 2H_2O$, where both J and J' have been determined by inelastic neutron scattering from the spin wave spectrum below T_c. The observed J' was about a factor of 2 smaller than given by (6.3) in terms of the known J and T_c.

NMR and ESR have been employed to estimate t_o and, in some cases, to compare the J' resulting from (6.2) with other measurements. In the intermediate range $\Gamma_{1d} t_o \sim 1$ the line shape will depend on t_o and therefore can be used to infer J'. This was done in ref. 17 for $CsMnCl_3 \cdot 2H_2O$ where the small departure from Lorentzian was fit to a theoretical line shape for which $|J'/J| \approx 2 \times 10^{-2}$, which is about 4 times larger than given by the neutron measurements.[25] The non-Lorentzian line reported in ref. 17 for $CuCl_2 \cdot 2NC_5H_5$ (CPC) was later shown to be in error,[20] and caused by misalignments within the bundle of needle-like crystals used.

The ESR width ΔH of a Lorentzian line can be written as

$$\Delta H = F_0(0) + F_1(\omega_o) + F_2(2\omega_o) \qquad (6.4)$$

where ω_o is the microwave frequency and $F_i(\omega)$ is the Fourier component of a spin correlation function such as (5.2). By judicious measurement of the angular and frequency dependence of ΔH it is sometimes possible to sort out the individual $F_i(\omega)$. The components at ω_o and $2\omega_o$ are adequately given by the short-time approximations, such as Gaussian, since the rapid modulation at ω_o or $2\omega_o$ washes out the long-time diffusive effects. If the broadening mechanism $\mathcal{K}'$ is known, one can then determine J from $F_1(\omega_o)$ and/or $F_2(2\omega_o)$. Next we note that $F_0(0)$ is proportional to $t_o^{1/2}$ ($\int_0^{t_o} t^{-1/2} dt \propto t_o^{1/2}$) if t_o is sufficiently long, and the constant of proportionality can be estimated from the short-time behavior involving the known $\mathcal{K}'$ and J as measured above. Thence t_o may be determined from a thorough study of the ESR width, even though the line is Lorentzian. This method has been used[26] for the salt $Cu(NH_3)_4PtCl_4$, and the reader is referred to that work for details of the calculation.

Both the NMR width and longitudinal relaxation rate $1/T_1$ are given by expressions similar to (6.4), except that the final term $F_2(2\omega_o)$ is conveniently absent. These expressions, of course, refer only to that part of the NMR relaxation which is caused by coupling of the nuclei to electron spins. Since the width is strongly influenced by nuclear dipole-dipole interactions as well, it is often difficult to study the electron spin correlations of interest by measuring the NMR linewidth. However, these nuclear-nuclear couplings do not influence T_1, and much useful information has been obtained from pulse measurements of the proton T_1 in 1d compounds. The dipolar coupling of a proton to electron spins is sufficiently weak that one should always be in the $\Gamma_{1d} t_o \ll 1$ region for NMR (remember that Γ_{1d} depends on the strength of $\mathcal{K}'$ as well as on J). Hence, the nuclear magnetization decays exponentially.

Because of the absence of $F_2(2\omega_o)$ and the appearance of only two-spin correlation functions, it is generally more straightforward to extract t_o from the frequency dependence of the NMR T_1 than from ESR linewidths. Recent studies[27-29] on TMMC and on the free radical Tanol[19] have successfully determined t_o.

Another, and quite direct, ESR method of measuring t_o is available in systems for which the chains are inequivalent owing to different orientations of the surrounding ligands. Suppose there are two chain types A and B whose principal axes of the anisotropic g-tensor make different angles with respect to the applied field, except when H is along symmetry directions. Then there will be

distinct frequencies ω_A and ω_B at which A and B would undergo resonance in the absence of any interaction between A and B type
chains. A J'_{AB} couples the chains and, if sufficiently strong,
merges the two lines into one with a width of the order of
$(\omega_A-\omega_B)^2 t_o$, where t_o is now the characteristic time for spin polarization to diffuse from type A to type B chains. Since $\omega_A-\omega_B$ is
proportional to H, the contribution of this mechanism to ΔH can
readily be identified by its strong field dependence. Further,
$\omega_A-\omega_B$ generally can be determined precisely from the average g-
tensor of the crystal plus symmetry considerations. Thus t_o can be
measured in a fairly direct manner. The method has recently been
applied[22] to $CuCl_2 \cdot 2NC_5H_5$(CPC).

In a quasi 2d system coupling between planes cuts off the
long time τ^{-1} decay of spin correlations in a manner completely
analogous to the above-mentioned effect of interchain coupling for
1d. The consequences of a finite cutoff t_o are not as marked in
2d as in 1d, simply because the low-dimensional anomalies are not
as drastic to begin with in 2d. The departure from a Lorentzian
line shape, though observable[13] in 2d, is nowhere near as great in
2d as in 1d, for example. Nonetheless, effects of t_o have recently
been documented[30] in the quasi-2d crystal $NaCrS_2$. A rather large
ratio of $|J'/J| = 1/6$ has been determined[31] from the high field
magnetization in the ordered state, which consists of ferromagnetic
planes antiferromagnetically coupled. (Here J' and J are inter-
and intraplane interactions, respectively.) It has been shown[30]
that a $|J'/J|$ this large alters the angular dependence of ΔH, and
indeed one observes a different angular variation in $NaCrS_2$ than
in[13] the nearly ideal 2d K_2MnF_4 and which agrees well with theory.[30]

VII. SPECIFIC FORMULAE FOR ESR LINEWIDTH

There is an obvious advantage to the relatively simple formula
(3.6) which, together with the definition (3.4) for ω_e, can be used
to predict the linewidth in 3d. The previous sections have pointed
to the anomalies and subtleties of resonance in 1d and in 2d which
make Eq. (3.6) quite erroneous. Unfortunately, it should also be
apparent by now that one cannot hope to have quite so compact an
expression. Indeed, a separate numerical computation may be
required for each compound in order to make meaningful comparisons
between theory and experiment. Perhaps it is out of despair over
the situation that several workers seem to persist in using (3.6),
even in 1d, for want of a replacement formula. In an effort to
give some guidelines to experimentalists who, rightly, do not wish
to take the pains of some of the complex calculations described in
the literature, we present in Table 7.1 some approximate formulae
which should be in the right ball park for the range of parameters
mentioned. The accompanying notes to the Table describe the limits
of validity and explain some of the symbols.

Table 7.1 Linewidth Formulae

System	General Formula	Numerical Values
1d, negligible interchain coupling	$$\frac{0.89\langle\Delta\omega^2\rangle^{2/3}_{k_z=0}}{(D/c^2)^{1/3}} \equiv 2.16\,\frac{\Gamma_{ld}}{\gamma}$$	$$1340|J|^{-1/3}(c/3)^{-4}\sqrt{S(S+1)}\left(\frac{g_c}{2}\right)^3$$
1d, appreciable interchain coupling ($\Gamma_{ld}t_o < 1$)	$$\sqrt{3}\,\Gamma_{ld}^{3/2}\,t_o^{\frac{1}{2}}$$	$$260(c/3)^{-6}[S(S+1)]^{11/6}|J|^{1/3}T_c^{-4/3}\left(\frac{g_c}{2}\right)^3$$
2d, negligible interplane coupling	$$\frac{2}{\sqrt{3}}\frac{\langle\Delta\omega^2\rangle_{k_\perp=0}}{8\pi D/a^2}\left\{\ln\left[\frac{(8\pi D/a^2)^2}{\langle\Delta\omega^2\rangle_{k_\perp=0}}\right]-1\right\}$$ $$\equiv\frac{2}{\sqrt{3}}\Gamma_{2d}\equiv\frac{2}{\sqrt{3}}\Gamma_o\left\{\ln\left(\frac{1}{\Gamma_o\tau_o}\right)-1\right\}$$	$$213\left(\frac{a}{3}\right)^6|J|^{-1}\sqrt{S(S+1)}\left(\frac{g_\parallel}{2}\right)^3$$ $$\times\{\ln[48\,J^2(a/3)^6(2/g_\parallel)^4]-1\}$$
2d, appreciable interplane coupling, $\Gamma_{2d}t_o < 1$	$$\frac{2}{\sqrt{3}}\Gamma_{2d}\left\{1-\frac{\Gamma_o}{\Gamma_{2d}}\left[\ln\left(\frac{1}{\Gamma_o\tau_o}\right)-1\right]\right\}$$ $$\Gamma_o\tau_o=\frac{\langle\Delta\omega^2\rangle_{k_\perp=0}}{(8\pi D/a^2)^2}\frac{8}{\pi}\left(\frac{J^2_Z}{J'^2_{Z'}}\right)$$	$$\Gamma_o\tau_o=3.1\times10^{-3}J^{-2}\left(\frac{a}{3}\right)^{-6}\left(\frac{g_\parallel}{2}\right)^4$$ $$\times(J^2_Z/J'^2_{Z'})$$

Notes for Table 7.1

a. Formulae for valid only for field H along chain axis (1d) or perpendicular to plane (2d) and at sufficiently high microwave frequencies that nonsecular contributions are negligible. Also, infinite temperature is assumed.

b. Except for bottom equation for $\Gamma_o t_o$, all formulae are for peak-to-peak linewidth ΔH_{pp}.

c. In last column J is in °K, lattice spacings a and c are in Å, T_c is in °K and ΔH_{pp} is in Gauss.

d. Formulae in last column for 1d (2d) assume intrachain dipolar coupling only (intraplane dipolar coupling only, square lattice).

e. $\langle \Delta \omega^2 \rangle_{k_z=0}$ is defined as follows. If the complete second moment is $\langle \Delta \omega^2 \rangle = N^{-1} \sum_k |F_k|^2 = N^{-1} \sum_{ij} |F(\vec{r}_{ij})|^2$, then

$$\langle \Delta \omega^2 \rangle_{k_z=0} = \left| F_{k_z=0} \right|^2 = N^{-2} \sum_{ij} \sum_{k\ell} F(\vec{r}_{ij}) F(\vec{r}_{k\ell}) \delta(\vec{r}_{ij} - \vec{r}_{k\ell})$$

where the Kronecker delta signifies that $\vec{r}_{ij}$ and $\vec{r}_{k\ell}$ must have the same components perpendicular to the chain. If only intrachain coupling is involved in $\mathcal{K}'$, then $F(\vec{r}_{ij})$ is zero unless $\vec{r}_{ij}$ is zero and the above reduces to

$$\left| F_{k_z=0} \right|^2 = N^{-2} \sum_{ij} F(z_{ij}) \sum_{k\ell} F(z_{k\ell}) = \sum_j F(z_{ij})^2 \,,$$

intrachain interactions only

The more general form is necessary if interchain dipolar coupling is included, but in compounds such as TMMC where the interchain distance between ions is much less than the intrachain separation, the above is sufficient. A similar definition holds for $\langle \Delta \omega^2 \rangle_{k_\perp=0}$, in which the Kronecker delta would be $\delta(z_{ij}-z_{k\ell})$ with z the direction perpendicular to the plane.

f. D is the diffusion coefficient.[32] For 1d we take the result obtained by computer studies of the motion of classical spins[33] and assume D is proportional to $\sqrt{S(S+1)}$ for general spin. Thus

$$D(1d) = 2.66 \, |J| \sqrt{S(S+1)} \, c^2/\hbar$$

(In ref. 33 the constant is 1.33, but the exchange coupling there is taken as $J\vec{S}_i \cdot \vec{S}_j$, rather than $2J\vec{S}_i \cdot \vec{S}_j$, as used here.)

For 2d we use[13]

$$D(2d) = \frac{1}{3} (2\pi)^{1/2} |J| a^2 \sqrt{S(S+1)} (Z/4)^{1/2}/\hbar$$

for Z neighbors at a distance a coupled by an intraplane exchange interaction J.

g. t_0 is cutoff time beyond which time correlation functions decay 3-dimensionally.

h. $\zeta = \langle \Delta\omega^2 \rangle_{k_\perp =0} / \langle \Delta\omega^2 \rangle = \left(\sum_j F_{ij} \right)^2 / \sum_j F_{ij}^2$, if only intraplane dipole coupling is considered.

i. Z and Z' are numbers of intra- and interplane neighbors.

j. g_c is g-factor parallel to chain axis; $g_{||}$ is g-factor along axis perpendicular to plane.

k. Formula involving T_c is for tetragonal lattice (equal interchain exchange coupling to 4 nearest interchain neighbors). T_c is related to t_0 by Eq. (6.3) and[17].

$$t_0 = \left[\frac{9}{32\, S(S+1)} \right]^{2/3} \left(\frac{2\pi D}{c^2} \right)^{1/3} \left(\frac{J'}{\hbar} \right)^{-4/3} \cdot 4^{-2/3}$$

which is $4^{2/3}$ smaller than the quantity defined as "t_0" in ref. 17, the difference being that the dipole correlations decay at the rate $4^{2/3}$ "t_0^{-1}" [see Eq. (39) of ref. 17]. The above assumes a single interchain exchange constant J' and 4 nearest interchain neighbors.

The following comments are in order regarding some of the results in the table.

1. The expression for 1d, negligible interchain coupling overestimates the linewidth in TMMC by about a factor of 1.8, a fact which must be taken into account when quantitative comparisons are made. However, it does give a reasonable estimate and has a definite prediction as to dependence on J and lattice spacing which can be checked against experiment, namely $\Delta H \propto |J|^{-1/3} c^{-4}$. For example,[34] $\Delta H |J|^{1/3} c^4$ differs by only 25% between TMMC and PyHMnCl$_3 \cdot$H$_2$O whereas $\Delta H |J| c^6$ differs by a factor of 5. PyHMnCl$_3 \cdot$H$_2$O has a non-Lorentzian line and other ESR features characteristic of a nearly-ideal linear chain.[34] If the 3d formula (3.6) were to apply, then the widths should scale as $|J|^{-1} c^{-6}$, which they obviously do not.

2. As an example of the use of the formula for 1d, appreciable interchain coupling, the numbers for CTS (c=5.33Å, J=3.15°K,

$T_c = 0.43°K$, $S = 1/2$, $g_c = 2.18$) give $\Delta H_{pp} = 29$ Gauss, compared with the observed[7] $\Delta H_{pp} = 35$ Gauss, which is remarkable, and likely fortuitous, agreement in view of the approximations made.

For CPC the agreement is nowhere near as good, a fact which may at least partly be due to the complex interchain interactions in that salt.[22]

3. The expressions for 2d are analytic approximations to the detailed calculations which match short- and long-time portions of the time correlation functions. For K_2MnF_4 (negligible interplane coupling) the formula in the table gives $\Delta H_{pp} = 156$ Gauss whereas the more detailed treatment of ref. 13 gave $\Delta H_{pp} = 127$ Gauss at infinite temperature.

4. For 2d the cutoff time t_o probably cannot be related to the 3d ordering temperature as it was for a quasi 1d system since very small Ising-like anisotropy within the plane is sufficient to induce 3d order, and, indeed, there is strong speculation that even a pure 2d Heisenberg magnet has a critical temperature.[35] In ref. 30 an anisotropic diffusion coefficient was considered, according to which we would roughly identify t_o by

$$c^2 = 4 \cdot 2 \, D' t_o$$

where D' is the coefficient for diffusion between planes separated by a distance c and where the factor of 2 is introduced because of the four-spin correlations, which are assumed to diffuse at twice the rate of the two-spin ones. D' is given by

$$D'/c^2 = 2(D/a^2)(J'^2 Z'/J^2 Z)$$

which is a generalization of Eqs. (11) and (12) of ref. 30 to the case of Z intraplane neighbors coupled by J and Z' interplane neighbors coupled by $J' \ll J$.

VIII. Discussion

We have shown that the consequences of exchange on the spin dynamics are to give rise to a coupled motion of the electron spins which is believed to be diffusive for long times in the paramagnetic state. This leads to an exchange narrowing of the ESR line in all dimensions, but the narrowing is less extreme in low dimensions, and there are anomalous features of the linewidth and line shape in 1d and 2d systems compared with 3d. We have discussed the nature of these effects and presented some approximate formulae which hopefully will serve as useful guides and replacements for the conventional 3d formula $\Delta H \propto \langle \Delta \omega^2 \rangle / \omega_e$, which simply is not applicable to low-dimensions.

Much has been learned of the nature of spin dynamics from magnetic resonance studies in 1d and 2d, but several questions remain, and much more useful information should be extractable, particularly in regard to the weak interchain or interplane exchange interactions and their relation to the chemical bonding. We would like to close these lecture notes by pointing out where future work is necessary and/or likely to be fruitful and that success of this work very much hinges on the ability of chemists to suggest and synthesize compounds with the desirable properties.

A major question is why the linewidth cannot be calculated more accurately in TMMC, which by all rights should be an ideal 1d system with no complications. Is the problem due to the failure of the factorization approximation for 4-spin correlation functions? This cannot be answered with certainty unless one knows for sure that the predominant broadening mechanism is the classical dipolar interaction. Studies of small amounts of Mn^{++} impurities in a diamagnetic isomorph of TMMC would be most helpful in this regard. The single-ion spectrum could tell one immediately whether a single-ion anisotropy of the size suggested by others[15] is present, and, of course, pair-spectra would pin down the spin-spin interaction. Unfortunately, such a crystal, Mn^{++} in a diamagnetic isomorph, has not yet been produced to our knowledge.

One of the more interesting features, we feel, of magnetic resonance in 1d is the way it is strongly influenced by only a relatively weak interchain coupling J'. This affords the possibility of measuring J' and making an independent check on the various theories which relate the ordering temperature T_c to J'. It would be particularly interesting, for example, to verify the dependence of ΔH upon T_c and other parameters suggested by the equation in Table 7.1. A good way to do this would be to have a series of compounds with the same, or nearly the same, intrachain properties but which differ in the interchain spacing. The linewidths and T_c's of the series could then be used as a reliable test of the theory.

Another direct measure of the cutoff time t_0 is possible in systems with magnetically inequivalent chains. One can then infer the interaction J'_{ie} between inequivalent chains. This information could in principle be used to check the relation between J' and T_c were it not for the fact that T_c depends on the interaction between equivalent chains J'_e as well as on J'_{ie}. Further, there may be more than a single exchange constant between like or unlike chains, as, for example, seems to be the case in CPC.[22] Clearly what is needed is a compound in which there is only one non-negligible interchain exchange constant, and that between inequivalent chains.

The above examples have been used to illustrate the importance of structural considerations and the desirability of obtaining data on new low-dimensional compounds which may be produced by clever chemists and/or crystal growers. Aside from the areas which have been treated in this paper, which has considered only the very high temperature limit, there are many interesting features of magnetic resonance concerned with the temperature dependence. These have been covered in other papers[13,36] and a review[10] and, as the effects mentioned above, they could be further clarified by experiments on more suitable crystals.

List of Text Books and Reviews

The following texts (T) and reviews (R) are general sources for exchange narrowed electron spin resonance and low dimensional systems.

T1. A. Abragam, Principles of Nuclear Magnetism (Oxford U. P., New York, 1961).

T2. C. P. Slichter, Principles of Magnetic Resonance (Harper and Row, New York, 1963).

T3. A. Carrington and A. D. McLachlan, "Introduction to Magnetic Resonance," (Harper and Row, New York, 1967).

R1. R. Kubo, in Fluctuation, Relaxation and Resonance in Magnetic Systems, edited by D. ter Haar (Plenum, New York, 1962).

R2. D. Hone, AIP Conf. Proc. 5, 413 (1972).

R3. P. M. Richards, in International School of Physics "Enrico Fermi," Course LIX, 1973 (to be published by Academic Press).

R4. D. W. Hone and P. M. Richards, Ann. Rev. Materials Sci. 4, 337 (1974).

R5. L. J. DeJongh and A. R. Miedema, Adv. Phys. 23, 1 (1974).

References Cited

1.   R. J. Birgeneau, J. Skalyo, Jr., and G. Shirane, Phys. Rev.
     B3, 1736 (1971).
2.   M. T. Hutchings, G. Shirane, R. J. Birgeneau, and S. L. Holt,
     Phys. Rev. B5, 1999 (1972).
3.   Y. Endoh, G. Shirane, R. J. Birgeneau, P. M. Richards and
     S. L. Holt, Phys. Rev. Lett. 32, 170 (1974).
4.   Text T1, pp. 501-506; Text T2, pp. 205-211; Review R1.
5.   R. Kubo and K. Tomita, J. Phys. Soc. Japan 9, 888 (1954).
6.   Review R1.
7.   G. F. Reiter and J. P. Boucher, "On the Theory of Exchange
     Narrowing in One and Two Dimensions," to be published.
8.   T. Morita, J. Math Phys. 12, 2062 (1971); A. C. Cook, Ph.D.
     Thesis, University of Kansas (1972, unpublished).
9.   J. E. Gulley, D. Hone, D. J. Scalapino, and B. G. Silbernagel,
     Phys. Rev. B1, 1020 (1970).
10. Reviews R2-R4.
11. R. E. Dietz, F. R. Merritt, R. Dingle, D. Hone,
     B. G. Silbernagel, and P. M. Richards, Phys. Rev. Lett. 26,
     1186 (1971).
12. R. R. Bartkowski and B. Morosin, Phys. Rev. B6, 4209 (1972).
13. P. M. Richards and M. B. Salamon, Phys. Rev. B9, 32 (1974).
14. R. Dingle, M. E. Lines, and S. L. Holt, Phys. Rev. 187, 643
     (1969).
15. K. Nagata and Y. Tazuke, J. Phys. Soc. Japan 32, 337 (1972).
16. Z. G. Soos, J. Chem. Phys. 44, 1729 (1966).
17. M. J. Hennessy, C. D. McElwee, and P. M. Richards, Phys. Rev.
     B7, 930 (1973).
18. G. F. Reiter, Phys. Rev. B8, 5311 (1973).
19. J. P. Boucher, F. Ferrieu, and M. Nechtschein, Phys. Rev. B9,
     3871 (1974).
20. W. Duffy, Jr., J. E. Venneman, D. L. Standburg and
     P. M. Richards, Phys. Rev. B9, 2220 (1974).
21. T. Oguchi, Phys. Rev. 133, A1098 (1964).
22. R. C. Hughes, B. Morosin, P. M. Richards, and W. Duffy, Jr.,
     "ESR and Structure of Magnetically Inequivalent Chains in
     $CuCl_2 \cdot 2NC_5H_5$," to be published.
23. P. M. Richards, "Critical Temperatures of Heisenberg Magnets
     by Second-Order Green Function Theory. . .," Phys. Rev. B, to
     be published.
24. Review R5.
25. J. Skalyo, Jr., G. Shirane, S. A. Friedberg, and H. Kobayashi,
     Phys. Rev. B2, 4632 (1970).
26. T. Z. Huang and Z. G. Soos, Phys. Rev. B9, 4981 (1974).
27. F. Borsa and M. Mali, Phys. Rev. B9, 2215 (1974).
28. M. Ahmed-Bakheit, Y. Barjhoux, F. Ferrieu, M. Nechtschein,
     and J. P. Boucher, Solid State Commun. 15, 25 (1974).

29. M. Villa, G. Bonera, and F. Borsa, "Proton Spin-Lattice Relaxation in the Rotating Frame in TMMC. . .," to be published.

30. P. M. Richards, K. A. Müller, H. R. Boesch, and F. Waldner, "ESR in a Two-Dimensional Compound with Appreciable Interplane Coupling: $NaCrS_2$," Phys. Rev. B, to be published.

31. K. W. Blazey and H. Rohrer, Phys. Rev. 185, 712 (1969).

32. See references contained in Ref. 33.

33. N. A. Lurie, D. L. Huber, and M. Blume, Phys. Rev. B9, 2171 (1974).

34. P. M. Richards, R. K. Quinn, and B. Morosin, J. Chem. Phys. 59, 4474 (1973).

35. Review R3.

EPR OF LOW-DIMENSIONAL SYSTEMS

Gerald F. Kokoszka

State University of New York, College of Arts and Science

Plattsburgh, New York 12901 U.S.A.

This paper will be broken down into four major sections: (1) a brief introduction to electron paramagnetic resonance (epr) studies in magnetically concentrated systems, (2) a discussion of epr studies of dimer, trimer and higher order clusters, particularly those involving transition metal ion complexes and the relationship of these studies to one and two dimensional systems, (3) epr studies of semi-dilute systems and finally, (4) a review of selected epr studies of one and two dimensional transition metal ion systems. The choice of topics is intended to provide not only an introduction to low dimensional epr studies for people of limited background in this area but also to provide some background materials for other speakers whose interests are related to these topics.

MAGNETICALLY CONCENTRATED MATERIALS

Early in the history of epr studies interest became focused on the wealth of fine and hyperfine structures which could be readily observed in magnetically dilute systems. Magnetic dilution was often achieved by studying the system of interest imbedded in a diamagnetic host lattice, in glasses, or a dilute solution in both the liquid and solid state. Magnetically concentrated materials, by contrast, are often characterized by extremely simple epr spectra. These normally consist of only a single line and usually of rather well defined lineshape.

There are several interactions that must be considered in characterizing the epr changes from the very dilute to the very

concentrated system and, for the time being, it is not necessary
or even desirable to restrict ourselves to the low dimensional
systems. At very low dilutions the spectrum of the individual
ions are the dominant feature but as the concentration of the
magnetic species increased, say up to the 5 or 10 percent level,
the spectra often exhibits characteristics due to the clustering
of magnetic ions. The specific details of these spectra will
reflect the crystal lattice geometry as well as the possible types
of interactions which may be present. This particular subject,
the epr of small clusters, will be covered in some detail in the
next section of this paper.

As the concentration increases still further, say up to the
30 to 60 percent level, the clusters become rather larger and begin
to link up. Short strands, sheets, or sort of crumpled sheets may
develop in three dimensional systems. This concentration range may
well see the development of rather long strands or trees of magnetic ions in two dimensional systems while in one dimensional
systems group or "runs" of magnetic ions will form. This region
is characterized for the most part by collapse of the structure
associated with single ions, pairs, and triads and the development
as a single structureless line. This topic will be pursued in the
third section on semi-dilute systems.

Finally, then, as the dilutent is eliminated the single line
whose width and shape are related to the variety and strength of
the various interactions present in the system of interest will be
the only observable spectral feature. We shall now proceed to discuss the types of interactions which may be present.

The simplest kind of interaction between paramagnetic ions is
the dipole-dipole interaction. The term may be divided in two
major parts, the secular term of the angular form $(3 \cos^2 \theta - 1)$
where θ is the angle between the magnetic field and the internuclear
vector and the non-secular terms, which with the secular terms, produce a $(1 + \cos^2 \theta)$ angular dependence. For systems which have no
other interactions between the ions, the secular terms are the only
ones contributing to the main absorption line and the second moment
of the line is given by[1]

$$M_{2d} = \langle \Delta \omega^2 \rangle = \frac{3}{4} \, r^4 \, h^2 I(I+1) \, \frac{1}{N} \sum_{j,k} \frac{(1 - 3 \cos^2 \theta_{jk})^2}{r^6_{jk}}$$

where the general moment expression is

$$\langle \Delta \omega^n \rangle = \frac{\int_0^\infty (\omega - \langle \omega \rangle)^n f(\omega) \, d\omega}{\int_0^\infty f(\omega) \, d\omega}$$

where $f(\omega)$ is the shape function of the line and n = 2. For purely dipolar interactions, the lineshape is usually gaussian. Under these circumstances the nonsecular terms produce satellite lines far removed from the main line. Their inclusion in the second moment would produce an error in the calculated second moment by 10/3 for simple lattices. The factor is large because, while the intensity of the lines is very small, the deviation from the center is quite large. This case is of little interest here.

When the interaction includes an exchange term, $H_e(\sim J/g\beta)$ as well as a dipolar field, H_d, then the exchange narrowing can occur[2,3]. In essence, the normal linewidth given by $H_d(\sim \sqrt{M_{2d}})$ is reduced by a factor H_d/H_e so that the final result is: M_{2d}/H_e. This equation holds subject to several simplifying assumptions and the condition $H_o > H_e$, where H_o is the observation field. If $H_e > H_o$ then the linewidth is $10\,M_{2d}/3H_e$ where the 10/3 factor is due to the inclusion of all terms in the dipolar Hamiltonian. Kubo and Tomita[4] derived the expressions to cover the whole range:

$$H = M_{2d}/H_e \left\{ 1 + 5/3\exp\left[-1/2\,(H_o/H_e)^2\right] + 2/3\exp\left[-2\,(H_o/H_e)^2\right] \right\}$$

It should be emphasized that this expression is subject to simplifying assumptions some of which will be discussed in detail later.

To summarize the discussion so far then, we note that when the exchange field is greater than the observation field the linewidth expression included all of the terms in the dipolar Hamiltonian and that the angular variation is of the form:

$$\sum (1 + \cos^2 \theta)$$

while for the exchange field less than the observation field the linewidth expression includes only the secular term and has an angular variation of the form:

$$\sum (3 \cos^2 \theta - 1)^2$$

In the former case the angular variation is nearly isotropic even for low symmetry lattices while in the latter case the angular variation can be quite pronounced even for high symmetry lattices[5,6]. Furthermore, in the passage from one extreme condition to the other a continuous variation is expected which is a function of several factors including the form of the electronic time correlation function.

The lineshape is related to the form of the correlation function of the local field by the relations:[7,8]

$$\psi(\mathcal{T}) = \frac{<\omega(\mathcal{T})\,\omega(0)>}{<\omega(0)>^2}$$

$$\omega(t) = \gamma H_{loc}(t)$$

$\psi(\mathcal{T})$ is related to the relaxation function, $\phi(t)$ by the expression:

$$\phi(t) = \exp -\left[<\omega(0)>^2 \int_0^t (t-\mathcal{T})\,\psi(\mathcal{T})\,d\right]$$

and the lineshape itself is related to $\phi(t)$ by

$$I(\omega - \omega_o) = (1/2\pi) \int_{-\infty}^{\infty} dt\, \phi(t) e^{i(\omega-\omega_o)t}$$

Now if $\psi(t)$ decays to a small value in a sufficiently fast time and if the first integral is finite then the $O(t)$ is exponential and the lineshape is Lorentzian. This is the common condition in three dimensional materials[9].

One of the major concerns of recent investigations in low dimensional systems is the effect of a non-convergence of the first integral due to the fact that $\psi(t)$ does not go to zero fast enough. Then the lineshape may be between Lorentzian and Gaussian. This case was first realized experimentally in TMMC. This very famous epr experiment and analysis was done by Dietz, Merritt, Dingle, Hone, Silbernagle and Richards[10], only three years ago.

An essential feature of their analysis is associated with the fact that the nonsecular terms in the dipolar Hamiltonian produce a contribution to $\psi(t)$ which decays sufficiently rapidly while the secular terms do not. Thus the final expression for $\phi(t)$ is:

$$\exp\left[-A_1(3\cos^2\theta - 1)^2 t^{3/2} - A_2 f(\theta)t\right]$$

where the first term is due to the secular parts while the second term in the exponential is due to the nonsecular terms. Further $f(\theta) = 0$ so the lineshape is truly diffusive only along $\theta = 0$ while at $\theta = 54°$, the lineshape is Lorentzian since the first term vanishes. The experimentsl results indicated a lineshape which nearly followed the secular contribution except in the region around $\theta = 54°$ thus confirming the analysis that the secular contributions dominate the lineshape over most of the unit sphere. The linewidth in one dimensional material is approximately given by:

$$(M_{2d})^{2/3}/H_e^{1/3}$$

A detailed set of equations for one, two, and three dimensions is given in the following paper by Richards.

It is worth noting that, for all H_e and M_2 being equal, the linewidth is narrowest in three dimensions and broadest in the one dimensional case.

So far we have limited ourselves to a consideration of only the dipolar contribution to the second moment. There are other interactions which may also be included. The most general bilinear spin-spin interaction term $\bar{S}_1:T:\bar{S}_2$ can be rewritten as the sum of a symmetrical and antisymmetrical part, T_A. The symmetrical part may, in turn, be written as a sum of an isotropic term which is the usual J and a traceless diagonal tensor, T_S. We then have

$$J\bar{S}_1 \cdot \bar{S}_2 + \bar{S}_1 : \tilde{T}_S : \bar{S}_2 + \bar{S}_1 : \tilde{T}_A : \bar{S}_2$$

the term involving T_A is of the proper form to be rewritten as a vector cross product:

$$\bar{C} \cdot (\bar{S}_1 \times \bar{S}_2)$$

These terms are usually not present because of symmetry requirements. For example, if the two ions are equivalent then $S_1 \times S_2$ must equal $S_2 \times S_1$ on physical grounds. But since the cross product changes signs on the interchange of the two vectors we require

$$\bar{C} \cdot \bar{S}_2 \times \bar{S}_1 = -\bar{C} \cdot \bar{S}_2 \times \bar{S}_1$$

which, in turn requires that C be identically zero. Inversion symmetry will thus eliminate this term from consideration. Moriya[11] has given a detailed analysis of this term and has shown, in systems which lack inversion symmetry, that it is usually of the order of $(\lambda / \Delta)J$, where λ is the spin-orbit coupling constant and Δ is of the order of the crystal field splitting. Therefore when it is present it may be an order of magnitude larger than the pseudodipolar term which will be discussed next.

The pseudodipolar exchange arises from the combined effects of an isotropic exchange interaction and spin-orbit coupling. This term was first discussed with regard to the pair spectrum of the dimer copper acetate monohydrate[12] and it could be of the same order of magnitude as the dipolar terms. This is of the order of $(\lambda / \Delta)^2 J$.

Date and coworkers[13] at Osaka have suggested that both the traceless symmetrical term (anisotropic exchange) in the bilinear spin-spin Hamiltonian and the antisymmetrical term may make a contribution to the linewidth. With an extension of the Kubo-Tomita theory[4] they show that these terms also contain secular and nonsecular components and that they can therefore produce a "10/3 effect", just as the purely dipolar term can. Yamada and Ikebe[14]

have also used this sort of treatment to discuss the high temperature epr behavior in K_2CuF_4 and we have used such a treatment for
the manganous formate system[15]. The latter two systems are two
dimensional while the copper benzoate system studied by Date, et
al. is one dimensional. None of these treatments consider the low
dimensional diffusive behavior as an alternative explanation of
the results. It should be noted that the diffusional model produces a similar angular dependence and so there are some unanswered
questions about these systems.

Richards and Salomon[16] have discussed diffusive spin dynamics
in the two dimensional system K_2MNF_4. Here because of the high
symmetry of the lattice and the presence of the S-state magnetic
ions, both anisotropic and antisymmetric effects should be minimal
or nonexistent and the results can be explained by a diffusional
model alone.

There are several other important interactions such as interchain interactions[9] in one dimensional systems or intersheet interactions in two dimensional systems as well as the effects of
diamagnetic and paramagnetic impurities. These and other topics
will be considered in later lectures in the magnetic section of this
conference.

EPR OF CLUSTERS

Epr studies of pairs, triads or higher order clusters in
magnetically semi-dilute materials or in naturally occurring
cluster complexes are of considerable importance in that they often
give detailed information about isotropic and anisotropic exchange
interactions as well as other interactions which may be present[17-19].
In a sense they are the simplest kind of low dimensional system.
The first investigation of this type was on copper acetate monohydrate[12] in which copper ions are some 2.6Å apart along a z axis
while four bridging carboxylate groups (O-C-O) form the edges of a
molecular framework. The magnetic properties of the single copper
ion in this lattice have been characterized by doping a small
amount of zinc (II) into the lattice[20]. The copper (II)-zinc (II)
pairs combine an $S = \frac{1}{2}$ species which has a $3d^9$, $2D$ electronic configuration with an $S = 0$ species which has a $3d^{10}$, $1S$ electronic
configuration. The magnetic behavior can be understood on the
basis of the following spin Hamiltonian:

$$H_i = g_z \beta H_z S_z + g_x \beta H_x S_x + g_y \beta H_y S_y + A_z S_z I_z + A_x S_x I_x +$$

$$A_y S_y I_y + Q I_z^2 + Q'(I_x^2 - I_y^2)$$

where $g_z = 2.344$, $g_x = 2.052$ and $g_y = 2.082$. The value for A_z is

-0.0147 cm^{-1}, A_x = +0.0010, A_y = 0.0013, Q = 0006 and Q' = 0.0000, respectively. When two such ions are exchange coupled together the resultant spin Hamiltonian is:

$$H = H_i + H_j + J S_i \cdot S_j + D_e(3S_{iz}S_{jz}) + E_e(S_{ix}S_{jx}) - S_{iy}S_{jy})$$

where J is the isotropic interaction while the anisotropic terms will include both an axial and rhomic term. Here E_e = 0 and there is no antisymmetric term due to the center of symmetry in the molecule. The anisotropic term contains both a dipolar and pseudo-dipolar contribution. The dipolar part is D_d = 00.12 cm^{-1} while the pseudodipole part may be estimated from a third order perturbation treatment involving spin orbit coupling and anisotropic exchange Hamilton ion. With f_1 and f_2 as the wave function of the orbital singlet ground states and e_1 and e_2 as the excited states the interaction has the form

$$\frac{\langle f_1f_2 \mid \lambda L_1 \cdot S_1 \mid e_1f_2 \rangle \langle e_1f_2 \mid H_{ex} \mid e_1f_2 \rangle \langle \lambda L_1 \cdot S_1 \quad f_1f_2 \rangle}{(E_e - E_f)^2}$$

In copper acetate this produces a contribution to the zero field splitting in the triplet:

$$D_E = -(1/12)\left[(\tfrac{1}{4})(g_z - 2)^2 J_z - (g_z - 2)^2 J_x \right]$$

where J_z is the exchange interaction between the d_{xy} excited state orbitals on both ions while J_x is the exchange interaction between the d_{xz} orbitals. The $d_{x^2} - y^2$ orbital is taken as the ground state in accordance with the single ion results. As a first approximation Bleaney and Bowers set $J_z = J_x = J_y$ = 315°K, and obtained a value D_E = +0.94 cm^{-1}. [17] When this is combined with the dipolar value -0.12 cm^{-1}, the predicted value is about 0.81 cm^{-1}, while the experimental result is $| 0.34 |$ cm^{-1}. This result suggests that the exchange terms in the excited state may be rather different from that in the ground but not by more than one order of magnitude[17,22].

The magnetic behavior of this dimer may be summarized as follows: The two $S = \tfrac{1}{2}$ species are exchange coupled by an anti-ferromagnetic coupling so that the lowest state is a magnetic singlet and the upper state is a triplet. The latter is split by the combined action of the dipolar and pseudodipolar terms with nearly axial symmetry so that two "allowed" transitions are observed at 25 GHz as well as a "forbidden" transition ($\triangle M = 2$) at about one half the g = 2 field. The g values are the same for the copper (II) monomer (the copper (II)-zinc (II) pair) as for the dimer. The A value for the dimer is half that found in the monomer following the general rule that the hyperfine pattern for a cluster of ions will have a separation of about A/n and a binomial distribution of intensities. The experimental values are summarized below:

Temperature	Cu$_2$Ac$_4$,2H$_2$O		ZnCuAc$_4$,2H$_2$O	
	T = 1 90°K	T = 1 300°K	S = ½ 77°K	S = ½ 60°K
g$_x$	2.08(3)	2.053(5)	2.052(7)	2.051(1)
g$_y$	2.08(3)	2.093(5)	2.082(7)	2.074(1)
g$_z$	2.42(3)	2.344(10)	2.344(5)	2.343(1)
D(cm^{-1})	0.34(3)	0.345(5)	-	
E(cm^{-1})	0.010(5)	0.005(3)	-	
A$_x$(cm^{-1})	<0.001	-	<0.0018	+0.00101(5)
A$_y$(cm^{-1})	<0.001	-	<0.0023	-0.00129(5)
A$_z$(cm^{-1})	0.008	-	0.0147(6)	-0.01469(1)Cu^{63}
	-	-	-	-0.01572(1)Cu^{65}
Q	-	-	-	+0.00064(15)
Q'	-	-	-	≈0
ref.	12	23	20	21

     As the cluster of ions gets larger the hyperfine pattern tends
to pull together. Neglecting transfer effects and given J$\gg$A, the
allowed transitions will contain $2nI_i + 1$ hyperfine lines for n
ions symmetrically exchange coupled and the spacing between lines
will be A/n. This has been discussed in some detail for larger
clusters by Ishikawa[24].

     One of the more important questions which may be answered in
part, by epr studies is the extent to which the electronic density
on one metal ion can interact with the neighboring metal nucleus in
a dimeric pair. Following Owen and Harris[17] we consider the
simplified hyperfine Hamiltonian:

$$H = J\bar{S}_i\cdot\bar{S}_j + A_i\bar{S}_i\cdot\bar{I}_i + A_j\bar{S}_j\cdot\bar{I}_j + a_i\bar{S}_i\cdot\bar{I}_j + a_j\bar{S}_j\cdot\bar{I}_i$$

where the first three terms have their usual meaning and where the
fourth term represents the interaction of spin "i" with nucleus
"j" while the fifth term is the interaction of spin "j" with
nucleus "i". If the two nuclei and the two spin are equivalent and
if J   A then this reduces to

$$H = \tfrac{1}{2}J\left[S(S+1) - S_i(S_i+1) - S_j(S_j+1)\right] + \tfrac{1}{2}(A+a)\bar{S}\cdot(\bar{I}_i+\bar{I}_j)$$

Thus the separation between adjacent hyperfine lines is now
expected to be (A+a)/2 rather than the A/2 produced by simple
theory. This transfer effect has been reported by May[25] for
V^{2+}-O^{2-}-V^{2+} pairs in MgO where the interaction is about 0.5%
greater than the A/2 for isolated ions. Duerst and Kokoszka[26]

reported a value of A/3 of 0.0034 cm^{-1} for Cu(S=$\frac{1}{2}$) - Ni(S=1) pairs
in a dimeric copper(II) complex. The A value, measured for
Cu(S=$\frac{1}{2}$) - M(S=0) pairs (M = Pb, Cd or Zn) is 0.0118 or a value
about 20% larger than three times the value of the Cu-Ni pairs.
Further they found a value of 0.0126 cm^{-1} for Cu-Ba pairs. These
appear to be the only two reports to date of such phenomena.

One of the more important aspects of the discussion so far
in this section is that the interacting ions remain fixed in their
positions in the crystal. However, there appear to be cases
where the energy levels can be significantly altered as a result
of the interacting ions changing their positions[17]. An interesting
and important example of this is found in the epr spectra of Mn^{+2}
pairs in MgO where the experimental results may be explained by a
Hamiltonian of the form:

$$J\vec{S}_i \cdot \vec{S}_j - j(\vec{S}_i \cdot \vec{S}_j)^2$$

However, Owen and Harris[17] believe that this result is due to exchange striction. They provide a simple explanation based on the
ability of the ions to distort along their internuclear vector.
With an exponential functional form for J

$$J = J_o \exp\left[-15r/r_o\right]$$

it is possible to successfully account[17] for the J values from
pair spectra in MgO and CaO as well as the value, from susceptibility measures of J in MnO.

Another topic which can be considered here briefly is the
behavior of inorganic cluster complexes which show some low level
electrical conductivity[27]. The dichloromonoaquopyridine-N-oxide
copper (II) Cu(pno) complex is known to be a dimer on the basis of
triplet state epr behavior and magnetic susceptibility behavior[28,29].
Although a detailed crystal structure has not been done on this
material, there is evidence for a structure which consists of a
linear array of donor dimers, and the possibility of a mobile triplet state in this material is still under investigation. The epr
spectrum could not be recorded at high frequencies, for example,
because the cavity Q was ruined by the insertion of a sample, in
much the same way as when a conductor is inserted. The discussion
of mobile triplet states in organic systems will be a topic covered
by Professor Soos[30].

SEMI-DILUTE SYSTEMS

One of the more interesting topics in recent epr studies is
the behavior of systems involving several metal ions. These

include (1) simple dilution studies, in which paramagnetic ions
are replaced by diamagnetic ions of nearly the same size, (2)
studies in which the paramagnetic ions are replaced by other paramagnetic ions which are strongly coupled to the lattice or, (3)
studies in which the paramagnetic ions are nearly replaced by
other paramagnetic ions thereby altering the local exchange field
and, in low dimensional diffusive systems, affecting the rate of
diffusion. Finally, we should consider as a separate but related
item the magnetic effects of increasing the size of the ligand
groups and thereby achieving a greater separation of chains in one
dimensional systems or sheets in two dimensional systems.

While several of these studies have been carried out on three
dimensional materials, the approach to full concentration of the
magnetic species involves intermediate stages which include the
formation of one or two dimensional clusters and ultimately groupings
which are interconnected across the entire lattice. There are, in
fact, bodies of mathematical information on these types of processes
and they are often discussed under the heading of percolation
theory or the theory of runs. Two reviews[31,32] of the former area
have recently appeared while the applications of the theory of
runs[33] to low dimensional systems does not seem to have gained
widespread attention. This may be because many of the results can
be derived from other statistical considerations. While these are
not major subject areas presently of interest to epr spectroscopists
there are at least one or two aspects which may be of some importance
and so a few comments on these areas seem appropriate here.

The basic problem in percolation theory is to describe the flow
of a fluid from certain source atoms along paths of similar atoms
in the lattice. Thus, at the outset, the resemblance to mixed
metal systems is clear. The percolation problem divides itself
into two categories: bond percolation and atom percolation. In the
former case the flow of the fluid is stopped by damming certain
bonds while in the atomic problem the flow is stopped by placing
blockages at the lattice site itself. Both aspects are of some
interest. In the former category we might envision a copper (II)
ion with unpaired electronic density in the xy plane serving as a
blockage to spin diffusion along the z axis but not in the xy plane,
although the rate of diffusion in the xy plane could be altered[34].
In the second category a diamagnetic dilutent might simply be
stopped or reflect the diffusing spin. This would be a site blockage.

There is a critical percolation probability (p_c) which depends
on the nature of the lattice and the number of nearest neighbors.
Below this value fluid emitting from a source, chosen at random,
will spread only locally while for $p > p_c$ a cluster of infinite
size will result. This, in turn, allows the fluid to spread

through the medium in percolation channels. Values of critical
percolation probabilities for some common two and three dimensional
lattices based on a particular method of calculation are given in
the table below. These are taken from Shante and Kirkpatrick[31] and
the interested reader is referred to their work for more detail.

While the application of this mathematical technique to transport theory in inhomogenous conductors or other areas of electrical
or thermal conductivity in binary mixtures may be more obvious, it
seems, to the present author that it may have applications in epr
studies as well. I will discuss this briefly later in this paper.

To return to a topic of more direct epr importance, let us
consider the variation of the second moment with magnetic concentration. Over twenty years ago Kittel and Abrahams[35] showed that the
second moment of a line is always proportional to the concentration,
while for the fourth moment a combination of first and second powers
of concentration contribute. However, for the case of strong exchange interactions it is believed that the quadratic power dependence
is dominant[36] with the expression:

$$\Delta H = \sqrt{M_2^3/M_4}$$

based on strong exchange, we should predict a $c^{\frac{1}{2}}$ dependence assuming
all other things are equal. The important point here is that even
accepting all of the other assumptions the real difficulty lies
with the hope that all other things will remain equal. As noted
earlier, the effects of small lattice distortions on the effective
exchange field can have such a large effect that the simple $c^{\frac{1}{2}}$
dependence might be obscured. Furthermore, there are changes in
lineshape which also spoil the hope of simple behavior[37]. In

Critical Probabilities for Common Lattices			
		Series Method	
Lattice	z	$p_c(b)$	$p_c(s)$
Honeycomb	3	0.6527 (exact)	0.700
Kagome	4	-	0.6527 (exact)
Square	4	0.5000 (exact)	0.590
Triangular	6	0.3473 (exact)	0.5000 (exact)
Diamond	4	0.388	0.425
s.c.	6	0.247	0.307
b.c.c.	8	0.178	0.243
f.c.c.	12	0.119	0.195
h.c.p.	12	-	-

addition there may be crystallographic changes which affect exchange pathways[38]. Other difficulties may also be envisioned.

The limited number of studies involving random dilution more often than not do point to linewidths which decrease with magnetic dilution, although a simple functional form is not usually obvious. In the series of compounds $Eu_xMe_{1-x}X$ where X = oxygen or sulfur and Me = Ca(II), Sr(II) and Ba(II). The linewidth varied in a similar way independently of X but decreased strongly for Ca(II), only mildly for Sr(II) and actually increased for Ba(II).

Two different sets of data are available on KMn_fMg_{1-f}[37,39] but the general trend is similar in both studies at the high concentration end. The general decrease in linewidth reflects both the decrease in the effective exchange field and the decrease in dipolar interaction but the latter has the dominating effect. Seehra's[37] analysis of this system is among the most complete discussion available to date, but a close comparison with the experiment was hindered by changes in lineshape.

There have been some semidilute studies on materials of geological interest[40,41]. In particular $Mn_xMg_{1-x}Al_2O_4$ was studied over a wide concentration range of X ($10^{-4} \leq x \leq 1.0$) and the decrease in linewidth, though not an $c^{\frac{1}{2}}$ decrease, was a noted feature.

In our laboratory the mixed valence material hexamminechromium (III) chlorocuprate (I,II) was studied as both a function of concentration and observation frequency. Because the exchange field was comparable to the observation fields (3000-12,000 gauss) over the full concentration range it was possible to observe the "10/3 effect" for all samples and to use this variation as a basis for evaluating the exchange field. While the exchange field fell off fairly rapidly with decreasing paramagnetic copper concentration the relatively slow decrease of the dipolar interaction produced an overall increase in linewidth with decreasing paramagnetic concentration. This is associated with the functional form of the second moment for the case of two interpenetrating magnetic lattices; one completely filled with paramagnetic ions (Type A) and one randomly populated with magnetic ions (Type B). The second moment in this case is given by:[42]

$$(1+f)^{-1} \left[M_{AA} + fM_{AB} + f^2M_{BB} \right]$$

where f is the fraction concentration on Type B ions, M_{AA}, M_{AB}, and M_{BB} are the second moments for the interaction of A ions with one another, A ions with B ions and B ions with one another respectively.

When we consider systems which are one or two dimensional before dilution then there are only two examples in the literature. Richards investigated 4% Cu(II) in TMMC and found a 50% increase in linewidth at $\theta = 0^\circ$ where the linewidth is dominated by diffusional apin dynamics and essentially no change at $0 = 54^\circ$ where the one dimensional effects are unimportant[34]. The cupric ions here seem to slow the diffusional motion but not stop it altogether. Since a full report is in the literature no further discussion will be given here.

The second case[15] of where there are measurable effects of magnetic dilution in a low dimensional system is the two dimensional system $Mn_xMg_{1-x}(HCOO)_2 \cdot 2H_2O$, where x = 1.0, 0.88 and 0.65. An analysis of the experimental results on this system based on a large contribution to the linewidth anisotropy from second nearest neighbor anisotropic exchange was shown to be consistent with the concentration dependence. However, after the paper had been accepted for publication, the two dimensional study by Richards and Salomon[16] was made available and provided a good example of the effects of spin diffusional motion in a two dimensional system. While a detailed reanalysis of the results of the semidilute system based on a diffusional model has not been accomplished due to the rather more complicated nature of the manganese formate system, two features do seem worth noting.

The first of these is that the angular variation of the x = 1 system can be consistent with a spin diffusional mechanism, even if the lineshape is very nearly Lorentzian. The second is that the suppression of spin diffusional dynamics would result, in the absence of anisotropic exchange, in a more isotropic angular variation. In fact this is observed and a substantially more isotropic linewidth is found in x $\approx$ 0.66. This value is only about 15% above the critical percolation probability for a square lattice and it is not clear if this result is completely accidental. In any event, the data from this magnetically semidilute system may lend themselves to a reinterpretation.

In the area of increasing ligand size to achieve greater isolation of sheets or chain, the efforts of Willett[43,44] on several two dimensional systems should be noted. In addition we call attention to a much earlier epr study of Abe and Shimada[45] on Mn-palmitate $Mn(C_{15}H_{31}COO)_2$ and Mn-stearate $Mn(C_{17}H_{35}COO)_2$ in which both exchange narrowing and a "10/3 effect" were observed. While the authors suggested the possibility of a one dimensional system, in this case, a two dimensional model with bridging carboxylate groups seems equally likely. A single crystal study would be most welcome on either of these materials because of the high probability of strong low dimensional character.

Finally an interesting preliminary report of three dimensional semidilute studies has recently appeared.

Cusumano and Troup[46] have studied the lineshape of DPPH in polystyrene and examined the lineshape as a function of concentration. When this was about 3×10^{20} spin/cm^3, they were able to explore the lineshape out to 20 times $\triangle H^{\frac{1}{2}}$ and they found an exponential decay in the wings. The center was still Lorentzian.

They were able to find $\psi(\tau)$ from $\phi(t)$ by integrating by parts and differentiating twice. They found:

$$\psi(\tau) = \frac{1}{<\omega(o)^2>} \left[\frac{\phi(\tau)\phi''(\tau) - [\phi'(\tau)]^2}{0(\tau)} \right]$$

where $0' = \dfrac{d\phi(\tau)}{d\tau}$ and $0'' = \dfrac{d^2\phi(\tau)}{d\tau^2}$

Now if $\phi(\tau)$ is a Lorentzian of the form $\dfrac{1}{1+\omega_x^2 t^2}$ where ω_x is an experimental parameter, then

$$\psi(\tau) = \frac{1-\omega_x^2 t^2}{1+\omega_x^2 t^2}$$

this leads to a spectrum for the random frequency modulation of the form $(X)\exp(-|X|)$

where $X = \omega/\omega_x$ since the Lorentzian function if valid only for small τ, then so is the expression listed above for $\psi(\tau)$. Therefore it is only in the wings that the expression listed above holds. They were further able to demonstrate that this behavior is consistent with the Blume-Hubbard correlation function in the high temperature limit of a Heisenberg ferromagnetic where exchange is the only process which produces fluctuations. It would appear that in the three dimensional case this model, which is Lorentzian in the center and exponential in the wings provides an excellent approximation to exchange narrowed lineshapes.

TWO DIMENSIONAL COMPLEXES

K_2MnF_4 is a two dimensional antiferromagnetic system which was studied by Richards and Salomon[16]. They provide a detailed analysis in terms of a theory based on a spin diffusive mechanism. In particular they successfully account for the linewidth, lineshape, and angular dependence as well as the temperature dependence. However, their model does not seem to be fully transferable to two dimensional ferromagnetic systems such as K_2CuF_4 [14] and $NiCl_2$ [47] where different temperature dependences are observed.

The angular dependence in K_2MnF_4 was roughly of the form $(3 \cos^2 \theta - 1)^2$ + constant. This functional form has also been used by Boesch and coworkers on $(CH_3NH_3)_2MnCl_4$ and $(C_2H_5NH_3)_2MnCl$.

The interest in the vector cross product, or Dzialoslinsky-Moriya, interaction term in epr spectroscopy dates back to 1968 when Seehra and Castner[48] showed that this produces a linear temperature dependence to the epr linewidth by means of a spin-phonon coupling. This has been observed in several two dimensional materials and is perhaps a general property of the A_2MnX_y salts[49]. Examples include the ethylammonium and propylammonium manganous bromide salts were synthesized and studied by Willett and Extine[43]. Evidence for the antisymmetric interaction was presented based on a linear decreasing temperature dependence. In addition, the angular variation was temperature dependent. The suggestion was made that the spin diffusional mechanism was operative at room temperature but not a dominant relaxation mechanism at 77°K. For several $(R NH_3)_2CuBr_4$ salts Willett and Extine[44] have found a simple $(3 \cos^2 \theta - 1)$ angular dependence. This result is not discussed in detail but it seems possible that the antisymmetric term may be an important factor here.

Yamada and Ikebe[14] have carried out a detailed investigation of K_2CuF_4. The study includes the high temperature angular variation of the linewidth as well as temperature and frequency dependence. The author indicated the strong effects of anisotropic exchange in this case in determining the linewidth.

Shia and Kokoszka[15] discuss the $Mn_xMg_{1-x}(HCOO)_2 \cdot 2H_2O$ system for x = 1, 0.88 and 0.65. They associate the angular dependence for x = 1.00 with anisotropic exchange associated with interlayer coupling. An alternate explanation of these results in terms of a diffusive mechanism also seems possible and here the loss of the anisotropic behavior for the x = 0.65 crystal may be associated with limiting the diffusive pathways.

ONE DIMENSIONAL COMPLEXES

In addition to the now famous TMMC salt[10] the pyridinium manganese(II) trichloride monohydrate (PMC) provides another example of a manganese(II) complex which has been found to have an epr spectrum which shows some characteristics of one dimensional exchange narrowing[50]. These include an angular variation which is quite similar to TMMC and a lineshape along the chain axis which is between the spin diffusive lineshape and a Lorentzian shape. The latter is associated with interchain coupling effects.

Apart from the organic materials and platinum complexes such as Magnus Green Salt[51] $Pt(NH_3)_4PtCl_4$ and $K_2Pt(CN)_4Br_{1/3} \cdot 3H_2O$ [52], which are discussed in other papers at the conference, most of the remaining one dimensional complexes incorporate the copper(II) ions.

The copper benzoate trihydrate was one of the first one dimensional copper(II) complexes studied[13]. From magnetic susceptibility data the value for J is $8.6 \pm 1K$ and no second order transition is found to 1.4K. The epr data on single crystals were taken at both 10 and 35 GHz and over a temperature range from 1.4 to 29°K. No frequency dependence was observed. The angular variation was not found to be in accord with the dipolar contribution alone but other effects had to be taken into account. The authors analyze these results in terms of an anisotropic exchange (vide supra) but a detailed analysis of the lineshape, especially for the magnetic field was not carried out. Thus it is possible that the effects of spin diffusion could be a factor in the angular variation of the high temperature data. Additional data and/or calculations should be welcome in this case.

In tetramine copper sulfate monohydrate[9] the value for $\triangle H^{\frac{1}{2}}$ at 8.68 GHz and room temperature is 40-45 gauss along the chain axis. The lineshape was strictly Lorentzian out to 10 halfwidths and deviated less than 1% at even $14 \triangle H^{\frac{1}{2}}$. The possible contributions of hyperfine fields to $H^{\frac{1}{2}}$ is about 26 gauss. It is possible that these complications such as anisotropic exchange could contribute here especially due to the non S-state of the copper(II) ions. The value of T_N here is 0.43°K and the antiferromagnetic coupling is J = 3.16K. A reasonable account of the data can be had for an interchain coupling of about $10^{-2}J$ or higher. The result is somewhat model dependent as are many of the estimates of this type.

In bis-(N-methylsalicylaldiminato) copper(II)[53,54] the planar molecules are stacked perpendicular to the C axis with a metal-metal distance of 3.3Å. The exchange is antiferromagnetic, J = 3.1°K and the ratio of intra- to interchain distances is nearly the same as in TMMC. The epr was done at 76°K with a 23.4 GHz spectrometer but the room temperature result was the same. The major experimental results are (1) a Lorentzian line at $\theta = 54°$ to C with a halfwidth at halfheight of 31 gauss and (2) a spin diffusion line profile at $\theta = 0°$ with $\triangle H^{\frac{1}{2}} = 120$ gauss.

These results were the first to show that the one dimensional spin diffusion effects are applicable for spin $\frac{1}{2}$ systems as well as higher spin systems when interchain coupling is small. This is particularly significant since it rules out a t^{-1} decay for the correlation function predicted by the rigorously solvable spin X-Y model. The latter would have given a lineshape which is the

fourier transform of exp (-t ln t) and be very nearly Lorentzian.
Some minor deviations from pure one dimensional behavior were
noted, however, and attributed to residual interchain effects.

$Cu(NH_3)_4PtCl_4$ is a one dimensional material in which the
tetramine copper(II) complexes lie perpendicular to the C axis.
The $PtCl_4^{2-}$ complex occurs between successive copper(II) complexes
and provide a measure of magnetic dilution which lead to comparable
hyperfine and dipolar fields as well as a value for H_e that is
between the observation fields at 9 and 36 GHz. This study is the
first detailed angular study of the "10/3 effect" and the results
confirm the general trend mentioned earlier, that the linewidth
becomes more isotropic for $H_o < H_e$ compared to $H_o > H_e$. [55]

Along the C axis the linewidth was 43.8 $\pm$ 1 gauss at X-band
and 38.5 $\pm$ 1 gauss at Q band. The linewidth decreased monotonically in all directions away from the C axis and did not exhibit
a minimum until the ab plane was reached. The Lorentzian lineshape out to about 7 linewidths indicates that the presence of
off-axial effects which have been discussed in greater detail by
Soos and Huang recently[56].

In dichlorobis (pyridine) copper(II) the square-planar four-
coordinate copper(II) complexes are stacked so that the major
axis for this system is perpendicular to the molecular plane. The
chlorides are trans and the crystal structure suggests Cu-Cl-
pyridine-Cu pathway as the dominant mechanism for interchain
interaction. The chloride ion is some 4Å away from the carbon
ortho to the coordinated nitrogen, while it is nearly 6Å from the
nitrogen itself. The interplanar metal-metal distance is 9.52Å.
The situation suggests only weak exchange pathways are possible.

While the epr lineshapes of these complexes were first
believed[53,9] to be non-Lorentzian a more recent reinvestigation of
this shows that the lineshape is Lorentzian out to ten halfwidths[57].
The earlier reports may have been less reliable due to problems
associated with the polycrystalline character of the sample. The
most recent study was on single crystals.

Copper pyrazine nitrate is a linear chain polymer which was
first synthesized by Reimann and coworkers who also examined the[58]
X-ray structure. The basic mode of the 6-fold coordination is a
square plane consisting of trans oxygen from the nitrate groups
and two nitrogens from the pyrazine molecules. Two other oxygens
from the nitrate ion are close enough to be considered to be coordinated but the angular deviation from octahedral is large. The
pyrazine acts as the bridging group and in spite of a 6.7Å separation between interacting copper(II) ions the J value derived from
the Bonner-Fisher model is about 5°K. This distance is not the

shortest copper-copper distance (which is 5.1Å) but is the most
likely pathway for exchange[59]. While any single crystal epr
study has not been carried out, a polycrystalline study[60] revealed
an additional discontinuity in the absorption curve between the
"parallel" and "perpendicular" positions. Viewed in retrospect,
this may have been a manifestation of a change in lineshape.
Additional single crystal studies of this system are in progress[61].

The magnetic behavior[62] of the quinoxalene complex of copper
(II) nitrate has also been studied in polycrystalline form. This
material is believed to possess a similar structure to the pyrazine
complexes.

Finally, calcium copper acetate hexahydrate has been recently
reinvestigated by epr[63]. This material is composed of alternate
metal ions stacked along the $\underline{c}$ axis and the calciums provide a
measure of magnetic dilution which is rather greater than that
found in the $Cu(NH_3)_4PtCl_4$ complexes discussed earlier[55]. Low
temperature magnetic susceptibility studies[64] (1.5 - 2°K) suggest
that the effective exchange field is about 1000 gauss. Under
these circumstances the non-secular components of the dipolar and
hyperfine fields would be nearly suppressed by the 10/3 effect
even in the absence of spin diffusion. As a result the linewidth
at the magic angle will not be dominated by non-secular effects
since they are nearly negligible so that a spin diffusive lineshape is observed over the full angular range[63]. It is possible
to account for the magnetic susceptibility and epr behavior in a
consistent way with no adjustable parameters by the one dimensional
model.

Acknowledgement is made to the donors of the Petroleum
Research Fund, administered by the American Chemical Society for
support during the preparation of this manuscript.

REFERENCES

1. C.P. Slichter. Principles of Magnetic Resonance (Chapter 3).
    Harper and Row, NY, 1963.
2. J.H. Van Vleck. Phys. Rev. 73, 679 (1969).
3. P.W. Anderson and P.R. Weiss. Rev. Mod. Phys. 25, 269 (1953).
4. R. Kubo and K. Tomita. J. Phys. Soc. Japan 9, 888 (1954).
5. A. Abragam. The Principles of Nuclear Magnetism (Chapter 4)
    Oxford, 1961.
6. B.R. Cooper and F. Keffer. Phys. Rev. 125, 896 (1962).
7. R. Kubo. in Fluctuations Relaxation and Resonance in Magnetic
    System. D. terHaar, Ed. Plenum Press, NY,1962.
8. J. Gulley, D. Hone, D. Scalapino and B. Silbernagel. Phys.
    Rev. 1B, 1020 (1970).

9. M.J. Hennessy, C.D. McElwee and P.M. Richards. Phys. Rev. $\underline{7B}$, 930 (1973).

10. R. Dietz, F. Merritt, R. Dingle, D. Hone, B. Silbernagel and P. Richards. Phys. Rev. Lett. $\underline{26}$, 1186 (1971).

11. T. Moriya. in Magnetism (Vol. 1). G. Rado and E. Suhl, Ed., Academic Press, NY, 1963.

12. B. Bleaney and K. Bowers. Proc. Roy. Soc. $\underline{A214}$, 451 (1952).

13. M. Date, H. Yamazaki, M. Motokawa and S. Tazawa. Prog. Theor. Phys. $\underline{465}$, 194 (1970).

14. I. Yamada and M. Ikebe. J. Phys. Soc. Japan. $\underline{33}$, 1334 (1972).

15. L. Shia and G. Kokoszka. J. Chem. Phys. $\underline{60}$, 1101 (1974).

16. P. Richards and M. Salomon. Phys. Rev. 00:000 (1974).

17. J. Owen and E.A. Harris. in Electron Paramagnetic Resonance (Chapter 6). S. Geschwind, Ed. Plenum Press, NY, 1972.

18. G. Kokoszka and R. Duerst. Coord. Chem. Rev. $\underline{5}$, 209 (1970).

19. G. Kokoszka and G. Gordon. Transition Metal Chemistry. $\underline{5}$, 181 (1969). R. Carlin, Ed. Dekker, 1969.

20. G. Kokoszka, H. Allen, Jr. and G. Gordon. J. Chem. Phys. $\underline{42}$, 3693 (1965).

21. A. Kawamori, S. Matsuura and H. Abe. J. Phys. Soc. Japan. $\underline{29}$, 1173 (1970).

22. G. Kokoszka, M. Linzer and G. Gordon. Inorg. Chem. $\underline{7}$, 1730 (1968).

23. H. Abe and J. Shimada. J. Phys. Soc. Japan. $\underline{12}$, 1255 (1957).

24. Y. Ishikawa. J. Phys. Soc. Japan. $\underline{21}$, 1473 (1966).

25. C.E.C. May. Thesis, Oxford (   ) quoted as reference 38 in Owen and Harris[17].

26. R.W. Duerst and G. F. Kokoszka. J. Chem. Phys. $\underline{51}$, 1673 (1969).

27. J. Lewis (Private communication)

28. G. Kokoszka, H.C. Allen, Jr. and G. Gordon. J. Chem. Phys. $\underline{46}$, 3013 (1967).

29. K.E. Hyde, G. Gordon and G. Kokoszka. J. Inorg. Nucl. Chem. $\underline{30}$, 2155 (1968).

30. Z. Soos (previous paper).

31. V.S. Shante and S. Kirkpatrick. Adv. in Phys. $\underline{20}$, 325 (1971).

32. J.W. Essam. Phase Transitions and Critical Phenomena. Vol. 2, p. 197. C. Domb and M. Green, Ed. Academic Press, NY, 1972.

33. W. Feller. An Introduction to Probability Theory and its Applications. Vol. 1 and 2. Wiley, NY, 1968.

34. P.M. Richards. Electron Spin Resonance . In The Impurity-Doped Heisenberg Linear Chain $(CH_3)_4NMnCl_3:Cu$. Phys. Rev. B (1974).

35. C. Kittel and E. Abrahams. Phys. Rev. $\underline{90}$, 238 (1953).

36. A. Samokhvalov, V. Babushkin, V. Bamburov and N. Lobacherskaya. Soviet Physics - Solid State. $\underline{13}$, 2530 (1972).

37. R.P. Gupta, M.S. Seehra and W.E. Vehse. Phys. Rev. $\underline{5}$, 92 (1972).

38. L. Shia, R. Scaringe and G. Kokoszka. EPR of Mixed Valence Complexes: Hexaaminecobalt(III) Chlorocuprates(I,II). J. Inorg. Nucl. Chem. (1975).

39. K. Horai and K. Saiki. J. Phys. Soc. Japan 21, 397 (1966) and
    private communication.
40. M. Stombler, H. Farach and C. Poole. Phys. Rev. B6, 40 (1972).
41. F. Gesmundo and C. DeOsmundis. J. Phys. Chem. Solids. 34, 1757
    (1973) and 34, 637 (1973).
42. R. Scaringe and G. Kokoszka. J. Chem. Phys. 60, 40 (1974).
43. R.D. Willett and M. Extine. Physics Lett. 44A, 503 (1973).
44. R.D. Willett and M. Extine. Chem. Phys. Lett. 23, 281 (1973).
45. H. Abe and J. Shimada. Nat. Sci. Report. Ochanomizie Univ.
    4, 77 (1953).
46. C. Cusumano and G.J. Group. Physics Lett. 44A, 441 (1973).
47. R.J. Birgeneau, L.W. Rupp, Jr., H.J. Guggenheim and P.A.
    Lindgard. Phys. Rev. Lett. 30, 1252 (1973).
48. M.S. Seehra and T.G. Castner, Jr. Phys. kondens Materie. 7,
    185 (1968).
49. B.C. Gerstein, K. Chang and R.D. Willett. J. Chem. Phys. 60,
    3454 (1974).
50. P.M. Richards, R.K. Quinn and B. Morosin. J. Chem. Phys. 59,
    4474 (1973).
51. F. Mehran and B. Scott. Phys. Rev. Lett. 31, 99 (1973).
52. F. Mehran and B. Scott. Phys. Rev. Lett. 31, 1347 (1973).
53. R.R. Bartkowski, M.J. Hennessy, B. Morosin and P.M. Richards.
    Solid State Commun. 11, 405 (1972).
54. R.R. Bartkowski and B. Morosin. Phys. Rev. B6, 4209 (1972).
55. Z.G. Soos, T.Z. Huang, J.S. Valentine, and R.C. Hughes. Phys.
    Rev. B9, 993 (1973).
56. T.Z. Huang and Z.G. Soos. Phys. Rev. B (1974) and B9, 4981
    (1974).
57. W. Duffy, Jr. J.E. Venneman, D.L. Strandburg and P.M. Richards.
    Phys. Rev. B9, 2220 (1974).
58. A. Santoro, A.D. Mighell and C.W. Reimann. Acta Cryst. B 26,
    979 (1970).
59. D.B. Losee, H.W. Richardson and W. Hatfield. J. Chem. Phys.
    59, 3600 (1973).
60. G.F. Kokoszka and C.W. Reimann. J. Inorg. Nucl. Chem. 32,
    3224 (1970).
61. W.E. Hatfield (private communication).
62. H.W. Richardson, W.E. Hatfield, H.J. Stoklosa and J.R. Wasson.
    Inorg. Chem. 12, 2051 (1973).
63. R. Adams, R. Gaura, R. Raczkowski and G. Kokoszka. Phys. Lett.
    A Sept. (1975).
64. J. McElearney, D.B. Losee, S. Merchant, and R. Carlin. J.
    Chem. Phys. 54, 4585 (1971).

MIXED VALENCE CHEMISTRY AND METAL CHAIN COMPOUNDS

P. Day

Oxford University, Inorganic Chemistry Laboratory
South Parks Road, Oxford, England

In 1896 Alfred Werner[1] began to investigate what, at that
time, were thought to be two isomeric forms of sodium bis-oxalato-
platinite. The curious feature which attracted Werner's
attention to these compounds was that while one was pale yellow,
the other was copper red. He was soon able to show that they
were not actually isomers at all, and that the yellow salt, which
had the formula $Na_2Pt(C_2O_4)_2 2H_2O$, could be converted into the red
one only by partly oxidizing it with chlorine or bromine water.
He also noted several other examples of platinum salts which
likewise existed in pairs, one member being a simple divalent
compound such as $(PtCl_3)K$ or $(PtCy_4)K_2$ – to use his nomenclature –
the other derivable from the first by partial oxidation. In
every case the divalent compound was pale in colour while the
partly oxidized one was very dark, frequently copper coloured,
with the appearance of a metallic reflector: Werner had discovered
the class of mixed valence platinum chain compounds about which so
much has been written in the last few years, and whose properties
form one of the principal topics of this Advanced Study Institute.

In fact Werner himself was not the first to prepare these
compounds, which had been known already for a number of years,[2]
but it was his insight which first connected the fact that they
were partly oxidized, i.e. lying intermediate between two formal
oxidation states, and their most immediately obvious physical
property, colour. In the same paper he remarks that it would be
surprising if other elements did not form analogous compounds,
and points to the example of the tungsten bronzes, though noting
that in that instance the metallic compound is formed by combining
a small amount of a lower oxidation state with an excess of a
higher, the opposite situation to that found in the platinum salts.

The final and, in the context of its era, quite extraordinary
imaginative leap made by Werner in this paper amounts to the
identification of similar charge transfer processes in organic
molecular complexes, for he noticed that formally, quinhydrone
is an addition compound between molecules in two different
oxidation states, quinone and hydroquinone, and that the deep
green, metallic reflecting crystals lose all colour when they
revert to their separated constituents in solution.   This latter
thought, too, initiated a line of enquiry which leads directly to
one of the other main topics of this Advanced Study Institute,
exemplified by the highly conducting TCNQ salts.

As Werner suspected, there are indeed a multitude of other
examples of partly oxidized, or mixed valence compounds, formed
by at least thirty-seven elements in the Periodic Table.   Not
all contain chains of metal ions, not all are deeply coloured
or conduct electricity.   The purpose of the present chapter
is to describe some representative compounds, where they occur
in the Periodic Table, and how their physical properties can be
predicted, at least in outline, from a knowledge of their
structures.   Since the topic of this Institute is one-dimensional
systems we shall naturally emphasise mixed valence compounds of
this type.

First, however, we must define what we mean by 'mixed valency'.
The presence of non-integral oxidation states in many compounds is
at once apparent from their formulae, e.g. in Fe_3O_4, if 'oxidation
state' is defined in the conventional way, by taking closed
electron configurations on the ligands: O^{2-} for instance.   Of
course, the definition of 'oxidation state' assumes that one can
identify the formal charge, and hence the number of valence
electrons, which a ligand would have in that particular ionic
configuration to which its actual configuration in the compound
most closely approximates.   Identifying the 'limiting' ionic
configuration in question can pose problems, either when the ligand
is one which can occur in an open shell configuration or when there
is extreme metal-ligand covalency.   An example of the former is
the ion $(NH_3)_5CoO_2Co(NH_3)_5$ $^{5+}$, whose e.p.r. spectrum[3] clearly
shows that the single unpaired electron is not localized on one
cobalt atom, or even shared equally between the two, but is
concentrated mainly on the oxygens.   Thus both cobalts are Co^{III}
and the oxygen is in the form of superoxide, O_2^-.   What, in a
molecular coordination complex, would be called metal-ligand
covalency would, in a continuous lattice solid, be described simply
by participation of more than one atom sublattice in the make-up
of the valence band.   Extreme examples in this category are
intermetallic compounds such as Hume-Rothery and Zintl phases,
whose formulae alone might lead one to think they contained ions
in non-integral oxidation states, e.g. Cu_3Si.   However, the
formulae of this type of compound are determined by the total

electron concentration in the lattice at which the Fermi surface approaches a critical point in the Brillouin zone, and the concept of oxidation state is quite inappropriate.

On the other hand some compounds which we know, from other evidence, to be of mixed valence type have formulae corresponding to oxidation states which are integral, but unusual for the element in question. Examples are $Pt(NH_3)_2Cl_3$ or Sb_2O_4. Such compounds certainly do not have the properties expected of Pt^{III} or Sb^{IV}, such as unpaired electrons, and are therefore taken as containing $Pt(II,IV)$ or $Sb(III,V)$. As we shall see later, oxidation state differences of two units are found towards the end of the transition series, and in the B subgroups.

To summarize then, we shall only be interested in those compounds in which integral closed shell oxidation states can be assigned to the ligands or anion sublattice so that, even if our description of the valence shell electrons is a collective, i.e. band, rather than a localized one, the basis functions of the valence band are primarily metal-based orbitals.

SIGNIFICANCE OF MIXED VALENCY FOR CONDUCTIVITY

It was first pointed out nearly forty years ago[4] that one could render simple binary transition metal oxides conducting by isomorphously substituting them with closed shell ions of different charge so that, to preserve the charge neutrality in the lattice as a whole, some of the transition metal ions had to change their oxidation states. In this way the number of valence electrons per metal ion becomes non-integral. The best known example is probably Li^+ replacing Ni^{2+} in the rocksalt structure of NiO, an equal number of other Ni^{2+} in the lattice being oxidized at the same time to Ni^{3+}. A 1% doping of Li^+ into NiO increases its specific conductivity by a factor of 10^{10}. Such 'controlled valency semiconductors'[5] are well known, but it is worth noticing that their conductivities are only high when the ions of differing oxidation state occupy sites of very similar geometry.

That it should be easier for charge to migrate through a lattice in which different cations have different charges follows from a very simple ionic argument. Consider a chain of ions such as Ni^{2+}. Transport of charge from one end of the chain to the other can only take place through ionic fluctuations, that is, by forming states of the type

$$...Ni^{2+}\ Ni^{2+}\ Ni^+\ Ni^{2+}\ Ni^{2+}\ Ni^{3+}\ Ni^{2+} ...$$

The energy required to create such a fluctuation depends on the ionization potential and electron affinity of Ni^{2+}, on the

attraction between the 'hole' on one cation and the excess
electron on the other, and on the polarisation energy resulting
from any changes in the ligand environments of the two cations
after the electron has been transferred.   Of course, the
electron affinity of Ni^{2+} is just minus the ionization potential
of Ni^+ so the amount of energy needed to produce separated holes
and electrons in the lattice is, in the ionic limit, related to
the difference between the second and third ionization potentials
of nickel.   This is a considerable amount of energy, about 18 eV
for ions in the gas phase.   In contrast, suppose we have a chain
containing both Ni^{2+} and Ni^{3+}.   Charge fluctuation can now occur
merely by exchanging electrons between the two oxidation states,
a process which, if the ligand environments of the ions were
identical, would cost no energy at all, since the electron
affinity of Ni^{3+} is equal and opposite to the ionization potential
of Ni^{2+}.

Two rather disparate cases which may prove useful analogies
to the situation we have been describing are the behaviour of
excess electrons injected into rare gas crystals, and the
π-electron excited states of long conjugated chains.   At first
sight, crystals of the rare gases might be thought of as extreme
examples of insulators, but excess electrons, either injected
from the electrodes or implanted by external bombardment, turn out
to have very high mobilities:[6] Ar^- exchanges very rapidly with
Ar^o.

Unsaturated chains of carbon atoms form two distinct classes,
called polyenes and cyanines.   In the former, which have the
general formulae $R_2C(CH)_{2n}CR_2$, the π-electron/atom ratio is one:
in our analogy they would be called single valence chain compounds.
Nevertheless, the bond lengths within the carbon chain are by no
means equal, even in extremely long chains such as β-carotene, but
alternate in such a way that to a first approximation one could
think of them as respectively single and double.   Valence bond
theory would write the ground state as approximating to the single
canonical structure

$$\cdots - CH = CH - CH = CH - CH = CH - \cdots$$

A solid state theorist interested in one-dimensional phenomena
might, on the other hand, see in the electron pairing and doubling
of the repeat distance a simple manifestation of a Peierls
distortion.[7]   Evidence that the ground state of even an infinitely
long chain of this kind would be insulating comes from the way in
which the energy of the first excited state varies with the chain
length.[8]   The transition, basically from the highest filled to the
lowest empty orbital, at first moves rapidly to lower energy as the
chain length increases, but finally reaches an asymptotic limit of
about 2.5 eV, i.e. we have an energy gap.   The reason why the

single neutral canonical structure written above is such a good approximation to the true polyene ground state is that, just as in the case of the single valence chain of nickel ions, structures in which there are charge fluctuations, e.g.

$$\overset{+}{\cdots -CH} = CH - \overset{-}{CH} - CH - CH = CH - \cdots$$

and $\overset{+}{\cdots -CH} = CH - CH - CH = CH - \overset{-}{CH} - \cdots$,

have very high energies.

The other class of conjugated carbon chains have π-electron/atom ratios different from unity. The simplest examples would be the hypothetical compounds $R_2C-(CH)_{2n+1}-CR_2^-$ with an odd number of conjugated atoms and an even number of π-electrons. In practice, the stablest derivatives have heteroatoms at the ends of the chains, e.g. $R_2N-(CH)_{2n+1}-NR_2^+$. The important point is that now we can no longer write a single valence bond canonical structure for the compounds. At the very least we must write an equal mixture of two:

$$R_2C = CH - CH = CH - CH = CH - \overset{-}{C}R_2$$

and $\overset{-}{R_2C} - CH = CH - CH = CH - CH = CR_2$,

with a further, smaller, admixture of others, such as

$$R_2C = CH - \overset{-}{C}H - CH = CH - CH = CR_2 .$$

The cyanines are therefore 'mixed valence' compounds. For comparable chain lengths the energies of the first excited states in these compounds are considerably lower than those of the polyenes.[9] For example when n = 1 the first transition of the cyanine chain lies at about 60% of the energy of the corresponding polyene, though it should be borne in mind that part of this difference comes from the fact that for equal n the cyanine chain is one atom longer. Nevertheless, when n = 6 (the longest chain for which data about the cyanine is available) the first transition in the cyanine has dropped to 40% of the energy of the lowest polyene transition. Extrapolating the curve of optical transition energy versus chain length for the cyanines one finds that in the asymptotic limit the energy gap is now close to zero. Consequently, without insisting in any way that the analogy between the carbon atom chains and our metal ion chains is a very precise one, we can see that making the number of valence shell electrons incommensurate with the number of atoms lowers the energy gap and stabilizes states in which charge fluctuations can propagate along the chain. Let us now look at the Periodic Table to see which elements are most likely to provide us with examples of mixed valence compounds.

MIXED VALENCE COMPOUNDS IN THE PERIODIC TABLE

Variable oxidation states, an obvious prerequisite for forming mixed valence compounds, are of course widespread in the d-block elements, and are also found at the beginning and end of the lanthanides, and on either side of gadolinium, where the $4f^7$ configuration is stabilized. In the actinides there are several examples near the beginning of the series, though increasing dominance of the +3 oxidation state after uranium makes it unlikely that many are going to be found near the end of the 5f block. In the post-transition series s- and p-blocks, particularly those following the 4d and 5d elements, the outer s-electrons are exceptionally firmly bound, as a result of the high effective nuclear charge which they experience. Consequently, in addition to the Group oxidation state N, e.g., +4 for Sn and Pb, many of these elements have a stable lower oxidation state N-2. Co-existence of these two oxidation states is a frequent source of mixed valence compounds in that part of the Periodic Table.

Table 1 lists those mixtures of oxidation states which have been found experimentally, with a few examples of the types of compound in which they occur. Obviously they reflect the broad variations of oxidation state stability in single valence compounds familiar to inorganic and coordination chemists. Several features of these oxidation states deserve comment. In the first transition series the combination (II,III) is the most frequent, while in the 4d and 5d blocks combinations of higher oxidation state are found, particularly in the earlier Groups, where one of the two states involved is the Group oxidation state. In the latter we are therefore dealing with the configurations (d^0, d'). Much of the single valence chemistry of the later 4d and 5d elements is dominated by the low spin d^6 configuration in octahedral compounds and low spin d^8 in square planar, so it is not surprising that both are found together in a lot of mixed valence compounds of Rh, Ir, Pd and Pt. Many of these are chain compounds, and are listed separately later.

SOME GENERAL TYPES OF MIXED VALENCE COMPOUND

In this section we list some of the main classes of material in which mixed valency is found. Not all, by any means, are chain compounds, but they serve to illustrate the extraordinary variety of chemical types available for study, or as starting points for the synthesis of further examples. The most comprehensive survey of mixed valence compounds published up till now is that of Robin and the present author,[10] to which reference should be made for more detailed information on many of the compounds listed here.

TABLE 1. <u>Elements for which mixed-valency compounds have been found</u>

Ti III,IV oxides oxyan-ions aq. soln.	**V** IV,V oxides oxyan-ions	**Cr** II,III hal-ides III,V oxides	**Mn** II,III oxide III,IV oxide	**Fe** II,III halide oxide salts	**Co** II,III oxide oxyan-ions amm-ines	**Ni** II,III oxide	**Cu** I,II hal-ides sulph-ide prot-eins	**Zn**	**Ga** I,III halide	**Ge**	**As**	**Se**	**Br**	**Kr**
Zr	**Nb** II,III hal-ides IV,V oxides	**Mo** V,VI oxides oxyan-ions	**Tc**	**Ru** III,IV amine II,III amm-ines	**Rh**	**Pd** II,IV hal-ides com-plexes	**Ag** 0,I halide I,III halide	**Cd** 0,II halide	**In** I,III halide	**Sn** II,IV aq. soln.	**Sb** III,V halide oxide	**Te**	**I** I_n^-	**Xe** VI,VIII oxyan-ion
Hf	**Ta** II,III hal-ides	**W** V,VI oxides oxyan-ions	**Re**	**Os** II,III amine N_2	**Ir** III,IV sulph-ates	**Pt** II,IV hal-ides com-plexes	**Au** I,III hal-ides	**Hg**	**Tl** I,III hal-ides salts	**Pb** II,IV halide oxide	**Bi** 0,III halide	**Po**	**At**	**Rn**

Lanthanides: Ce: III,IV Oxide
                Pr: III,IV oxide
                Tb: III,IV oxide
                Eu: II,III oxide

Actinides: Pa: IV,V   oxide
                U: IV,V   oxide
                    IV,VI   oxide
                Np: V,VI   oxide
                Pu: III,IV oxide

The most important structural division is between those compounds in which the mixed-valence interaction occurs between ions in a crystal lattice, and those in which the chromophore is a distinct polynuclear species.   In the former the effects of the interaction would naturally disappear if the lattice were separated into its components whilst the latter, which, in the solid state form molecular crystals, would still exhibit, for example, their mixed valence colours in solution.   One might think that the discrete molecular mixed valence compounds would be less interesting in the context of one-dimensional behaviour, but they can sometimes serve as useful models for portions of chains.

1.   Oxides

(a)   Rocksalt, MO. $A_x^I M_{1-x}^{II,III} O$ (A=Li;   M= Mn, Fe, Co, Ni)

(b)   Spinel, $A^{II} B_2^{III} O_4$.   Face-centred cubic lattice of O^{2-} with cations occupying octahedral and tetrahedral holes.

   Normal spinels, $(A_{tet}^{II})(B_{oct}^{III})_2 O_4$ ($Mn_3 O_4$, $Co_3 O_4$)

   Inverted spinels, $(B_{tet}^{III})(A_{oct}^{II})(B_{oct}^{III}) O_4$ ($Fe_3 O_4$)

(c)   Perovskite, ABO_3.   Corner sharing BO_6 octahedra with 12 coordinate A ions.

   e.g. $La_x Sr_{1-x} MO_3$ (M = Mn, Fe, Co)

(d)   Bronzes, $A_x^I B_y O_z$, where $B_y O_z$ is the formula of a normal valence oxide and $0 < x < y$.

   e.g. $Na_x TiO_2$, $Na_x V_2 O_5$, $Na_x MoO_3$, $Na_x WO_3$.

   The cubic structure is analogous to that of perovskite, but the hexagonal, stable at lower x, contains 'tunnels' of Group IA cations.

(e)   Magneli phases, $M_n O_{qn-1}$.   Based on ordered arrangements of excess interstitial ions in 'shear planes' within the lattice of a single valence d^0 or f^0 oxide.

   e.g. $Ti_n O_{2n-1}$, $4 < n < 9$

   $V_n O_{2n-1}$, $4 < n < 8$

   $Nb_{3n+1} O_{8n-2}$, $n = 5,7$ (n = 6,8 have slightly different structure)

$$Mo_nO_{3n-1}, \quad 8 < n < 14$$

$$Pr_nO_{2n-1}$$

2. Halides

Continuous lattice mixed valence halides are rare, and no general structure types are apparent. Some examples of binary halides are:

$GaCl_2$, $(Ga(GaCl_4))$

Tl_2Cl_3, $(Tl_3(TlCl_6))$

$Bi_{12}Cl_{14}$, $((Bi_9^{5+})(Bi_2Cl_9)(BiCl_5))$

Cr_2F_5, Ag_2F

A number of lattices contain mixed valence halogeno-anions:

$Co(NH_3)_6^{3+}(Cu^{II}Cl_5)_x^{3-}(Cu^ICl_4)_{1-x}^{3-}$: rocksalt lattice

$Co(NH_5)_6^{3+}(Pb^{IV}Cl_6)_{0.5}^{2-}(Pb^{II}Cl_6)_{0.5}^{4-}$: rocksalt lattice

$A_2^I(Sb^{III}X_6)_{0.5}^{3-}(Sb^VCl_6)_{0.5}^-$: antifluorite lattice (K_2PtCl_6)

$A_2^I(Au^ICl_2)(Au^{III}Cl_4)$

3. Chain compounds

(a) Directly interacting metal ions (Figure 1(a)). Formed by stacking square planar d^{8-x} molecules. Partial oxidation (x) is achieved either by removing Group IA cations or incorporating excess anions, e.g.

$$K_{1.6}Pt(C_2O_4)_2 \cdot 2.5H_2O$$

$$K_2Pt(CN)_4Br_{0.30} \cdot 3H_2O \text{ (KCP)}$$

(b) Metal ions interacting through anion bridges (Figure 1(b) and (c)). Either alternating square planar d^8 and octahedral d^6 molecules, e.g.

$$\left[Pt^{II}(EtNH_2)_4\right]\left[Pt^{IV}(EtNH_2)_4X_2\right]X_4 \cdot 2H_2O: \text{ Wolfram's red salt}$$

or alternating square planar d^8 and linear d^{10} molecules, e.g.

$$\left[Cu^{II}(NH_3)_4\right]\left[Cu^IX_2\right]_2$$

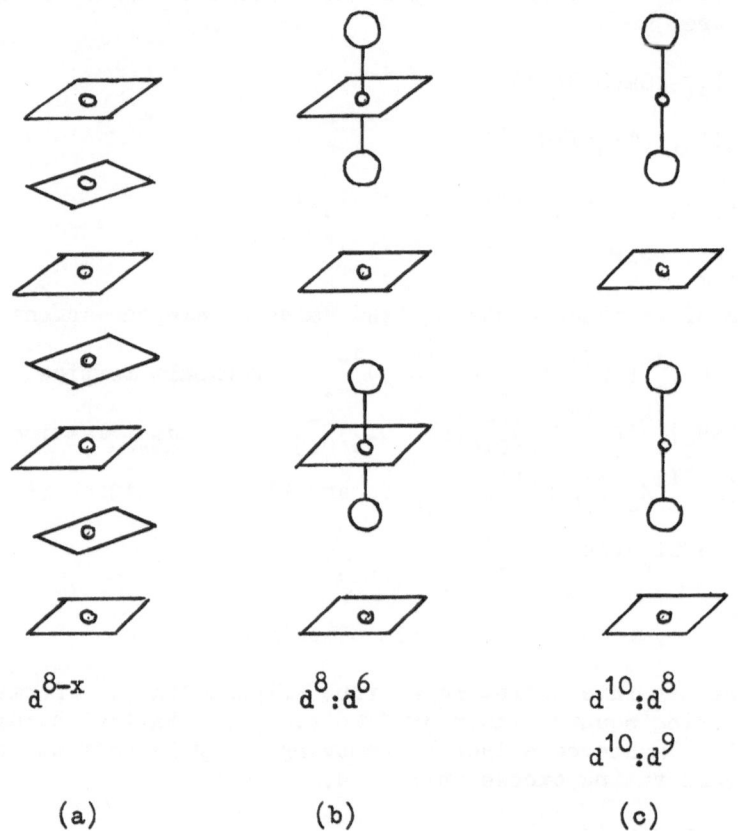

Figure 1.   Three types of mixed-valence metal atom chains.

4. Cyanides

The bridging properties of CN^- lead to its incorporation in a variety of mixed valence compounds, e.g.

$$Fe_4^{III}(Fe^{II}(CN)_6)_3 \cdot 14H_2O \ : \ \text{Prussian blue}$$

5. Discrete molecular compounds

(a) Directly interacting metal ions. Apparently lower valent compounds in the first half of the 4d and 5d blocks sometimes contain discrete mixed valence metal atom clusters. Examples are $M_6X_{12}^{n+}$ (M = Nb, Ta; X = Cl, Br, I; n = 2, 3, 4), and $Tc_2Cl_8^{3-}$. The former are cubic clusters based on regular octahedra of metal ions, the latter has two planar $TcCl_4$ groups placed plane to plane in an eclipsed arrangement.

(b) Metal ions bridged through anions and molecules.

Oxide: 'Heteropoly-blues' are formed when the d^o polynuclear oxyanions of Mo^{VI} and W^{VI} are reduced by 1-6 electrons, e.g. $(PW_{12}O_{40})^{n-}$. 'Ruthenium red' is a linear trinuclear cation in which Ru atoms are bridged by O^{2-}, their octahedral coordination being completed by NH_5 molecules: $(NH_3)_5RuORu(NH_3)_4ORu(NH_3)_5^{6+}$.

Neutral molecules: Taube[11] has synthesised a number of Ru and Os complexes with formulae $(NH_3)_5MXM(NH_3)_5^{n+}$ where X is N_2, pyrazine, 4,4'-dipyridyl etc. Organometallic sandwich compounds based on the bicyclo-pentadienyl ligand have also been prepared recently.[12]

CLASSIFICATION OF PROPERTIES OF MIXED VALENCE COMPOUNDS

Among the compounds whose structures we have just listed are metals, semiconductors and insulators, ferromagnets, antiferro-magnets and diamagnetic materials, and compounds of all colours, including white and black. We therefore seek some means of classifying (and if possible predicting) these physical properties. In 1967 Robin and Day[10] developed a classification scheme for mixed-valency compounds, based on the degree to which the sites occupied by the ions of differing valency could be distinguished in the ground state, and the consequent ease or difficulty of transferring an electron from one site to another. Consider the following simple model. A Ti^{III} ion, with

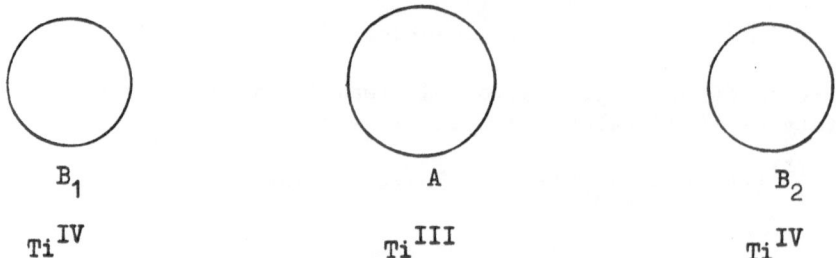

$$B_1 \qquad\qquad A \qquad\qquad B_2$$

$$Ti^{IV} \qquad\qquad Ti^{III} \qquad\qquad Ti^{IV}$$

d^1 configuration has Ti^{IV} ions (d^0) on either side of it.   The Ti^{III} and Ti^{IV} sites are distinguishable (A & B), either through being surrounded by anions in a different symmetry (e.g. one octahedral, the other tetrahedral) or by having different Ti-X bond lengths.   The anions are not shown in the simple picture above.

The zeroth-order ground-state wave function is just the product

$$\Psi_o = \psi_A^{III}\,\psi_{B_1}^{IV}\psi_{B_2}^{IV}$$

where ψ_A^{III} is the wave function of a trivalent ion on the A site, etc.   With a certain expenditure of energy the single d-electron could be transferred from the A to either of the B sites, but as the two B-sites are indistinguishable we have to make the two linear combinations

$$\Phi_{\pm} = \psi_{B_1}^{III}\,\psi_{B_2}^{IV} \pm \psi_{B_1}^{IV}\,\psi_{B_2}^{III}$$

The excited-state wave functions are therefore

$$\Psi_{\pm} = \psi_A^{IV}\,\Phi_{\pm}$$

Of these, Ψ_+ has the same symmetry as the ground-state Ψ_o, i.e. it transforms under the same representation, so the true ground-state is the linear combination

$$\Psi_o' = (1 - \alpha^2)^{\frac{1}{2}}\Psi_o + \alpha\,\Psi_+$$

where α is determined by solving the secular equation, and depends on the energy difference $E_+ - E_o$ and on the integral

$$\int \Psi_+^* \; H\,\Psi_o \; dx$$

where H is the hamiltonian of the system.   We call α the valence delocalization coefficient.

Since the ionization potential of Ti^{III} is equal and opposite to the electron affinity of Ti^{IV}, any energy difference between Ψ_0 and Ψ_+ can only come from the different ligand-fields around the sites A and B. From the structures of mixed valency compounds we should therefore be able to predict whether α will be large or small.

When A and B are <u>very different</u> (e.g. one octahedral and one tetrahedral), $E_+ - E_0$ is large, is small and Ψ_0 alone is a good description of the ground state, i.e. the valencies are firmly trapped. Then the properties of the compound are, to a good approximation, the sum of those of the individual ions. These we call <u>Class I</u> compounds. Good examples are $GaCl_2$ (50% of the Ga sites are tetrahedral, and 50% dodecahedral) and $[Co(NH_3)_6]_2$ $[CoCl_4]_3$ (high-spin tetrahedral Co^{II} and low-spin octahedral Co^{III}).

If the A and B sites are <u>identical</u> the electron can be transferred from one to the other with no expenditure of energy. Hence, if the lattice is continuous the material will be a metal. As the valences of the individual ions are 'smeared out', we do not expect to see any of the characteristic properties of the single valence states. Materials of this kind are called Class IIIB. Examples are the bronzes. If the lattice is not completely continuous, delocalization only takes place within a finite cluster of equivalent ions. Then, although none of the individual ion properties are seen, the lattice does not conduct. This type of substance, of which the cluster ions $Nb_6X_{12}^{2+}$ are good examples, we call <u>Class IIIA</u>.

Intermediate between the two extremes are compounds in which the A and B sites are <u>similar but distinguishable</u>, (e.g. both octahedral, but with slightly different metal-ligand distances). Then E_+ lies only a small distance above E_0 and though the valences are substantially trapped in the ground-state (α is small but finite) they can exchange with only a small expenditure of energy. Such materials still exhibit the single ion properties characteristic of the ground-state, but have low-energy charge transfer bands, are semiconductors, etc. Examples are Prussian blue, ruthenium red, etc.

The physical properties which we expect to find associated with each class of compound are summarized in Table 2.

The model for mixed valence behaviour we have been describing is essentially a static one, and takes no account of the dynamical behaviour either of the electrons or the nuclei. It is therefore important to notice that even when the mixed valence sites are surrounded by equal numbers of similar ligand atoms, we nevertheless only get what was described above as Class III behaviour

TABLE 2.   Physical Properties of the Classes of
Mixed-Valency Compounds

	Class I	Class II	Class IIIA (clusters)	Class IIIB (infinite lattice)
Optical	No mixed-valence transitions in visible	One mixed-valence transition in visible; low intensity	One or more mixed-valency transitions in visible; high intensity	Opaque; metallic reflection in visible
Electrical	Insulator; resistivity greater than 10^{12} ohm cm.	Semiconductor; resistivity $10-10^8$ ohm cm.	Insulator or high-resistance semiconductor	Metallic conductor; resistivity $10^{-2}-10^{-6}$ ohm cm.
Magnetic	Diamagnetic or paramagnetic to very low temperature	Magnetically dilute; either ferro- or anti-ferromagnetic at low temperatures	Magnetically dilute	Pauli paramagnetism or ferromagnetic with high Curie temperature
Other properties (IR,UV, Mossbauer,etc.)	Spectra of constituent ions at normal frequencies	Spectra of constituent ions close to normal frequencies	No spectra of constituent ions	No spectra of constituent ions

if electrons can be transferred from site to site more rapidly than the ligand nuclei can relax to equilibrium positions corresponding to the individual trapped valences. In continuous lattice compounds this criterion has been recognized for some years as the one which distinguishes the realms of validity of the band and polaron hopping models of electron transport, for example in Li doped NiO.[13] Only a year or two ago, on the other hand, were the first examples found among discrete molecules which forced a consideration of this question. There, the most famous instance has been the compound $(NH_3)_5Ru(pyrazine)Ru(NH_3)_5{}^{5+}$. Both Ru atoms are six-coordinate, five nitrogen donor atoms coming from the ammonia molecules and the sixth from the pyrazine.[11] At first glance the compound ought to reside in our Class IIIA, yet the evidence of numerous physical techniques (e.g. Mossbauer and electronic spectroscopy) is that the valences are trapped, and that it properly belongs in Class II. To see how this situation comes about, we refer to Figure 2, wiich shows two identical potential energy surfaces, one for the state described to a first approximation as $Ru_A^{II}Ru_B^{III}$, the other as $Ru_A^{III}Ru_B^{II}$, where A and B label the two atoms of the metal dimer. The amount of energy required for the adiabatic transfer of an electron from A to B is given by E_{Ad} and the energy of a Franck-Condon transition (that is, the maximum of the mixed valence optical absorption band) by E_{FC}. If the two surfaces are harmonic along the vibrational coordinate in question, there is a simple relation between these two energies:

$$E_{Ad} = E_{FC}/4,$$

a point first made by Hush.[14] At the crossing point of the two configuration coordinate curves the complex is in a delocalized state consisting of equal combinations of the two valence structures $Ru_A^{II}Ru_B^{III}$ and $Ru_A^{III}Ru_B^{II}$. If there is a resonance interaction between the two structures two new surfaces are formed, at energies higher and lower than the original crossing point, as indicated by the dotted lines. Whether the energy of the dimer is minimized in the delocalized (Class IIIA) or one of the localized (Class II) states then depends on the magnitude of the resonance integral H_{res} compared with E_{Ad}: if $H_{res} \gg E_{Ad}$ we have complete delocalization, but if $H_{res} \ll E_{Ad}$ the valences are trapped.[15]

Another way of looking at this criterion is to note that if the 'excess' electron is placed on atom A it will oscillate back and forth between A and B at a frequency

$$\nu = 2H_{res}/h.$$

The question then becomes one of assessing the magnitude of ν in comparison to the frequency of the vibrational mode which would convert the geometry of the A site into that of the B. Strictly speaking, of course, there is no reason to suppose that it is safe

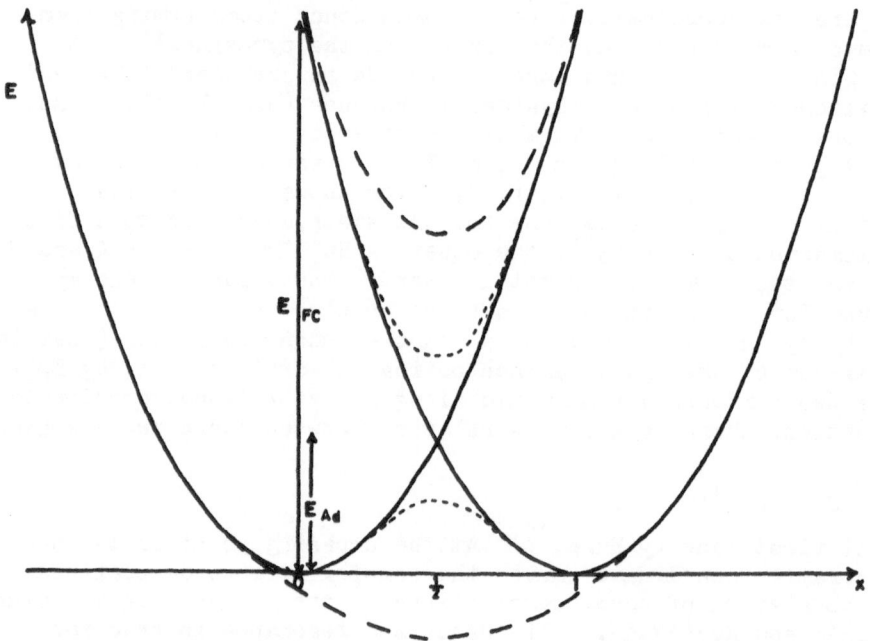

Figure 2. Potential energy surfaces for the states $Ru_A^{II}Ru_B^{III}$
and $Ru_A^{III}Ru_B^{II}$ (full lines). The dotted line represents the
surface for H_{res} E_{Ad} and the dashed line the surface for
H_{res} E_{Ad}. Reproduced by permission from J. Amer. Chem. Soc. <u>94</u>,
2885 (1972).

to apply the Born-Oppenheimer approximation to this type of process, since the wavefunction of the mixed valence electron is certainly a function of the atomic coordinates. However, no treatment of this question as a full vibronic problem ever appears to have been made. If it were, it would be the molecular analogue of the Holstem model of the small polaron.[16]

MIXED VALENCE METAL CHAIN COMPOUNDS

Amongst the specific examples of mixed valence compounds which we cited earlier are quite a large number in whose structures chains of metal atoms can be distinguished. These include not only molecular lattices, but many continuous lattice oxides and halides. For instance in the simple rutile (TiO_2) structure columns of TiO_6 octahedra share opposite edges, the columns being joined together through their corners. Metal atom chain compounds provide us with examples falling in all three of the classes of mixed-valence behaviour postulated by our classification scheme, and can therefore be used to show how the scheme works out in practice.

Class I

As examples of Class I behaviour in metal chain compounds we take the B-subgroup oxides Pb_3O_4 and Sb_2O_4. The structure of the former[17] consists of chains of PbO_6 octahedra sharing opposite edges, in which the Pb-O distances are all equal to 2.15 A. These chains are joined together through further Pb atoms which only have three nearest neighbour oxygens, at about 2.2-2.3 A. Thus there are two very distinct Pb sites, in agreement with the assignment of valences at $Pb_2^{II}Pb^{IV}O_4$. In polycrystalline form the compound is a good insulator, and its colour (red) is simply the sum of the colours of PbO and PbO_2, both also red. Likewise in Sb_2O_4[18] one finds two very different modes of Sb coordination: octahedral, with bond lengths between 1.96 and 1.99 A, and distorted tetrahedral, with two pairs of bond lengths at 2.02 and 2.22 A. The latter sites form chains through the lattice. In fact there are two polymorphs of Sb_2O_4, but in contrast to the blue Sb(III,V) halide salts, in which both valence states are found in octahedral coordination, the two forms of the mixed valence oxide are colourless.

Class II

Good examples of Class II mixed valence behaviour are those halide bridged platinum chain compounds, such as $PtenCl_3$ (en = ethylenediamine) and Wolffram's red salt, which consist of

alternate square planar and tetragonally distorted octahedral
molecules (Figure 1(b)). A firm indication that these crystals
are to be considered as trapped valence Pt(II), Pt(IV) in their
ground states comes from, among other things, their X-ray photoelectron (XPS) spectra.[19] In the Pt 4f region the XPS spectra of
the single valence constituent complexes of $PtenCl_3$, namely $PtenCl_2$
and $PtenCl_4$, each consist of two peaks, corresponding to ionization,
from $4f_{5/2}$ and $4f_{7/2}$ spin-orbit levels, although it is important
to notice in passing that the correct XPS spectra of these and
other complexes which lend themselves to easy oxidation and
reduction are only obtained after very carefully cleaning the
surface of the sample. The XPS spectrum in the Pt 4f region of
the mixed valence complex itself, on the other hand, consists of
three distinct peaks, which can be deconvoluted into two partly
overlapping pairs, i.e. $4f_{5/2}$ and $4f_{7/2}$ of Pt^{II} and Pt^{IV}. Likewise the infrared spectrum of $PtenBr_3$ is a superposition of the
spectra of $PtenBr_2$ and $PtenBr_4$.[20]

We designate this type of compound as Class II from a
structural point of view because although two Pt sites of differing
geometry can be distinguished, one might imagine them as being
mutually interconvertible simply by moving pairs of axial halide
ions in phase away from the octahedrally coordinated Pt atoms
towards the square planar. It is a pity that no-one has ever
searched experimentally for evidence of the longitudinal acoustic
phonon modes which accomplish this interconversion, for example by
inelastic neutron scattering. If they had, knowledge of the
behaviour of these modes might open up a fascinating new field of
vibronic interactions leading to valency inversion.

The optical properties of the halide bridged Pt chain
compounds agree with our assignment of them to Class II because,
in addition to absorption in the near ultraviolet which can be
assigned to local ligand field excitations of the constituents,
again through comparison with the single valence molecules, they
all have additional low energy absorption, polarised parallel to
the chains, of which no trace appears in the spectra of the
constituents. The additional bands are assigned to electron
transfer from Pt^{II} to Pt^{IV}, specifically, from the filled z^2 orbital
of the former to the empty z^2 of the latter. Much effort has
been expended over the last few years in determining the order of
the filled and empty orbitals in square planar d^8 and tetragonally
distorted d^6 complexes,[21] and extrapolating the results of those
studies enables us to construct the diagram for the mixed valence
Pt compounds shown in Figure 3.

Most of the work on the optical properties of these compounds
dates back to the 1950's, and was performed at room temperature
with a rather simple microspectrophotometric setup.[22] The main
experimental obstacle to extending the spectra to low temperatures
and higher resolution is the intensity of the transitions, although

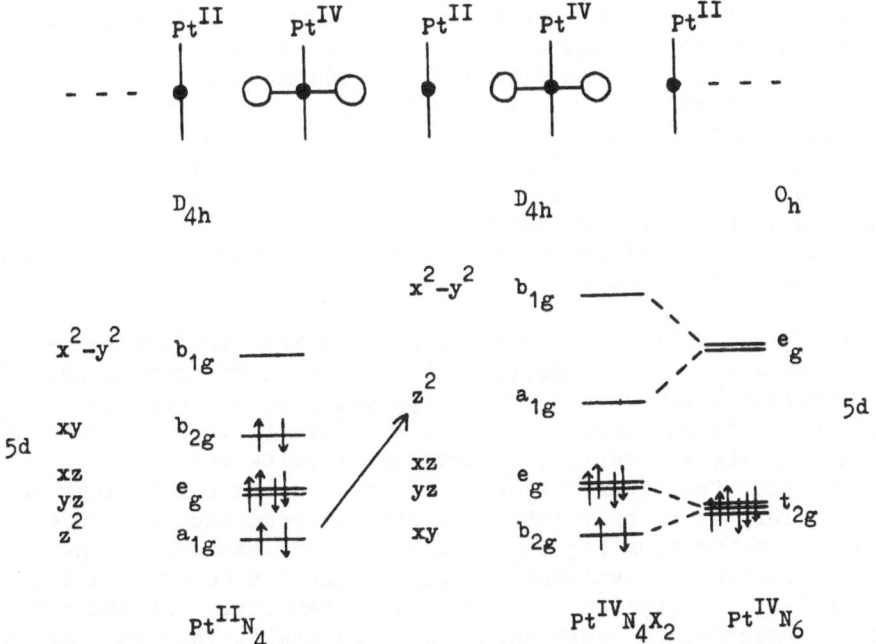

Figure 3.   Energy levels and transitions in halide bridged
Class II Pt(II,IV) chains.

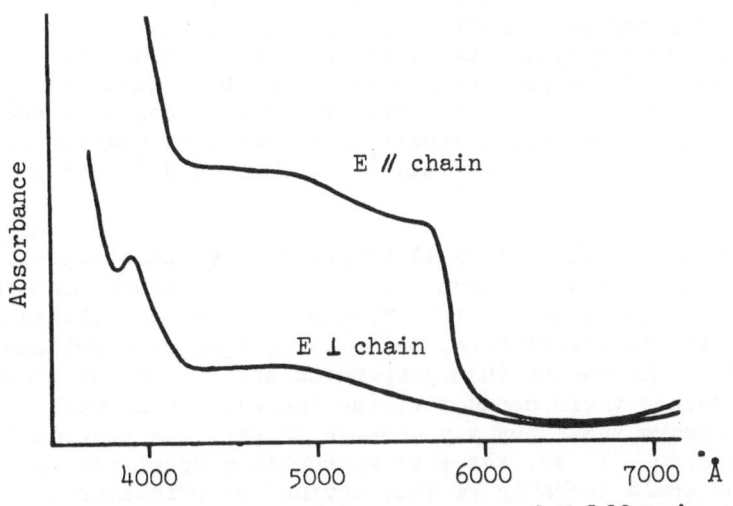

Figure 4.   Polarised absorption spectrum of Wolffram's red salt
at 4°K.

we ourselves recently succeeded in obtaining a 4°K spectrum of
Wolffram's red salt (Figure 4) by using an extremely small (0.1 mm^2)
crystal and replacing the standard 1P28 photomultiplier of the
Cary 14 spectrophotometer with one of much greater sensitivity.
The most notable feature of this spectrum is the way the mixed
valence absorption band shifts and changes shape from room
temperature to 4°K.   An accurate examination of such variations
as this will provide much useful information about changes in the
potential energy surfaces as the electron is transferred from one
metal atom to the other.

     At first sight it may seem surprising that the metal-to-
metal charge transfer transitions in this type of crystal should
be so intense because in PtenBr$_3$, for example, the nearest
neighbour Pt-Pt distance is 5.60 A.[23]   Direct overlap between 5d
orbitals on the two centres cannot therefore be very great.
Nevertheless, following the model of mixed valence behaviour we
gave earlier, it is only through mixing between the zero order
ground state function $Pt_A^{II}Pt_B^{IV}$ and $Pt^{II} Pt^{IV}$ excited charge
transfer states that valence delocalization can occur, and the
mixed valence absorption thus gain its intensity.   If the direct
integral which could cause such mixing is small, then an indirect
mechanism involving Pt_A^{II} Br and Br Pt_B^{IV} excited states must be
the most important, rather as in the Anderson model of superexchange.[24]   This idea can be written out in a second-order
perturbation formalism and, if the energies of the metal-to-ligand
and ligand-to-metal charge transfer states can be found, either by
identifying them in the spectra of the mixed valence compound or
related chromophores, and the metal-ligand resonance integrals
estimated from reliable wavefunctions, then valence delocalization
coefficients for Class II compounds can be calculated.   From the
coefficients one can in turn calculate the transition moment of the
mixed valence absorption band, for comparison with the observed
intensity.   At the present time we do not have sufficient data to
carry out this kind of calculation for the Pt chain compounds, but
it has been applied successfully to a number of simpler model
systems, such as Prussian blue,[25] and Taube's $Ru^{II,III}$ pyrazine
dimer.[26]

     From their other Class II properties we would expect, and
indeed find, that the halide bridged Pt chain compounds behave as
high resistance semiconductors rather than metals, although the
conductivity is anisotropic, being two orders of magnitude larger
parallel to the chains than perpendicular.[27]   On the other hand,
under pressure their conductivities increase by as much as 10^6,
and it appears that they may undergo transitions towards Class III
structures.[28]   If so, the most appropriate model for their
behaviour would probably be that devised by Drickamer and Slichter[29]
to rationalize spin-state and oxidation state changes in single
valence Fe^{II} and Fe^{III} compounds under pressure.   In this approach

one considers two potential energy surfaces, one for each state of
the system, whose separation in energy and space coordinates varies
as the pressure changes.

Class III

The most famous examples of Class III mixed valence chain
compounds are of course the Pt cyanides and oxalates first
prepared by Knop[2] and Werner,[1] and more recently revived and
investigated by Krogmann and his colleagues.[30]   Other lecturers,
particularly Professor Krogmann himself, will give more detailed
accounts of the structures and properties of these compounds.
Suffice it to say here that the structures of salts such as
$K_2Pt(CN)_4Br_{0.30}3H_2O$ (KCP) and $K_{1.6}Pt(C_2O_4)_2.2.5H_2O$ are based on
plane-to plane stacking of the square planar molecular units
(Figure 1(a)) and that, the Pt sites being all identical, we
expect from Table 2 that they will behave as metallic conductors.
Because the distance between adjacent chains is so large (9.87 A
in KCP) the metallic conductivity will be essentially one-
dimensional.   Further, if the valences are entirely smeared out
on the time-scale of the measurement, no single ion excitations
should be observable.

Broadly speaking, these simple predictions from the mixed
valence classification scheme are indeed borne out by the observed
properties.   Neither X-ray photoelectron[31] nor Mossbauer[32] spectra,
methods spanning the widest difference in time-scale (10^{-16} and
10^{-7} seconds respectively), reveal any dissimilarity between the
Pt sites.   At least near room temperature, the conductivity
parallel to the Pt chains is metallic, increasing with decreasing
temperature, and with the incident electric vector parallel to
the chains a metallic plasma edge also appears in the optical
reflectivity,[33] rather reminiscent of the one found in tungsten
bronzes.   On the other hand, below room temperature the behaviour
of KCP begins to diverge in a most interesting way from the simple
Class III prediction.   For example, on lowering the temperature
below about 200 K the temperature variation of the conductivity
gradually changes from that of a metal to the activated variation
expected of a semiconductor so that, when $4°K$ is reached, the
specific conductivity in the direction of the chains is no more
than 10^{-12} ohm^{-1}cm^{-1}.[34]   This change in the character of the
conductivity is most plausibly connected with 'softening' of the
longitudinal acoustic phonon mode of the Pt chain which has a
wavelength between six and seven Pt-Pt spacings.[35]   Were the mode
to soften completely, and a phase transition to occur, the new
phase would have a trapped valence Class II structure and the
change to an activated conductivity would be explained.   It is
surely significant that the latest estimate of the wavelength of
the phonon instability (6.67 Pt-Pt spacings[36]) is close to the

periodicity along the chain required by valence localization to the most common Pt oxidation states, Pt^{II} and Pt^{IV}. There now seems general agreement that the limiting formula of KCP is the one we have written above, in particular that the Br:Pt ratio is 0.30 and not 0.33, and that in spite of difficulties caused by efflorescence, its limiting H_2O:Pt is 3. The formal Pt^{II}:Pt^{IV} ratio in the compound would thus be 20/3:1.

In terms of our simple model, KCP at low temperature thus becomes a weakly trapped Class II system. That the lowest common ratio of Pt^{IV} to Pt^{II} is apparently a non-integral number need not worry us unduly: it simply means that the repeat distance along the c-axis is much longer than has previously been assumed. In fact evidence is already accumulating[37,38] that the KCP structure is considerably more complicated than was originally thought.[30] Starting with the shear and block structures of the Magneli compounds,[39] and the tunnel structures of bronzes,[40] inorganic chemists have become increasingly familiar over the last few years with the idea that many apparently non-stoichiometric compounds are actually stoichiometric, though with very large unit cells and elaborate formulae. $Nb_{12}O_{29}$[41] and $Mo_{17}O_{47}$[42] are just two examples from the very large number now known. Perhaps the mixed valence Pt chain compounds will prove to be further additions to the list. At any rate, it will be clear from this chapter that the story of mixed valency compounds opened by Werner is still far from complete and that this fascinating group of materials will continue to provide us with puzzles in relating structures to electronic properties for a long time to come.

References

1.   A. Werner, Z. anorg. Chem. 12, 46 (1896).

2.   W. Knop, Ann. Chem. 43, 111 (1842).

3.   I. Bernal, E.A.V. Ebsworth and J.A. Weil, Proc. Chem. Soc., 57 (1959).

4.   J.H. de Boer and E.J.W. Verwey, Rec. Trav. Chim. 55, 541 (1936).

5.   E.J.W. Verwey, P.W. Haaijmann, F.C. Romeijn and G.W. van Oosterhout, Philips Res. Rep. 5, 173 (1950).

6.   L.S. Miller, S. Howe and W.E. Spear, Phys. Rev. 166, 871 (1968).

7.   R.E. Peierls, 'Quantum Theory of Solids, Oxford University Press, London, 1955, p. 108.

8. F. Sondheimer, D.A. Ben-Efraim and R. Wolovsky, J. Amer. Chem. Soc. 83, 1675 (1961).

9. S.S. Malhotra and M.C. Whiting, J. Chem. Soc. 3812 (1960).

10. M.B. Robin and P. Day, Adv. Inorg. Chem. and Radiochem. 10, 247 (1967).

11. C. Creutz and H. Taube, J. Amer. Chem. Soc. 91, 3988 (1969).

12. D.O. Cowan, C. LeVanda, J. Park and F. Kaufman, Acc. Chem. Res. 6, 1 (1973); U.T. Mueller-Westerhoff and P. Eilbracht, J. Amer. Chem. Soc. 94, 9272 (1972).

13. I.G. Austin and N.F. Mott, Science 168, 71 (1970).

14. N.S. Hush, Prog. Inorg. Chem. 8, 391 (1967).

15. B. Mayoh and P. Day, J. Amer. Chem. Soc. 94, 2885 (1972).

16. T. Holstein, Ann. Phys. (N.Y.), 8, 343 (1959).

17. M.K. Fayek and J. Leciejewicz, Z. anorg. Chem. 336, 104 (1965).

18. D. Rogers and A.C. Skapski, Proc. Chem. Soc., 400 (1964).

19. D. Cahen and J.E. Lester, Chem. Phys. Lett. 18, 108 (1973); J. McGilp, Part II Thesis, Oxford, 1973.

20. G.W. Watt and R.E. McCarley, J. Amer. Chem. Soc. 79, 4585 (1957).

21. For a review see D.S. Martin, Inorg. Chim. Acta Rev. 5, 107 (1971).

22. S. Yamada and R. Tsuchida, Bull. Chem. Soc. Japan, 29, 894 (1956).

23. T.D. Ryan and R.E. Rundle, J. Amer. Chem. Soc. 83, 2814 (1961).

24. P.W. Anderson, Phys. Rev. 79, 350 (1950).

25. B. Mayoh and P. Day, J.C.S. Dalton Trans. 846 (1974).

26. B. Mayoh and P. Day, Inorg. Chem. 13, 2273 (1974).

27. T.W. Thomas and A.E. Underhill, J. Chem. Soc. A, 512 (1971).

28. L.V. Interante, K.W. Browall and F.P. Bundy,

29. H.G. Drickamer, C.W. Frank and C.P. Slichter, Proc. Nat.
     Acad. Sci. U.S.A., $\underline{69}$, 933 (1972).

30. K. Krogmann and H.D. Hausen, Z. anorg. Chem. $\underline{358}$, 67 (1968).

31. M.A. Butler, D.L. Rousseau and D.W.E. Buchanan, Phys. Rev. B
     $\underline{7}$, 61 (1973).

32. W. Ruegy, D. Kuse and H.R. Zeller, Phys. Rev. B $\underline{8}$, 952 (1973).

33. D. Kuse and H.R. Zeller, Phys. Rev. Lett. $\underline{27}$, 1060 (1971).

34. H.R. Zeller, Festkorperprobleme, $\underline{13}$, 31 (1973).

35. B. Renker, H. Reitschel, L. Pintschovius, W. Glaser,
     P. Bruesch, D. Kuse and H.R. Zeller, Phys. Rev. Lett. $\underline{30}$,
     1144 (1973).

36. B. Renker, L. Pintschovius, W. Glaser, H. Rietschel, R. Comes,
     L. Liebert and W. Drexel, Phys. Rev. Lett. $\underline{32}$, 836 (1974).

37. J.H. Dieseroth and H. Schultz, Phys. Rev. Lett. $\underline{33}$, 963 (1974).

38. D. Griffiths, P. Day, C.J. Sampson and F.A. Wedgwood, Solid
     State Comm., in press.

39. A. Magneli, Acta Cryst. $\underline{6}$, 495 (1953).

40. L. Kihlborg, Adv. Chem. Ser. $\underline{39}$, 37 (1963).

41. R. Norin, Acta Chem. Scand. $\underline{20}$, 871 (1966).

42. L. Kihlborg, Arkiv. Kemi $\underline{21}$, 471 (1964).

ELECTRICAL TRANSPORT AND SPECTROSCOPICAL STUDIES OF THE

PEIERLS TRANSITION IN $K_2[Pt(CN)_4]Br_{0.30} \cdot 3(H_2O)$

H. R. Zeller

Brown Boveri Research Centre

CH - 5400 Baden Switzerland

Abstract

A review is given of the optical properties and electrical transport studies on $K_2[Pt(CN)_4]Br_{0.30} \cdot 3(H_2O)$. The material undergoes a very broad transition from a metallic into an insulating state which was shown by diffuse x-ray and inelastic neutron scattering experiments to be caused by a Peierls instability. Due to the one-dimensional character of the compound fluctuations dominate in a very large temperature interval and impede the successful application of convential phase transition concepts. It is shown, however, that a simple description of the phase transition is possible.

I. INTRODUCTION

In many respects $K_2[Pt(CN)_4]Br_{0.30} \cdot 3(H_2O)$ (KCP) is a unique model system for the experimental study of dimensionality effects[1]. Large single crystals can be grown allowing the use of virtually all solid state experimental techniques. The electronic structure of KCP is exceedingly simple with a parabolic free electron like conduction band[2] which is almost ideally one-dimensional. The heavy mass of Pt with respect to the other constituent ions of the crystal makes x-ray studies of minute distortions of the Pt chain possible[3,4] and also leads to relatively soft longitudinal acoustical modes in the

Pt chain which can be studied by inelastic neutron scattering[5],[6]. As a result KCP is a unique model system for the study of dimensionality effects and is the best understood one-dimensional (1-D) conductor today.

Chemistry and structural properties[7],[8] of KCP are discussed in other papers of this volume and will not be reviewed here. Also we will just briefly sketch the theoretical aspects of the Peierls instability and concentrate on the temperature and frequency dependence of the dielectric properties of KCP.

II. THE PEIERLS TRANSITION (THEORY)

In a 1-D metal the Fermi surface consists of two points located at $\pm k_F$. Energy conserving two body interaction are thus restricted to a momentum change of $\Delta k \approx 0$ or $\Delta k \approx 2k_F$. The corresponding response functions show logarithmic singularities at $2k_F$ resulting in an instability of the 1-D metallic system[9],[10],[11]. Depending on the interaction parameters a charge density wave, spin density wave, singlet or triplet superconducting ground state results[12]. Up to now all such theories are weak coupling theories and neglect the presence of a crystal lattice. Contact with experimental systems is unfortunately still impossible and in order to discuss the Peierls transition[13] which is a special case of a charge density wave instability we may as well adopt a free electron model. Within a nearly free electron model the energy of the electronic system in a 1-D metal has a logarithmic singularity as a function of a lattice distortion with wave vector $2k_F$[14],[15]. The elastic energy on the other hand is quadratic in the distortion and as a consequence no matter how stiff the crystal is, the total energy of the system can always be lowered by a distortion. The ratio of elastic to electronic energy change determines the equilibrium value in the distorted state. A distortion with wavevector $2k_F$ creates an energy gap at the Fermi energy and hence transforms the material into a semiconductor.

It is a well known truism that in a stictly 1-D system with short range interactions no phase transition can occur at $T \neq 0$. Even in the absence of a true phase transition one would expect that fluctuations strongly affect physical properties in a sizable temperature interval. The residual 3-D coupling necessarily present in a real system is expected to lead to a finite transition temperature[17] but also to smoothen out the singularities which lead to the transition itself. Generally speaking it is a nontrivial question whether the dimensionality effects[1] which show up in 1-D mathematical models do in fact survive the nonideality of experimental systems. As we will see the Peierls transition survives in KCP due to the fact that the electronic system is almost ideally one-dimensional. This is true to a much lesser degree for the elastic properties[6] and 3-D coupling is introduced into the electronic system mainly by electron-phonon interaction, impurities and disorder.

Optical experiments provide relatively easy access to the parameters governing the Peierls transition. The visible and ultraviolet spectrum yields information about the band structure, transitions across the Peierls gap are found in the infrared and the charge density wave is observed in far infrared from which the electron-phonon coupling constant can be deduced.

III. BAND STRUCTURE OF KCP

Based on standard arguments a conduction band deduced from an atomic d_z^2 orbital should exhibit a large effective mass and a tight binding scheme is expected to be appropriate. This is not born out by experiment. Fig. 1 shows the reflectivity as a function of wavelength at $T = 40°K$ and room temperature. The salient feature is a rather sharp plasma edge in the visible and the absence of interband transitions at higher energies. From the observed plasma frequency[1,18] assuming 1.7 electrons per Pt we deduce an effective mass of $m* = m_e$, i.e. free electron behaviour[2].

An even stronger proof (the above argument is based on a Drude model) comes from the conductivity sum rule which states that the properly normalized integral[18] $\int_0^\omega \sigma(\omega') \, d\omega'$ asymptotically approaches the total number of electrons per atom. The only quantum mechanical system in which there are no interband transitions is the free electron system. As a consequence the sum rule will be exhausted for the intraband transition at $\omega \approx 0$. In the case of KCP $\sigma(\omega)$ can be determined from a Kramers Kronig analysis of the reflectivity[2,18]. The resulting integral $n(\omega)$ is shown in Fig. 2.

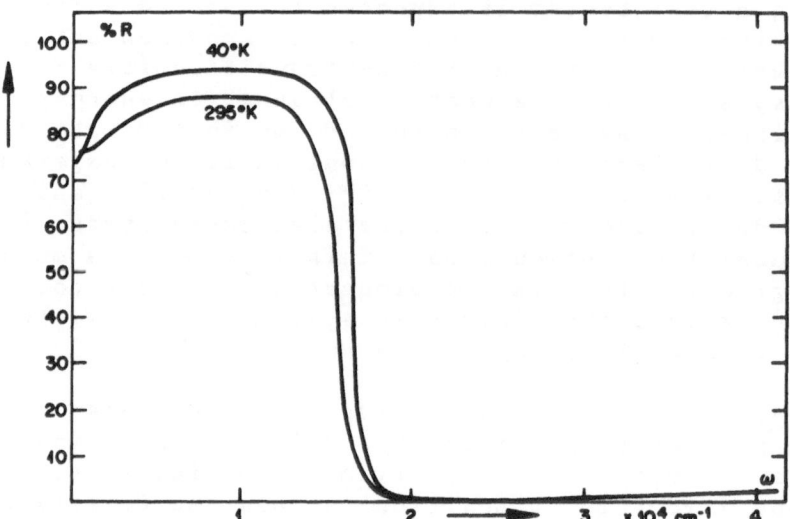

Fig. 1   Reflectivity of single crystal KCP for light
          polarized parallel to the conducting axis. Note
          the sharpening of the plasma edge and the struc-
          ture in the infrared with decreasing temperature.

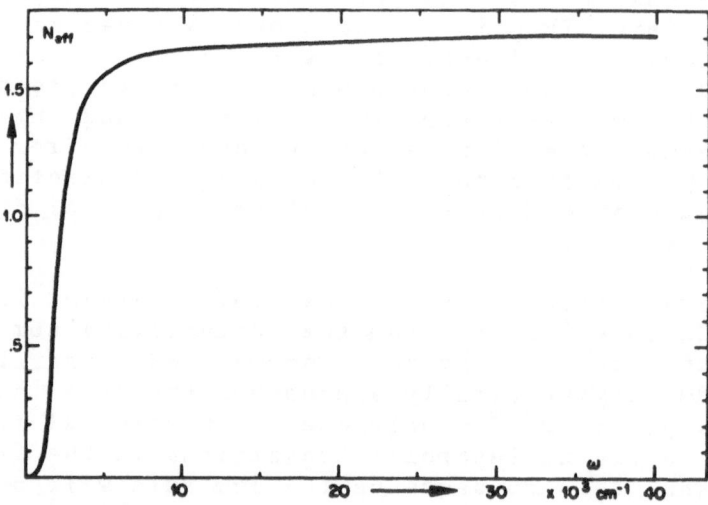

Fig. 2   Conductivity or oscillator strength sum rule.
          For free electrons n_{eff} reaches a plateau of
          1.7 e/Pt above the intraband transition at
          $\omega \approx 0$. (T = 40oK)

n (ω) reaches its asymptotical value of 1.7 electrons per Pt already at ω ≈ 10'000 cm⁻¹. This implies that above 10'000 cm⁻¹ the electrons are free and no longer feel the periodic potential of the lattice. It will be shown later that the deviations from free electron behaviour at lower frequencies are connected with a Peierls gap.

We thus arrive at the band structure given in Fig. 3 which assumes a parabolic band with effective mass $m^* = m_e$ and zone boundaries introduced by the lattice periodicity. (Each Pt is assumed energetically equivalent). From Fig. 3 it is possible to calculate the energy of the first interband transition without any free parameter. For ideally free electrons this transition should have zero oscillator strength. Hence both, oscillator strength and energy of this transition give information about deviations from ideally free electron behaviour.

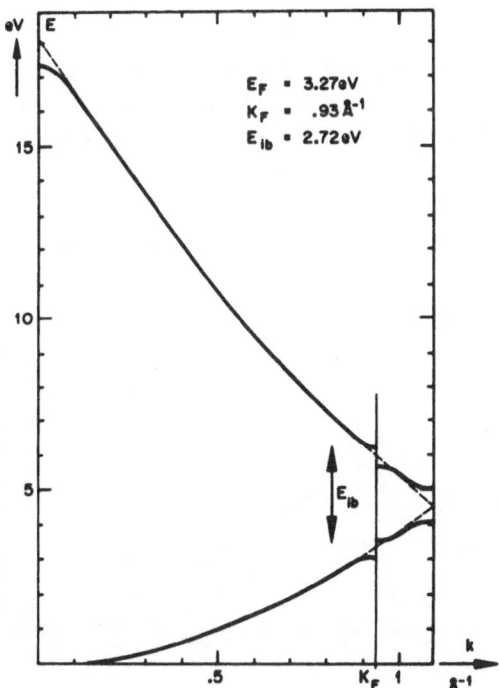

Fig. 3   Free electron band structure of KCP (Note the absence of any free parameters).

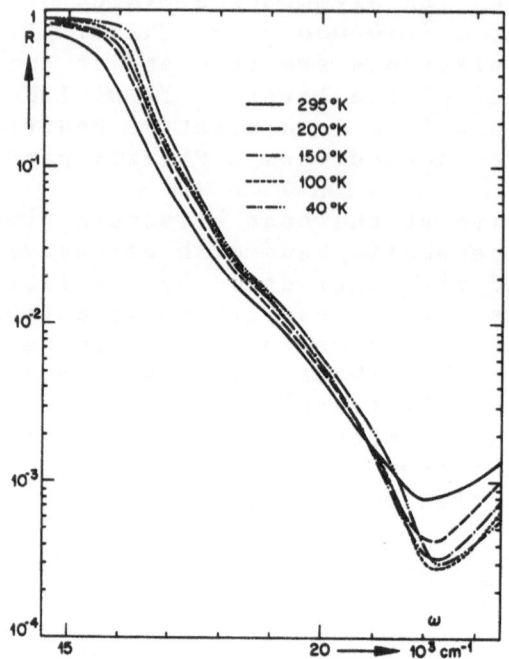

Fig. 4   Logarithmic plot of the plasma edge at various
         temperature. The weak structure is interpreted as
         due to the interband transition indicated in Fig. 3.

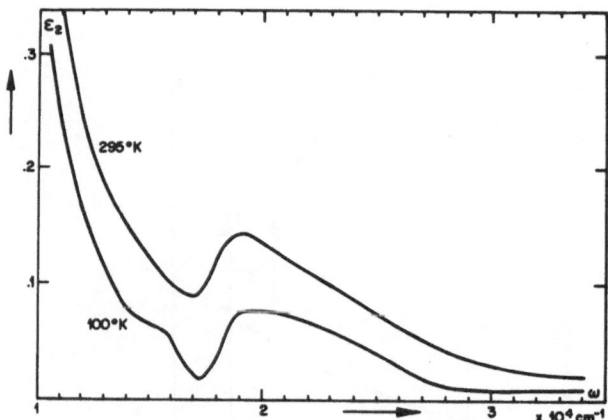

Fig. 5   Dielectric function in the region of the plasma
         edge.

Fig. 4 shows detailed measurements of the plasma edge which exhibits a weak structure[2,19]. A Kramers Kronig analysis yields a nearly temperature independent structure superimposed on a temperature dependent background (Fig. 5). The oscillator strength of this structure is of the order of 1 % of the total and its position is about 10 % below the predicted value again supporting the free electron picture.

From the optical experiments on KCP alone it cannot be completely excluded that the band structure is complex with bands overlapping at E_F and that both the plateau in n (ω) at 1.7 e and the free electron plasma frequency are accidental and do not reflect free electron behaviour. However, this latter possibility can be easily discarded. First with bands overlapping at E_F the simple relation between chemical band filling and Fermi momentum observed by Comès et al[3] would break down. Second if the above properties would not reflect a sum rule but be accidental then we would expect the plasma frequency to depend strongly on Pt-Pt distance. High pressure experiments on KCP and comparison with related compounds shows that this is not the case[2].

The explanation for this at first sight strange behaviour is connected with the dimensionality of the system. It is well known that in 3-D nearly free electrons can be obtained by a sufficient overlap of s-wave functions such as for instance in an alkali metal. For d-electrons this is not possible. A d-orbital in a crystal is directional resulting in a strong spatial dependence of electron density. Free electrons on the other hand imply a constant electron density. As a result d-bands in 3-D have large effective masses and intraband transitions with small oscillator strength. In 1-D systems we just ask for free electron behaviour along one axis. A nearly constant electron density along the c-axis can by easily achieved by overlapping s hybridized d orbitals[20]. In this respect in 1-D systems there is no difference between s,p,d etc. bands simply due to the fact that in 1-D "directionality" of a bond has no meaning.

The simplicity of the experimentally found band structure not only simplifies theoretical calculations of the Peierls transition but also the rather large band width renders electron-electron correlation effects relatively unimportant.

IV PEIERLS TRANSITION IN KCP

The first direct experimental evidence for a Peierls instability in KCP came from diffuse x-ray experiments by Comès et al[3],[4]. Later the diffuse x-ray results were confirmed and extended by inelastic neutron scattering techniques[5],[6]. Both experiments are discussed in detail in other papers of this volume. The existence of a sinusoïdal lattice distortion with the theoretically predicted period of about 6.7 Pt-Pt spacings is clearly established. It turns out that the simple mean field picture of a lattice mode going soft and reaching zero frequency at the metal to semiconductor transition T_p[15],[21] completely fails. At room temperature instead of a well defined soft mode a broad structure is found[6] in the structure factor coexistent with a pronounced central peak. Also by lowering the temperature the structure hardens rather than softens despite the fact that the dc conductivity above 200°K has a metallic temperature coefficient[22]. This clearly indicates that the physical properties are fluctuation dominated as expected for a 1-D system. It is thus of paramount importance to obtain information not only on the elastic properties as determined by inelastic neutron scattering but also on the electronic properties of KCP. A simple and powerful tool to study electronic properties is optical spectroscopy.

We have thus performed a detailed study of the optical properties of KCP in the infrared as a function of temperature. The prominent feature, a structure in the reflectivity[18] is shown in Fig. 6. As can be seen the structure becomes sharper with decreasing temperature. A Kramers Kronig analysis of the data shows already at room temperature a strong resonance in $\varepsilon_2(\omega)$ at 1600 cm^{-1} (Fig.7). With decreasing temperature this resonance becomes sharper and sharper (Fig. 8).

The structure at low temperature is interpreted as to arise from transitions across the Peierls gap implying a gap value of $2\Delta \cong \cdot 2$ eV. From mean field theory we would then predict a transition temperature T_p^{MF} of about 800°K. In a 1-D system T_p^{MF} is clearly a fictional number but its physical implication is that we have to expect strong fluctuations and precursor effects at $T \lesssim \sqrt{2} \ T_p^{MF}$. Furthermore below $T \lesssim \sqrt{2} \ T_p^{MF}$ the fluctuating order parameter will be essentially the zero temperature order parameter. Physically this means that starting from a constant

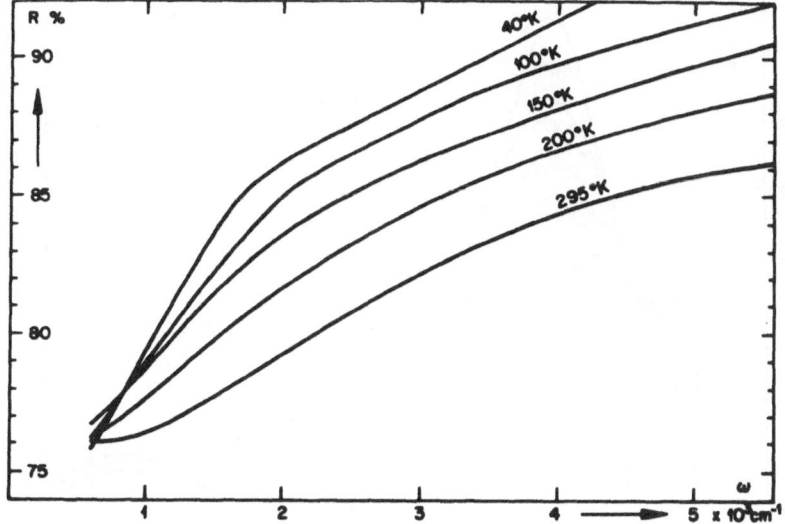

Fig. 6   Temperature dependence of the infrared reflectivity of KCP for light polarized parallel to the
conducting axis.

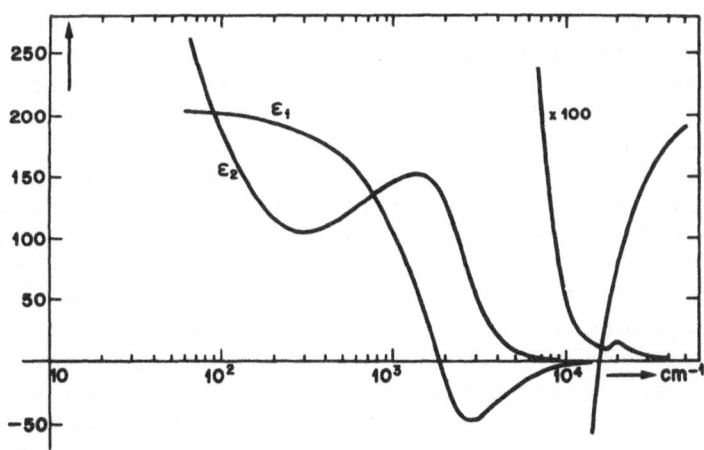

Fig. 7   Overall dielectric properties of KCP at room
temperature.

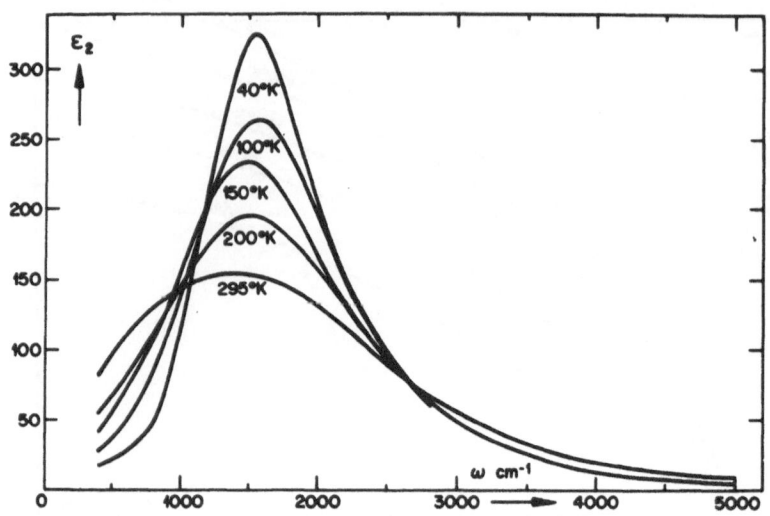

Fig. 8   Temperature dependence of ε_2 in the infrared.

density of states at high temperature a pseudogap will gradually form and eventually become a real gap but there will be no temperature dependence of the optical gap. As Fig. 8 shows this is born out by experiment.

In an ideal case one should be able to determine 2Δ also from the thermally activated dc conductivity. Fig. 9 shows the dc conductivity[1],[22],[23] as a function of temperature for KCP. At low temperatures an activation energy corresponding to a gap of $2\Delta \approx \cdot 115$ eV is deduced compared to $2\Delta \approx \cdot 2$ eV from optical measurements. The discrepancy is not really surprising. Due to the random occupation of available Br^- sites[7] the singularities in the density of states at $E_F \pm \Delta$ are smeared out to some extent with tails extending into the forbidden zone. Nuclear magnetic resonance measurements show that the tails do not extend all the way down to E_F. A qualitative picture of the smearing of the singularity can be obtained from the optical experiments (Fig. 8) consistent with the observed activation energy of the dc conductivity.

Nuclear magnetic resonance measurements[24] show that at room temperature the density of states at the Fermi energy $N(E_F)$ is more than 50 % of the high temperature asymptotic value implying that there is only a very weak

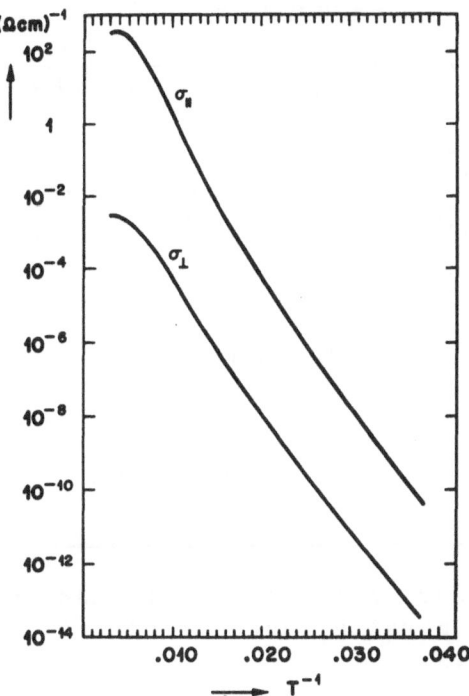

Fig. 9   DC-conductivity of KCP as a function of temperature for parallel and perpendicular directions.

pseudogap. Nevertheless a strong structure in $\varepsilon_2(\omega)$ is observed at the gap value (Fig. 8) and the dc conductivity
is more than an order of magnitude smaller than expected
from a Drude extrapolation of the high frequency conductivity[25]. This clearly shows that mobility or lifetime effects
are important and in fact model calculations by Strässler
and Schneider[26] show that the electron phonon interaction
strongly reduces the mobility in the energy range 2Δ.
Hence the general picture is that with decreasing temperature fluctuations strongly reduce the mobility even
before they significantly affect the density of states.
The Peierls transition also shows up in the thermopower
data[27,28], (Conflicting data have been reported in the
literature[29]. The discrepancies can be explained by the
fact that KCP is a solid electrolyte and that electro-
chemical reactions occur with most contact materials).
At room temperature the thermopower along the c-axis of
KCP is small, nearly temperature independent and ne-

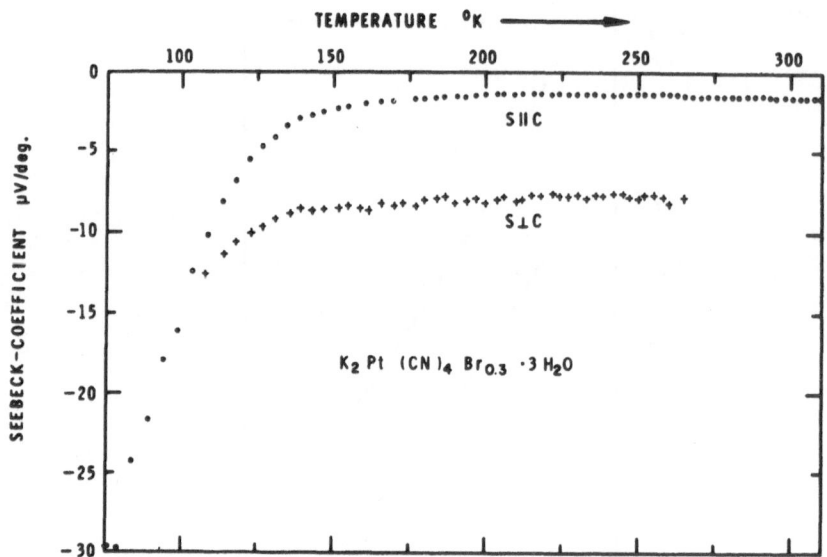

Fig. 10   Thermopower of KCP for field and temperature
          gradient parallel and perpendicular to the
          Pt-Pt axis.

gative (n-type) (Fig. 10). The value of - 1.5 µV/°K[28]
is not too far from - 2.2 µV/°K predicted by the free
electron model. For an ideal metal with no energy dependence of the scattering time the thermopower is proportional to T. In the fluctuation region of a 1-D Peierls
transition a strongly energy dependent τ is expected which
may explain the observed high temperature behaviour. Consistent with the NMR Knight shift[24] measurements the rapidly decreasing density of states N (E_F) causes a rapid increase in the thermopower with decreasing temperature.
The transverse thermopower never is metallic and shows
that the carriers are localized transverse to the strands.

V FROEHLICH MODE

Fröhlich[14] has argued that a 1-D conductor in its
Peierls distorted state should be a superconductor. This
is due to the fact that within a continuum model there is
no restoring force for the charge density wave. As a consequence the charge density wave can be freely shifted
through the crystal carrying electrons along. Scattering
processes are effectively suppressed by the presence of

an energy gap. Bardeen[30] and coworkers[31] have recently
pointed out that although the Fröhlich model does not
lead to true superconductivity in a real system it may
explain the unusual conductivity features[32] of TTF TCNQ.

Lee Rice and Anderson[33] have studied the Fröhlich
collective mode in a real system. The charge density wave
established by the Peierls distortion has a complex order
parameter which can be described by an amplitude and a
phase (this is not true for the half filled band case
where phase and amplitude cannot be distinguished).
Correspondingly there exist two types of fluctuations of
the order parameter involving fluctuations of the phase
or of the amplitude. The amplitude mode is not optically
active and has a finite frequency while the phase mode
is optically active and has within an idealized model
zero frequency i.e. it is the mode which in the Fröhlich
model leads to superconductivity. However, in a real crystal restoring forces are present also for the phase mode.
They can be provided by disorder, impurities, commensura-
bility of the charge density wave with the lattice or
3-D coupling. As a result the phase mode will have a small
but finite frequency. Even in an idealized model damping
of the phase mode is only nonexistent with regard to electron-phonon interaction. In a real crystal phonon lifetime
effects and impurity scattering introduce damping and the
phase mode of the pinned charge density wave can be thought
of as one of the two zone center LO phonons originating
from the $2k_F$ LA mode in the high temperature phase[34].

In KCP the phase mode and its temperature dependence
have been studied by far infrared reflection[34] (Fig. 11).
At 4.2°K a pronounced structure exist which can be fitted
with moderate success by a single oscillator with a longitudinal frequency of about 58 cm^{-1} and a transverse frequency of about 25 cm^{-1}. The data suggest that there is
a distribution of transverse or pinning frequencies and
that the unusually high low frequency dielectric constant
$\varepsilon_1 \gtrsim 3000$[38] is caused by the low frequency end of the distribution. The oscillator strength of the low temperature
structure is about 10^{-3} corresponding to an effective mass
of 10^3. Based on this the electron phonon coupling constant
λ can be estimated[34,35] to be about $\lambda \approx .2 - .3$.

We expect that neither the oscillator strength nor
the pinning force constant are very much temperature dependent at room temperature and below (T $\lesssim \sqrt{2}$ T_p^{MF}).

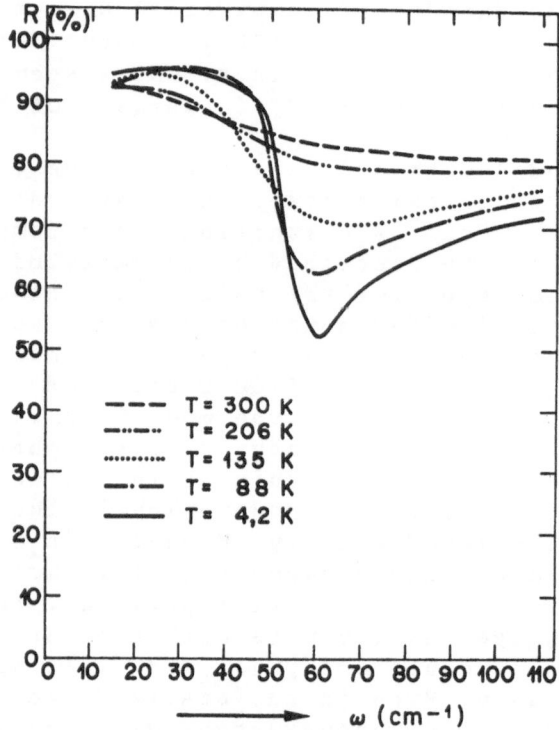

Fig. 11   Temperature dependence of the far infrared
          reflectivity.

In fact virtually all of the observed temperature dependence can be accounted for by a temperature dependent
damping. (Below 100°K the effect of conduction electrons
is negligible but at higher temperature single particle
exitations significantly contribute to the imaginary part
of the dielectric constant and cause a smearing of the
observed structure in addition to the intrinsic damping of
the mode.)

Fig. 12 shows the optical conductivity of KCP at
40°K. The large and broad peak at $\approx$ 1600 cm^{-1} $\hat{=}$ ·2 eV
corresponds to exitations across the Peierls gap, the
small structure at $\approx$ 20 cm^{-1} is due to the pinned Fröhlich mode.

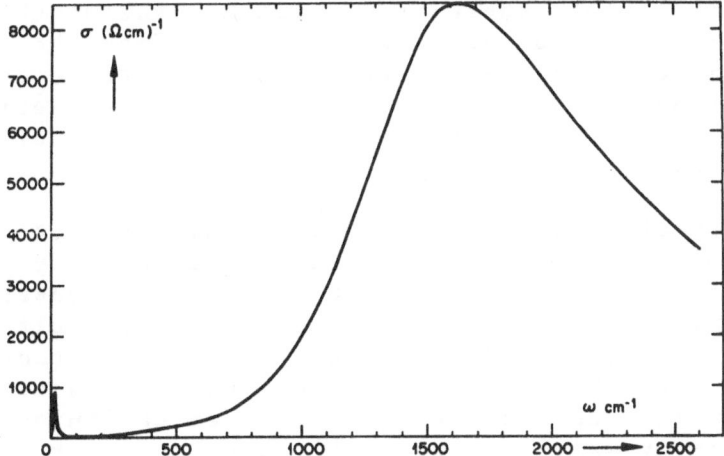

Fig. 12   Frequency dependence of the conductivity at
          T = 40°K from a Kramers Kronig analysis of the
          reflectivity data.

It was suggested by Bardeen[30] that the Fröhlich mode
could be thermally unpinned at higher temperatures and
could lead to an enhanced dc conductivity (paraconductivity) in a temperature interval near the pinning temperature. A subsequent phenomenological model calculation[31]
was supported by a microscopic calculation by Strässler
and Toombs[35]. On the other hand calculations by Fukuyama,
Rice and Varma[36] show that the fluctuations are resistive rather than conductive and suppress rather than enhance the conductivity. The main difference between the
calculations of Fukuyama et al and Bardeen et al resp.
Strässler and Toombs is in the momentum relaxation. In
35 it is assumed that the scattering rate of electron-
hole pairs is essentially proportional to the occupation number of $2k_F$ phonons and the phonons in turn
get rid of the momentum by unspecified channels. The
Fröhlich mechanism which corresponds to strong phonon drag
will then effectively suppress the scattering rate and
cause an increase in the conductivity. In 36 on the other
hand it is assumed that the only momentum relaxation chan-
nel of the coupled electron-phonon system is impurity
scattering. As a consequence the existence of a finite
pinning frequency (due to impurities) below T_p leads to
a divergent scattering rate at T_p, i.e. the fluctuations
are resistive.

Experimentally there is no evidence for a substantial contribution to the dc conductivity from the unpinned
Fröhlich mode at any temperature. In order to be observable
i.e. to produce a detectable effect in the reflectivity
the contribution would have to be at least 50 % of the
total conductivity.

In principle the charge density wave can not only
become unpinned by thermal excitation but also by an
electric field. At 4.2°K we have observed the onset of
strong nonlinearities[37] in the current versus voltage
characteristic (Fig. 13) at about 2000 V/cm. Although
more work would be required to prove that the nonlinea-
rity is in fact due to the field induced unpinning of the
charge density wave, a crude estimate shows that the
observed voltage is not unreasonable[35].

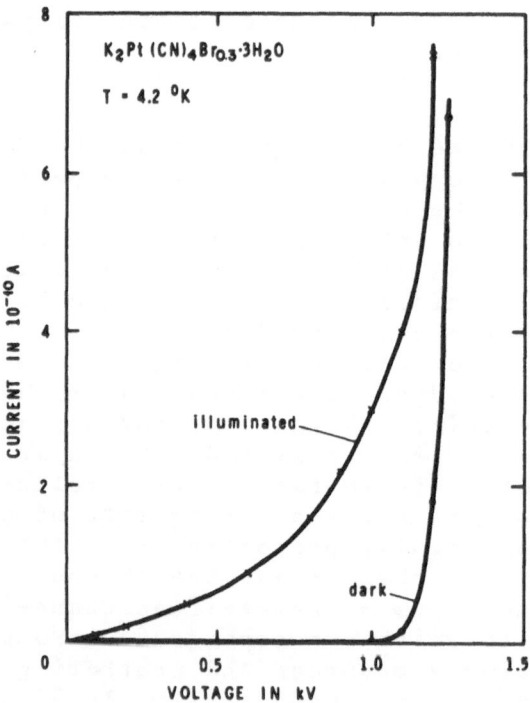

Fig. 13   Current versus voltage characteristic of KCP at
          4.2°K. Crystal dimensions were approx. 3x3x5 mm,
          i.e. the onset of the strong nonlinearity in the
          dark state is at about 2000 V/cm. The "illumina-
          ted" curve corresponds to illumination with unfil-
          tered light of a tungsten lamp.

VII SUMMARY

From the large number of independent experiments on KCP a fairly consistent picture has emerged. The occurrence of a Peierls instability has been directly demonstrated by diffuse x-ray and inelastic neutron scattering results. All other experiments can be successfully incorporated into this picture by properly taking into account fluctuations.

Physically we have to distinguish between two temperatures, a temperature below which a charge density wave gets established and a temperature at which the charge density wave gets pinned. Or in other words the two relevant parameters are $\sqrt{<Q^2>}$ and $<Q>$ where Q is the (complex) amplitude of the charge density wave. In the true metallic region which for KCP is not experimentally accessible $\sqrt{<Q^2>}$ and $<Q> = 0$. The intermediate region with $\sqrt{<Q^2>} \neq 0$ and $<Q> = 0$ is still metallic in the sense that there is no true energy gap but is governed by fluctuation effects which strongly depress the dc conductivity and cause a central peak in the inelastic neutron data. At low temperature (below 100°K for KCP) $<Q> \neq 0$, the charge density wave is pinned and semiconductor properties are found.

The picture is greatly simplified by the fact that in the experimentally accessible temperature region $\sqrt{<Q^2>}$ has already essentially reached its zero temperature value. As a consequence the optically observed pseudogap will occur at the zero temperature gap value. Even at room temperature where fluctuations cause a strong depression of the low frequency conductivity this occurs in the energy region of the zero temperature gap.

The work on KCP has clearly established the importance of two typical concepts of 1-D physics namely the Peierls instability and fluctuations. KCP is sufficiently one dimensional to exhibit a Peierls instability and in an extremely wide temperature region the physical properties are governed by fluctuations. Residual three dimensionality is responsible for the pinning of the charge density wave below about 100°K.

Optical spectroscopy and transport studies have turned out to be an essential and valuable tool in the study of the electrodynamics of the charge density wave of a 1-D electron system.

It is a great pleasure to thank P. Brüesch, R. Comès, P. Fulde, H. Launois, M. J. Rice, T. M. Rice and S. Strässler for making unpublished work available and for numerous stimulating discussions.

References

1   H.R. Zeller, Advances in Solid State Physics, Vol. XIII p. 31 (1973).
2   H.R. Zeller and P. Brüesch, to be published in Phys. Status Solidi (b) 65, issue 2 (1974).
3   R. Comès, M. Lambert, H. Launois and H.R. Zeller, Phys. Rev. B8, 571 (1973).
4   R. Comès, M. Lambert and H.R. Zeller, Phys. Status Solidi (b) 58, 587 (1973).
5   B. Renker, H. Rietschel, L. Pintschovins, W. Gläser, P. Brüesch, D. Kuse and M.J. Rice, Phys. Rev. Lett. 30, 1144 (1973).
6   B. Renker, L. Pintschovins, W. Gläser, H. Rietschel, R. Comès, L. Liebert and W. Drexel, Phys. Rev. Lett. 32, 836 (1974).
7   K. Krogmann, Angew. Chem. Int. Ed. Engl. 8, 35 (1969).
8   H.J. Deiseroth and H. Schulz, to be published.
9   Yu.A. Bychkov, L.P. Gor'kov and I.E. Dzyaloshinsky, Zh. Ekdsperim. Teor. Fiz. 50, 738 (1966). (Soviet Phys. JETP 23, 489 (1966)).
10 N. Menyhard and J. Solyom, J. Low Temp. Phys. 12, 529 (1973).
11 J. Solyom, J. Low Temp. Phys. 12, 546 (1973).
12 S.T. Chui, T.M. Rice and C.M. Varma, to be published.
13 R. Peierls, Ann. Phys. 4, 121 (1930) and "Quantum Theory of Solids" (Oxford University Press Oxford 1955).
14 H. Fröhlich, Proc. Roy. Soc. A 223, 296 (1954).
15 M.J. Rice and S. Strässler, Solid State Commun. 13, 125 (1973).
16 P.A. Lee, T.M. Rice and P.W. Anderson, Phys. Rev. Lett. 31, 462 (1973).
17 M.J. Rice and S. Strässler, Solid State Comm. 13, 1389 (1973).
18 J. Bernasconi, P. Brüesch, D. Kuse and H.R. Zeller, J. Phys. Chem. Solids, 35, 145 (1974).
19 H. Wagner, H.P. Geserich, R.v. Baltz and K. Krogmann, Solid State Commun. 13, 659 (1973)) attribute the structure to plasmon excitations. Their assumption

of a hole like band is shown to be incorrect in the main text.

20 The above arguments originated in a series of discussions with P. Fulde and S. Strässler.

21 M.J. Rice and S. Strässler, Solid State Commun. 13, 1931 (1973).

22 H.R. Zeller and A. Beck, J. Phys. Chem. Solids, 35, 77 (1974).

23 A.S. Berenblyum, L.I. Buravov, M.D. Khidekel', I.F. Shchegolev and E.B. Yakimov, Pis'ma Zh. Eksp. Teor. Fiz. 13, 619 (1971).
    (JETP Lett. 13, 440 (1971)).

24 H. Niedoba, H. Launois, D. Brinkmann and H.U. Keller, to be published.

25 D. Kuse and H.R. Zeller, Phys. Rev. Lett. 27, 1060 (1971).

26 S. Strässler and W.R. Scheider, personal communication.

27 I.F. Shchegolev, Phys. Stat. Solidi (a) 12, 9 (1972).

28 D. Kuse and H.R. Zeller, Solid State Commun. 11, 355 (1972).

29 For instance M.J. Minot and J.H. Perlstein, Phys. Rev. Lett. 26, 371 (1971).

30 J. Bardeen, Solid State Commun. 13, 357 (1973).

31 D. Allender, J.W. Bray and J. Bardeen, Phys. Rev. B 9, 119 (1974).

32 L.B. Coleman, M.J. Cohen, D.J. Sandman, F.G. Yamagishi, A.F. Garito and A.J. Heeger, Solid State Commun. 12, 1125 (1973).

33 P.A. Lee, T.M. Rice and P.W. Anderson, Solid State Commun. 14, 703 (1974).

34 P. Brüesch and H.R. Zeller, Solid State Commun. 14, 1037 (1974).
    P. Brüesch in "Proceedings of the German Physical Society Conference on One-Dimensional Conductors, Saarbrücken (1974).

35 M.J. Rice, S. Strässler and W.R. Schneider, ibid.
    S. Stässler and G.A. Toombs, Physics Lett. 46 A, 321 (1974).

36 H. Fukuyama, T.M. Rice and C.M. Varma, Phys. Rev. Lett. 33, 305 (1974).

37 D. Kuse and H.R. Zeller, unpublished results.

38 R.C. Jacklevic and R.B. Saillant, Solid State Commun. 15, 307 (1974).

X-RAY AND NEUTRON DIFFRACTION STUDIES

OF THE PEIERLS INSTABILITY

B. Renker and R. Comès[*]

Angew.Kernphysik, Kernforschungszentrum Karlsruhe, Germany

Labor.de Physique des Solides, Université Paris-Sud, Orsay,France[*]

1. INTRODUCTION

The influence of a one dimensional [1d] electron band over the lattice leads to the well known Peierls distortion and the enhanced Kohn anomaly [1]. Fig. 1a shows the occuring of a Peierls distortion for the special case of a half filled band. A distortion with a periodicity of 2c creates a gap at k_F which causes the electrons to flow from occupied higher states to lower ones [Fig. 1b]. Thus the kinetic energy of the electronic system will be lowered. As the distortion will happen on the expense of elastic energy there will be an equilibrium state with a finite distortion which is the Peierls state. By a Peierls distortion we get a zone boundary at k_F or a new reciprocal lattice point at $2k_F$ and thus this effect is observable at wave vectors $q = 2k_F$ in the elastic scattering.

Looking at the dynamical properties of a metal it is well known that the ions due to the existence of the electron gas move in a screened field which is given by the unscreened interaction potential divided by the dielectric function.

$$V_{eff} = \frac{V}{\epsilon [q, \omega]}$$

Thus electronic properties will influence the lattice dynamics. $\epsilon [q, \omega]$ can be calculated within the RPA where we consider the simplest polarisation graph:

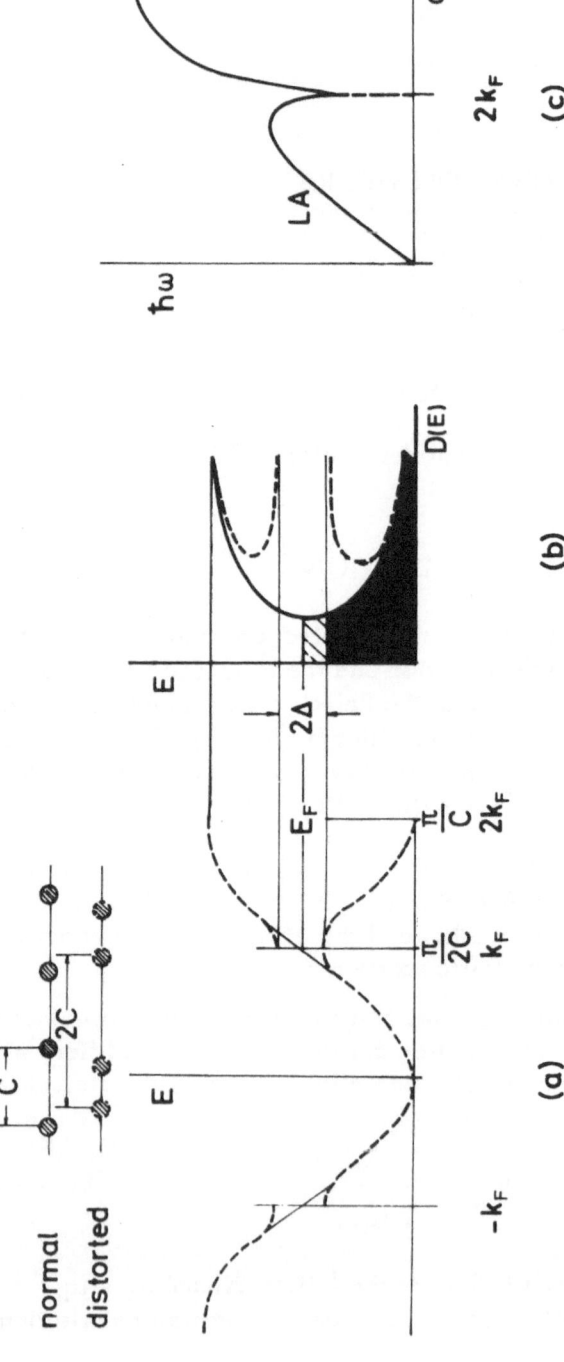

Fig. 1 Characteristic anomalies caused by a 1d-conduction band

a] The Peierls distortion for the special case of a half filled band. A doubling of the unit cell parameter
c in real space leads to an electronic gap at k_F.

b] The related electronic density of states. The Fermi energy is lowered by an amount of Δ due to the
Peierls distortion.

c] A giant Kohn anomaly appears at $q = 2k_F$ in the LA branch of the phonon dispersion due to an
anomaly in the screening of the ion-ion interaction by the electrons at $q = 2k_F$.

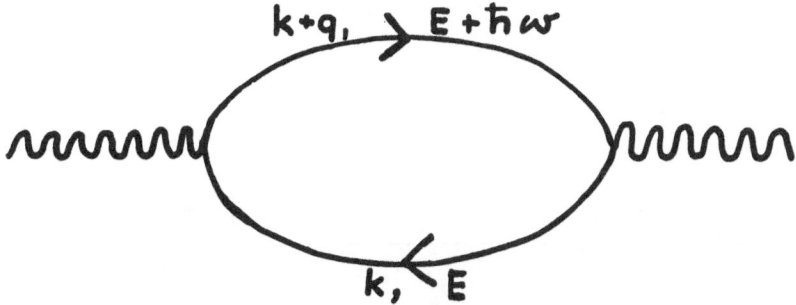

It is found that for a 1d-band ε [q,o] pocesses a logarithmic singularity at
$q = 2k_F$ which by the electron-ion-interaction affects the phonon dispersion
at this wave vector [see Fig. 1c]. Due to a smearing out of the Fermi edge
at finite temperatures this effect is less pronounced at higher temperatures.
In principle the metal insulator transition in a quasi 1d-system is expected
to proceed as a second order phase transition via the softening of the $2k_F$
phonon which condenses into the Peierls state. But it has been shown that
this transition may be largely effected by other effects such as fluctuations
/2/ and the Peierls transition will coincide with a general 3d-phase transition /3-4/.

The Kohn anomaly as well as the Peierls distortion is due to an inherent
instability of a 1d-system against distortions with a wave vector $2k_F$.
Whereas the Kohn anomaly is dynamical and is observed in inelastic scattering at energy transfers of some meV, the Peierls distortion is of static
origin and will be observed in the elastic scattering. Both effects are related to the wave vector $2k_F$ which is of the order of the reciprocal lattice
spacings.

Fig. 2 shows various experimental techniques. Momentum and energy
variables are plotted on logarithmic scales and the axes intersect in a region
which corresponds to the reciprocal spacings and energy levels found in
condensed matter. Thus it can be seen that neutrons are the only form of
radiation for which, both the momentum and the energy transfer are in the
desired region. The energy of X-rays is of the order of some 10 keV which
is far too large for a measurement of phonon energies. Thus the scattering of
X-rays involves an integration over the energy transfers and provides information as a function of the momentum variable only.

Light scattering and γ-ray scattering [Moessbauer scattering] are on
the periphery of this diagram and are limited for the investigation of special
problems. Thus X-ray and thermal neutron scattering are the most suitable
tools for a general investigation of our problem.

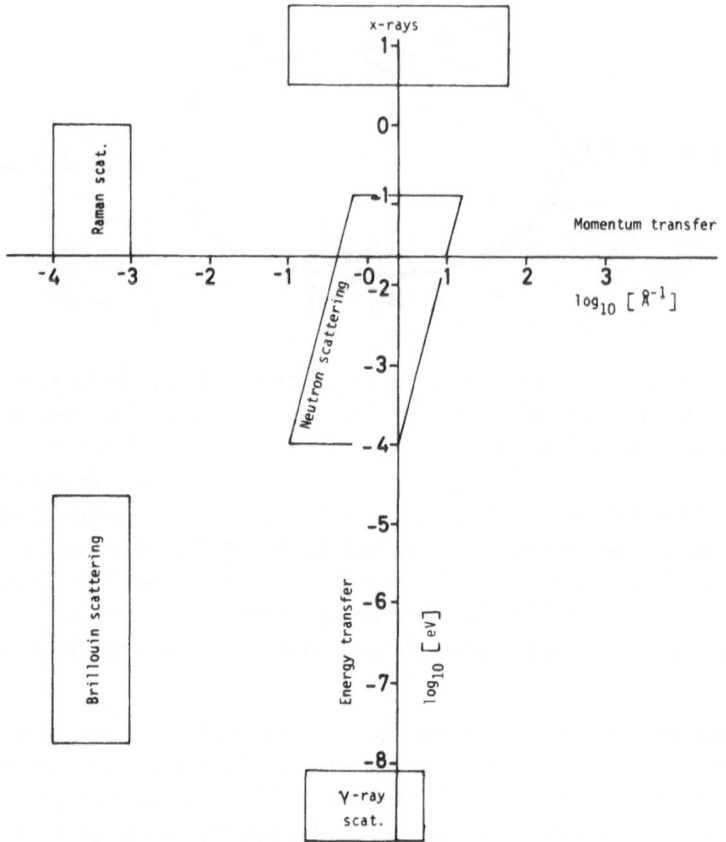

Fig. 2   Momentum-energy-space diagramm showing the regions covered by
         different types of radiation. The axes intersect in a region which
         corresponds to the reciprocal spacings and energy levels found in
         condensed matter.

2. SCATTERING OF X-RAYS

2.1 Experiment

The conventional Laue technique which employs a white beam of X-rays
together with a single crystalline sample gives a characteristic pattern of
points if we scatter at a 3d-object. Fig. 3 shows how these results change
if we scatter at a 1d-object.

Corresponding rays which after scattering have a certain phase shift
against one another are located on cones around each scatterer. It can be

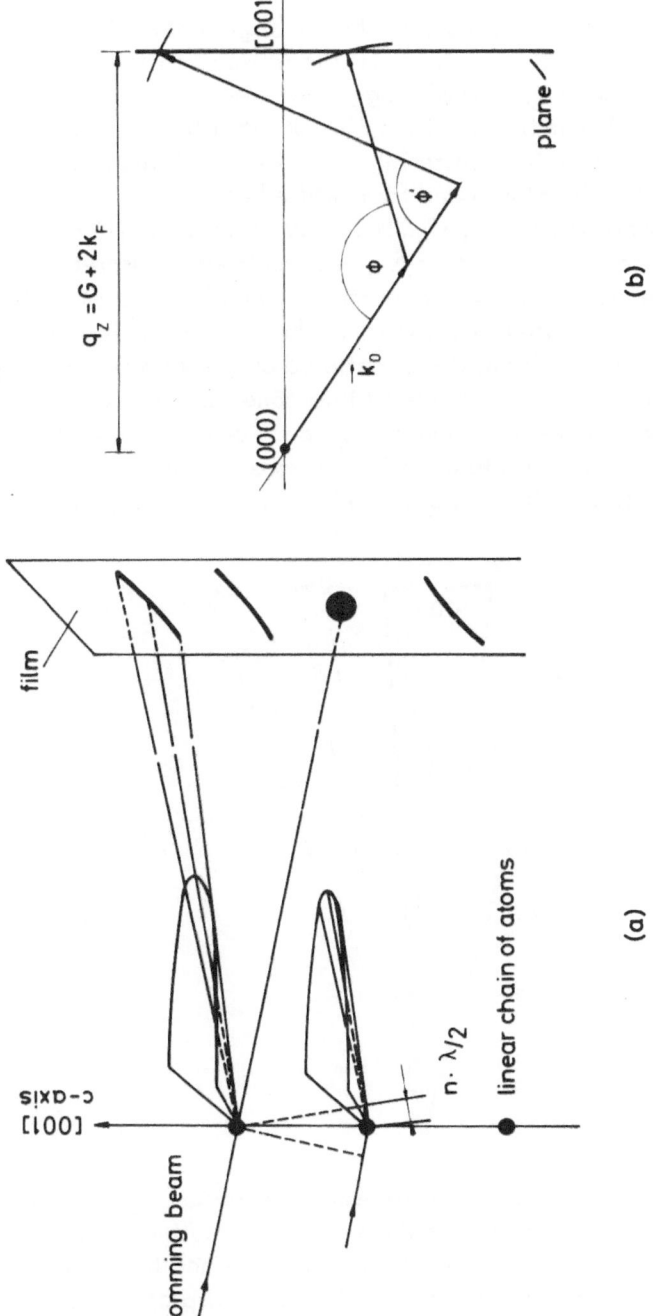

(b)

(a)

Fig. 3   Scattering of X-rays from a linear chain of atoms. Possible points for Bragg scattering are on planes in reciprocal space.

seen that the projection on a film gives a set of slightly curved lines. Since the system possesses periodicity only in chain direction, the only condition for coherent elastic scattering is, that the z-component of the momentum transfer is a multiple of $2\pi/c$, if c is the lattice constant in z-direction. Thus in reciprocal space instead of Bragg points we have planes perpendicular to the z^+-axis. It can be seen from Fig. 3b that for every direction of the incoming beam there is a scattering angle for which the Bragg condition is fulfilled. On the other hand for a white beam of incoming X-rays, 1 d-Bragg scattering will appear simultanously at all scattering angles. Thus only a beam of monochromatic X-rays gives a pattern of discrete lines shown in Fig. 3a. Therefore the normal Laue analysis which uses a white beam is unable to detect a 1d scattering. There may be another difficulty. Compared to a 3d-system where the intensity is scattered only into a few Bragg points, in a 1d-system the intensity is scattered into cones and hence is distributed over a much larger volume in momentum space. For this reason the intensity is much weaker for a 1d-structure. It is therefore extremely difficult to observe weak 1d-scattering in the neighbourhood of strong Bragg points.

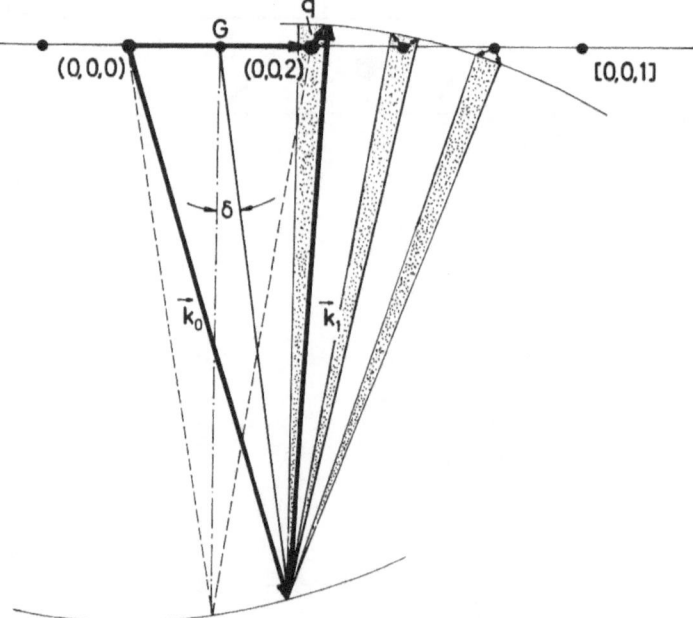

Fig. 4 Diffuse scattering of X-rays in reciprocal space. By rotating the sample by an amount of δ coherent elastic scattering can be avoided and inelastic scattering will be observed. Since the cross section $\sigma[\omega, q]$ is proportional to $1/\omega^2$ and ω is proportional to the wave vector q for acoustical phonons with small values of q diffuse scattering occurs.

Fortunately by rotating the sample [Fig. 4] it is possible to find a position where any bragg reflexion is avoided. It can be seen from Fig. 4 that in this case we will be able to observe weak scattering. In the case of phonons the scattering cross section is proportional to $1/\omega^2$ [at higher temperatures] and we get some remarkable intensity only from very low frequency phonons. For acoustical phonons we have $\omega \sim q$ and thus instead of Bragg points we will observe diffuse spots. These diffuse spots are comparable in intensity with the 1d-lines and therefore by this technique it is possible to observe the 1d-scattering together with the normal 3d-one.

In the actual experiment the primary X-ray beam from a molybdenum anode was first reflected from a doubly bent lithium fluoride crystal. The [200] Bragg reflection was used to provide a monochromatic beam of $\lambda = 0.709 \,\overset{\circ}{A}$ and the scattered intensity was measured on a flat camera. By focusing on the photographic plate it was possible to obtain good patterns of the superstructure with high resolution during an exposure of about 12 hours.

2.2 Analysis and Discussion

Fig. 5 shows a pattern of $K_2 Pt [CN]_4 Br_{0.3} \cdot 3H_2O$ [KCP] as observed at 300 K /5-6/. The unit cell of KCP is tetragonal. The cell dimensions are $a \times a \times 2c$ with $a = 9.87 \overset{\circ}{A}$ and $c = 2.88 \,\overset{\circ}{A}$ where c is in the direction of the Pt-chains and measures the distance between successive planes of Pt atoms /7-8/. Several conclusions can be drawn from the pattern shown in Fig. 5:

[i]   The intense layers of diffuse spots are perpendicular to the c^+-axis. As their mutual distance is proportional to $2\pi/c$ they are due to the scattering from the Pt-planes. Diffuse spots which correspond to the true lattice parameter 2c are rarely to see since usually the structure factors are very small.

The thin continuous lines which appear symmetric to the rows of diffuse spots are due to the 1d-superstructure. From the fact that their intensity increases with increasing scattering vector it can be stated that the observed superstructure is due to a distortion of the lattice and is not caused by a superlattice resulting from an ordering of particular atoms or molecules such as bromine and water. Therefore the atoms involved in the distortion must consequently in the average be separated by the same distance of $2.88 \,\overset{\circ}{A}$. It can be seen from Fig. 5 that the distance between the continuous line and the neighbouring row of diffuse spots is roughly 1/6-th of the

distance between the rows of the diffuse spots. Consequently there
exists a distortion parallel to the c-axis with a superperiod of
6×2.88 Å.

From the stoichiometry of KCP we conclude that each Pt-atom
can contribute 2 electrons to the conduction band. If by partial
oxydation of bromine we remove roughly 1/3 of an electron from
each Pt-atom we get an energy band which is filled up to 5/6.
The Peierls splitting into 5 filled bands and one empty band leads
to a superstructure of 6 times the Pt-Pt distance and hence is in
excellent agreement with the experimental observation.

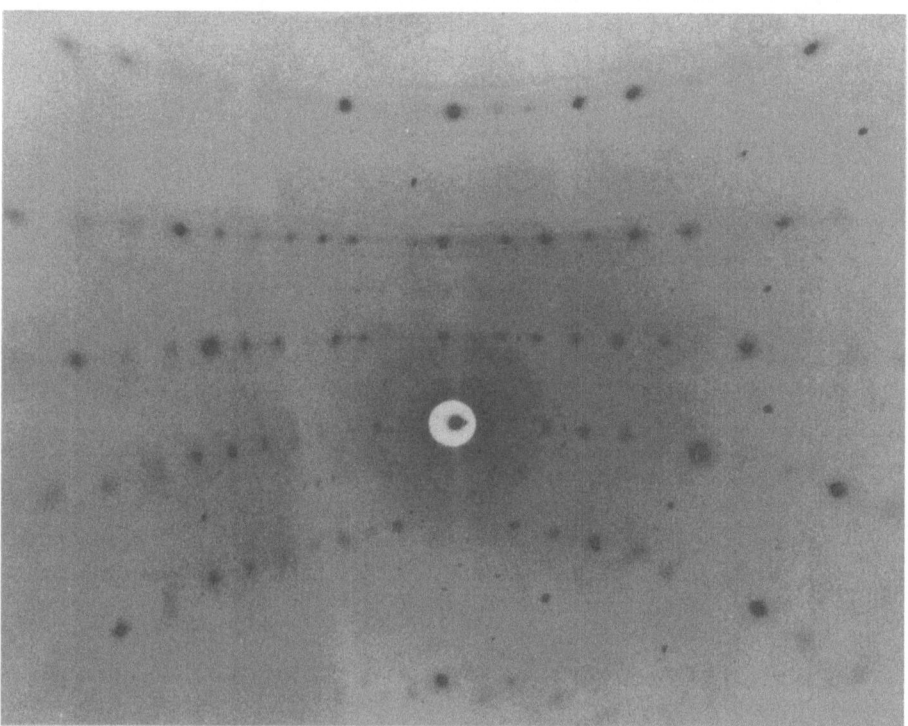

Fig. 5 Scattering of X-rays from $K_2Pt[CN]_4 Br_{0.3} \cdot 3H_2O$ at room
temperature. The diffuse spots are due to the 3d-structure and the
thin continous lines reflect the 1d-superstructure.

[ii]   The diffuse sheets appear only as satellite lines symmetric to the layers of diffuse spots. Thus, since one observes one Fourier component only, it can be concluded that the distortion is sinusoidal. A calculation of structure factors for a linear chain with sinusoidal distortion yields the selection rules given in Fig. 6. Taking the magnitude of the distortion δ equal to 0.5 percent, which seems to be a realistic value, we find for a superperiod of 6c that the even levels [except l = 6, 12 etc.] have at least three orders of magnitude lower intensities than those corresponding to the diffuse sheets and will therefore be unobservable in an experiment.

We shall see that these selection rules are fully confirmed by the neutron scattering results and we may use these formulas to calculate the magnitude of the distortion.

[iii] The diffuse sheets are sharp and well defined. Assuming that the observed linewidth is not affected by the resolution of the spectrometer we further can establish a lower limit for the coherence length of a distorted chain, which is found to be larger than 170 $\mathring{A}$ or 60 Pt-Pt distances.

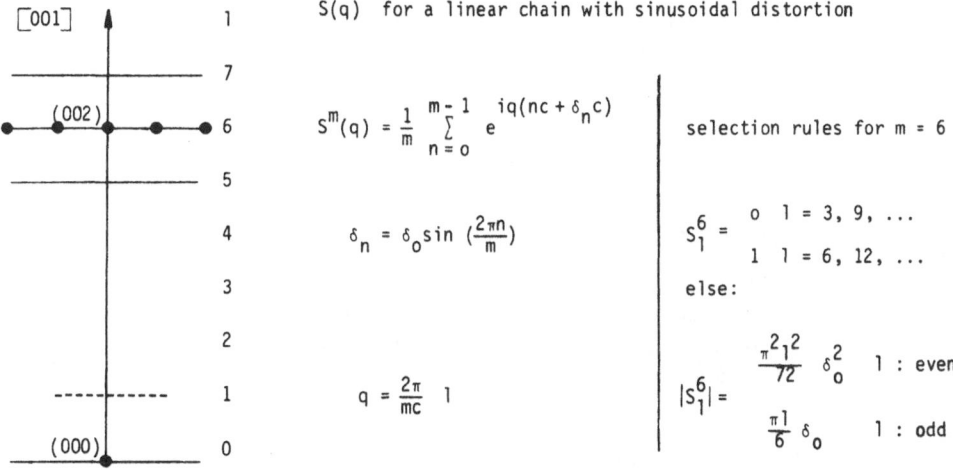

Fig. 6 Calculated structure factors for a linear chain of atoms with sinusoidal distortion.

Fig. 7 visualizes this kind of a distortion. The sinusoidal distortion along a Pt-chain and the 3d-arangement of different chains in the lattice are shown. It is important to realize that there is no correlation in phase between the distortions on different chains. This results from the fact that the single diffuse sheets do not show any structure.

A more accurate evaluation shows that instead of $6 \times 2.88\,\text{Å}$ the true period of the superstructure is $6.67 \times 2.88\,\text{Å}$. Thus the true superperiod is incommensurable with the lattice, but it coincides exactly with $2k_F$. It has already been discussed that both effects, the Peierls distortion and the Kohn anomaly, appear at $2k_F$. Since X-ray scattering cannot distinguish between elastic and inelastic scattering the observed superstructure may also be due to a giant Kohn anomaly with very low frequencies. Thus from the present experiment we cannot find out whether the observed superstructure is static or dynamic.

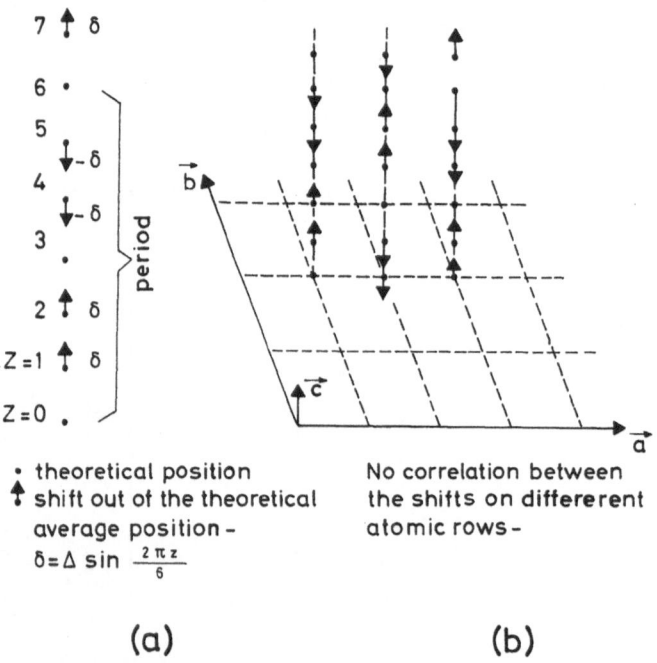

(a)   (b)

Fig. 7 Sinusoidal distortion of a single chain of atoms and the spacious arrangement of different chains showing no correlation concerning the phase of the distortion.

For a check of our interpretation it is important to look at another Pt-compound with different oxydation number. Fig. 8 shows the diffraction pattern for the rubidium compound [Rb$_2$ Pt [CN]$_4$ Br$_{.23} \cdot$ x H$_2$O] in comparison with that of KCP. Again the patterns look very similar. Due to the different oxydation number we observe the diffuse sheets at a different position. Since the concentration of bromine is not exactly one quarter we have a superstructure only of about 8 times the Pt-Pt spacings. Again the exact value coincides with 2k$_F$.

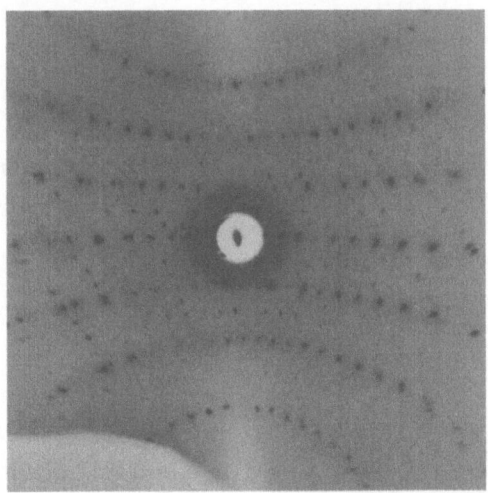

Fig. 8 Rb$_2$Pt [CN]$_4$ Br$_{0.23} \cdot$ x H$_2$O at room temperature. The sheets which corresponds to the 1d-superstructure are observed at different positions compared to Fig. 5.

It is known from conductivity measurements that these substances become insulating at low temperatures /9/, thus temperature dependent measurements are important. Fig. 9 shows a comparison of patterns of KCP at 300 K and 77 K /5/. It is seen that the intensity in the diffuse sheets starts to concentrate on particular points at lower temperatures, which indicates that some correlation between the distortions in neighbouring chains starts to develop. The reciprocal points which correspond to the new diffuse maxima

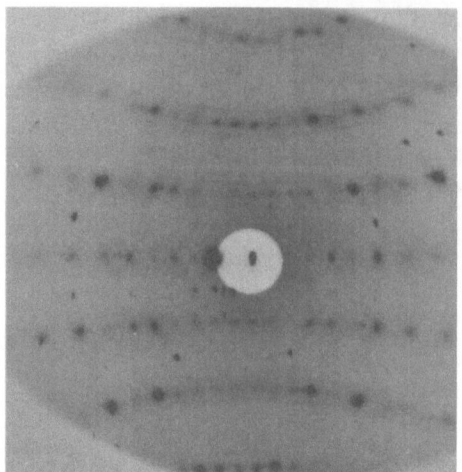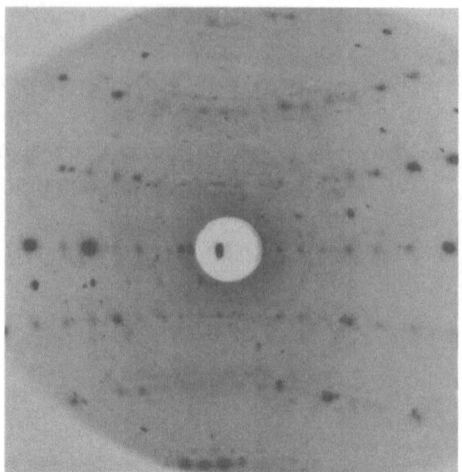

Fig. 9   Comparison of patterns of $K_2Pt[CN]_4Br_{0.3} \cdot 3H_2O$ at room
temperature [left]   and at 77 K   [right]. The intensity on the
continuous lines of the 1d-superstructure concentrates at points
[$a^*/2$, $a^*/2$, $c^*/3$] at lower temperatures. a^* and c^* are
reciprocal lattice parameters.

are found to be [$\frac{a^*}{2}$, $\frac{a^*}{2}$, $\frac{c^*}{3}$] where a^* and c^* are the reciprocal
lattice parameters. In real space a larger unit cell of 2a x 2a x 3c corre-
sponds to these values. Since the diffuse spots are still part of a line we
conclude that a real phase transition including long range order has not been
achieved at 77 K. To account for the localisation of the new diffuse maxima
neighbouring chains of Pt-atoms must have been distorted with opposite
phases. Fig. 10 visualizes this kind of distortion.

3. SCATTERING OF NEUTRONS

3.1 Experiment

By this technique we easily can distinguish between elastic and inelastic
scattering and thereby gain essentially new information. Fig. 11a shows

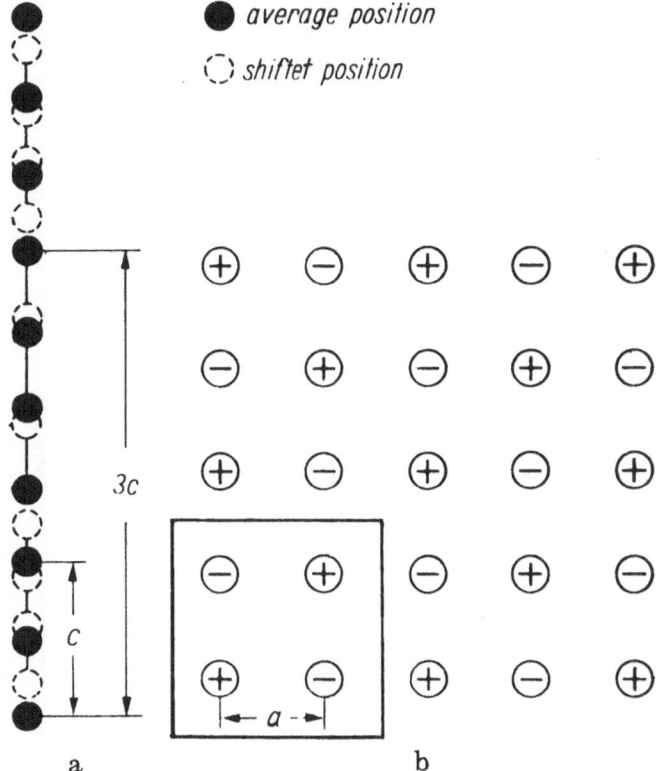

Fig. 10 Ordering in the phases of the distortions at different chains and low temperatures [schematic]. The a-parameters of the unit cell are doubled.

schematically the setup of a triple axis spectrometer which is used for such a kind of measurements. In difference to the scattering of X-rays an additional single crystal is used to determine the energy of the scattered neutrons by aid of a second Bragg reflection. If the sample is a single crystal and its orientation relative to the incoming beam is known, we can relate the measured energy and momentum transfer of the neutrons directly to the values ω and q of a phonon. The experimental parameters which are the incident energy, the orientation of the sample, the scattering angle and the

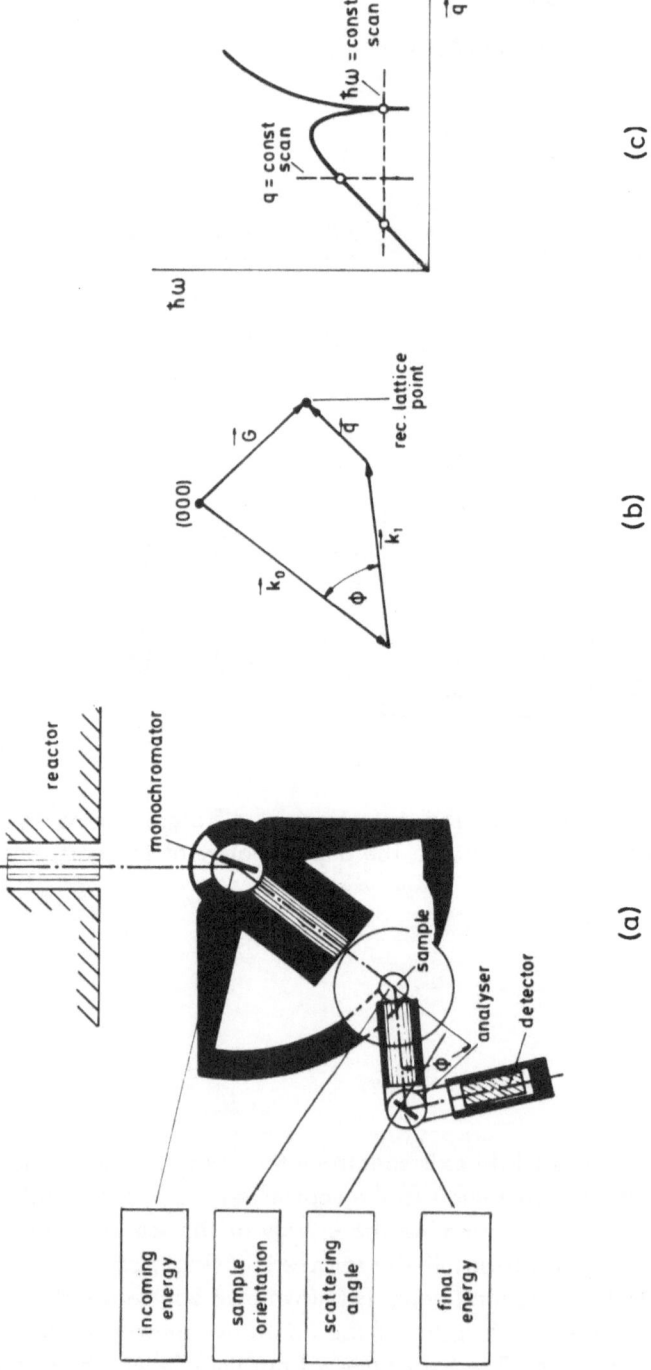

(a)   (b)   (c)

Fig. 11   Neutron spectroscopy with a triple axis spectrometer
a]   Schematic drawing of the spectrometer. Experimental parameters are the incoming energy, the
     sample orientation, the scattering angle and the final energy.
b]   The measurement in reciprocal space.
c]   Special scans in the $\hbar\omega$-q plane. The points where the scans cross the dispersion line
     correspond to maxima in the scattered neutron intensity.

final energy may be chosen such that we fulfil the conditions in energy and momentum transfer for the excitation of a special phonon as shown in reciprocal space by Fig. 11b. An advantage of this technique is that the parameters may be varied such, that one measures along definite lines in the $\hbar\omega, q$ plane. This is shown in Fig. 11c for the special cases of a "const. ω" - and a "const. q"-scan. In the present experiments the experimental resolution was about $0.04\,\mathring{A}^{-1}$ in q and about 0.6 meV in $\hbar\omega$. A disadvantage of this technique is, that in order to get good intensities large single crystals are required for samples which is a problem for the linear conductors. The size of the KCP crystals used was about 0.4 cm^3. Another difficulty arises from the fact that even perfect single crystals may cause an amount of incoherent scattering. Thus the presence of hydrogen with its extremely high incoherent scattering cross section is particularly unfavourable for coherent neutron scattering. Therefore deuterated samples are used when ever available.

3.2 Results and Discussion

3.21 Inelastic measurements at room temperature

Fig. 12 shows some acoustical phonon branches which have been measured by inelastic neutron scattering at room temperature /10-11/. It is seen that an enhanced Kohn anomaly appears in the LA branch at $[0,0,0.3\,\frac{\pi}{c}]$. It is easy to show that the value $0.3\,\frac{\pi}{c}$ corresponds exactly to $2k_F$ if in agreement with the chemical formula we assume that each Pt-atom contributes 1·.7 electrons to the conduction band [actually $2k_F$ comes into the first Brillouin zone by the Umkapp process: $\frac{2\pi}{c} - 1.7\,\frac{\pi}{c}$].

The branch along the zone boundary in $[0.5, 0.5, \zeta]$ direction is also polarized parallel to the chain direction and shows the same enhanced anomaly. Thus the Kohn anomaly appears as a deep depression along the line $q_z = 2k_F$ = const. which proves directly that the anomaly is due to a 1d-property of the sample. It was possible to resolve the dip by const. ω-scans at 1.0 and 0.75 THz. Thus outside a narrow region of about $\pm\,0.02\,\mathring{A}^{-1}$ around $2k_F$ [which is our present resolution] it seems that we still observe single phonons and the dispersion can be well described by simple models /10/. There are strong indications that in the very near neighbourhood of $2k_F$ the phonons are no longer well defined and instead of single collective excitations we should discuss the dynamical structure factor /12,13/.

The TA branch in chain direction shows no indication of an anomaly which states that the TA phonons are not influenced by the conduction electrons. The TA branch in $[\zeta, \zeta, 0]$ direction gives an impression of the coupling between the chains and indicates that except the region around $2k_F$ the phonon spectrum is quite 3-dimensional.

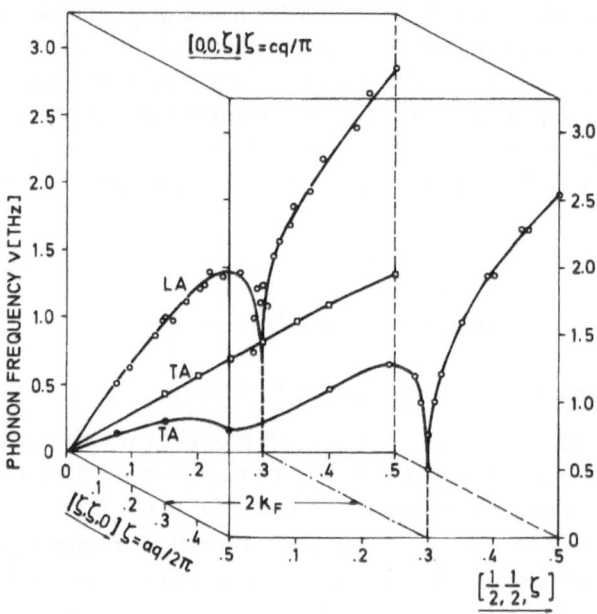

Fig. 12 Acoustical phonon branches in $K_2Pt[CN]_4Br_{0.3} \cdot 3D_2O$ at room temperature. An enhanced Kohn anomaly appears along the line $q = 2k_F = const.$

3.22 Elastic Measurements /14/

Fig. 13 shows the result of a const. q scan at $[0,0,2k_F]$ and room temperature. The hump between two and seven meV is due to the contribution of the soft phonons in the anomaly. In this scan, due to the limited resolution [about 0.8 meV in this case], we integrate over a large part of the anomaly. The quasielastic part shows a pronounced peak around $\hbar\omega = 0$. Since a const. ω-scan at $\hbar\omega = 0$ again shows a well defined peak at $[0,0,2k_F]$ the observed peak is due to a superstructure and cannot be caused by incoherent elastic scattering only. Additional measurements show that this peak occurs along lines with $q_z = G \pm 2k_F$ in reciprocal space [G is a reciprocal lattice vector of the type [0,0,2l]] thus the corresponding superstructure is 1d. This result seems surprising at a first view because it seems to prove existence of both, the enhanced Kohn anomaly and the Peierls distortion, at room temperature.

In principle this observation may be explained in two ways:

i] As long as the Peierls gap is comparable with the phonon energies on Fermi surface scattering, which gives the most important contributions to the dielectric polarisation will be diminished but not completely suppressed. Thus the Kohn anomaly will be reduced in magnitude but it will still be observable.

ii] On the other hand the observed scattering around $\hbar\omega = 0$ must not be truly elastic. In this case we interpret the scattering as quasielastic and caused by critical fluctuations preceding the real Peierls transition. There would be an analogy to other structural phase transitions and in principle we should be able to measure a natural linewidth of the central peak in $\hbar\omega$. The present results do not show such a broadening but the true linewidth may be much smaller than the instrumental resolution [about 0.8 meV in this scan].

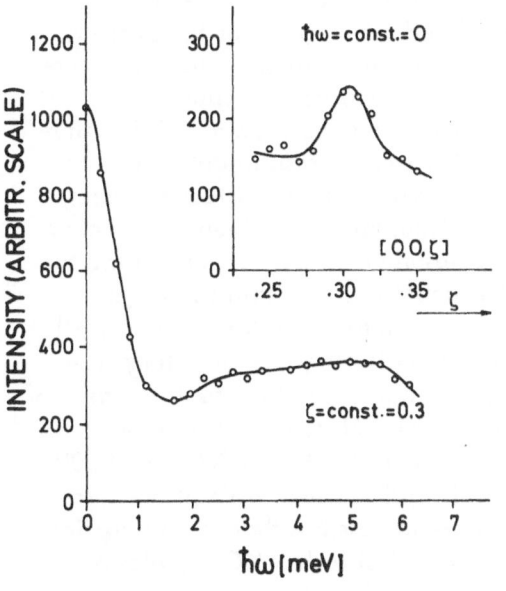

Fig. 13
Const. q-scan at $[0,0,2k_F]$.
Besides the soft phonon contributions in the inelastic part a central component at $\hbar\omega = 0$ is observed.

If we look at this peak at different reciprocal lattice points we confirm the X-ray result of a sinusoidal distortion. As shown above in fig. 6 from the ratio of the intensities at $\pm\,2k_F$ beneath the [0,0,2] and the [0,0,4] reciprocal lattice points relative to the intensities of the bragg reflexions we can calculate the magnitude of the distortion which we find to be about 0.5 % of the Pt-Pt distance at 300 K. This calculation in which we compare very intense bragg peaks to very weak scattering gives nevertheless only a crude estimation.

Fig. 14 shows the temperature dependence of this elastic intensity measured at $[0,0,2k_F]$ and $[\frac{\pi}{a},\frac{\pi}{a},2k_F]$ corrected for incoherent background scattering. One should note that the scales for the two curves are different. Thus we see that the intensities are about the same at both points between 315 K and about 200 K. Consequently in this region the distortions impressed on the chains are uncorrelated and 1d. Additionally it can be seen that the intensities are almost temperature independent within this region. Then below 160 K the intensity at the superlattice point increases rather rapidly. It is known from the X-ray measurements that this increase corresponds to a 3d-ordering of the distortions along the single chains. Finally the intensity saturates at about 40 K and does not show any remarkable change down to 6 K.

From this observation we may draw the following conclusion: The fact that the distortions are uncorrelated at higher temperatures implies that a Peierls state if it exists at room temperature cannot be stabilized by an interaction between the chains. By stabilized we mean, that the phase of a static distortion must be fixed relative to the lattice in order to make sense. But if the distortions are pinned by defects or impurities already at room temperature, it is hard to explain the 3d-ordering observed at 100 K. Thus although the present observations due to the limited resolution do not provide a direct evidence for critical fluctuations their existence is most likely. Independent evidence for critical fluctuations is given from recent theoretical work /3,15,16/ and NMR measurements /17/. Concerning the X-ray and neutron results there are still problems to explain the large coherence length ξ of the fluctuations at room temperature and the small temperature dependence of the quasielastic scattering at higher temperatures. Prediction of Lee and Anderson /3/ concerning the absolute value of the coherence length [about 7 Pt-Pt distances] and its temperature dependence [ξ increases by a factor of more than 2 between 300 K and 200 K] are in disagreement with our experimental results. These estimations were derived for a real order parameter but it seems possible that the assumption of a complex order parameter which is more suitable for KCP yields a better agreement with the experimental results /15/.

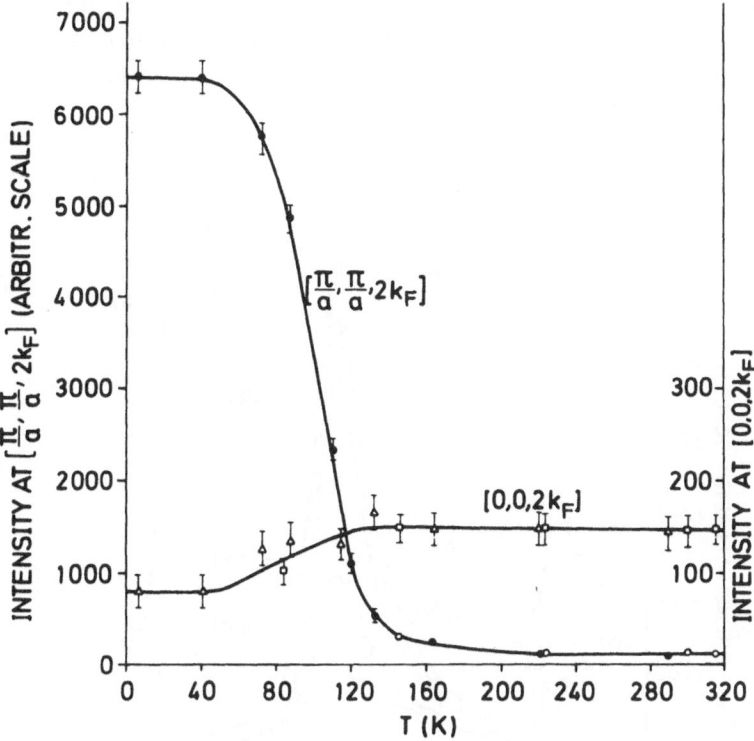

Fig. 14 Temperature dependence of the elastic intensities measured at
$[0,0,2k_F]$ and $[\frac{\pi}{a}, \frac{\pi}{a}, 2k_F]$. A transition of the type disorder-
order is observed at 100 K at $[\frac{\pi}{a}, \frac{\pi}{a}, 2k_F]$. As the
intensity at $[0,0,2k_F]$ does not go to zero no long range order is
established perpendicular to the chains.

The intensity measured at $[0,0,2k_F]$ [Fig. 14] decreases as soon as the
3d-ordering starts. Thus we observe that the 1d-intensity which at higher
temperatures is uniformly distributed over a plane in reciprocal space concentrates at lower temperatures at the new reciprocal lattice points. This
increase is interpreted first as critical scattering and then as Bragg scattering.
But we see that even below 40 K where the 3d-ordering saturates the 1d-
intensity does not go to zero as it would be expected for a real long range
ordered structure. Therefore we conclude that only short range order is
established perpendicular to the chains. Fig. 15 gives direct evidence for
the missing of any long range order. It shows a series of elastic scans along
the $[\zeta, \zeta, 2k_F]$ direction for different temperatures. The scans are centered

at $[0.5, 0.5, 2k_F]$. Again we see the increase in intensity as the temperature is lowered. Additionally it is seen that the linewidth is considerably larger than the resolution of the spectrometer which is marked below. This is the case even at 6 K.

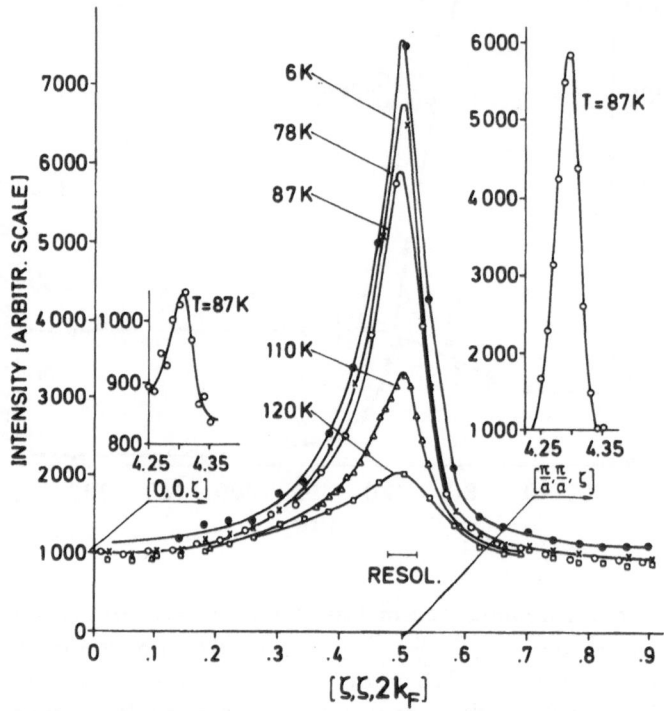

Fig. 15   Scans along $[\zeta\ \zeta\ 0]$ in the plane $q = 2k_F = $ const. The line-width observed in scans along the $[00\ \zeta]$ direction is equal to the resolution width of the spectrometer. From this the lower bound of 20 Pt-Pt spacings can be deduced for the correlation length ζ. Perpendicular to that direction an enhanced line width is observed.

Scans parallel to the chain direction which are shown by the inserts do not show any broadening compared with the instrumental resolution. From this one can derive a lower limit for the correlation length of a distorted chain. If we define a correlation length by the inverse of the natural line width in q, we find as a lower limit 20 Pt-Pt spacings [the somewhat better resolution of the X-ray experiment yields 60 Pt-Pt spacings.]

The evaluation of the linewidth perpendicular to the chain direction is shown in Fig. 16. Extrapolating the line width linearly down to zero we find a transition temperature of~100 K. It is seen that the correlation length does not diverge at 100 K but it saturates at a value of 33 Å which corresponds to about three interchain distances. Thus for KCP the 3d-ordering is restricted to the formation of domains which are large in chain direction but which are small perpendicular to that direction. The true reason for this is unknown but it may be possible that imperfections such as impurities or chain interruptions are responsible for this behaviour.

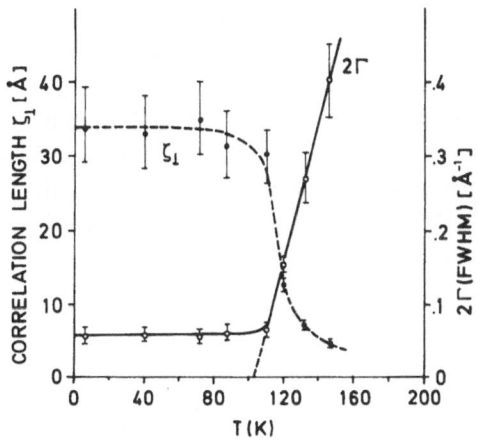

Fig. 16
Evaluation of the line width of scans along [ζ, ζ,0] shown in Fig. 14. The exptrapolation yields the transition temperature $T_o \sim 100$ k. The phase transition is incomplete and ζ saturates at 33 $Å^{-1}$ or about 3 inter chain distances.

3.3 Temperature Dependence of the Soft Mode

As discussed the central mode in KCP reflects the instability of the system against a Peierls distortion. It is caused by the 1d-character of the conduction band and consequently it is observed on a plane with $q = 2k_F = $ const. The soft mode which constitutes of the phonons in the Kohn anomaly shows the same 1d-property but a $\bar{q}$ dependence of the soft mode frequency is observed at room temperature: Whereas in chain direction a clear separation from the central mode is observed [Fig. 12], const q-scans performed at the low temperature superlattice point [$\frac{\pi}{a}$, $\frac{\pi}{a}$, $2k_F$] do not show a separation between the soft phonon part and the central component within our resolution [0.8 meV]. The fact that the phonon anomaly at $[\frac{\pi}{a}, \frac{\pi}{a}, 2k_F]$ reaches down to lower frequencies than in the remainder part of the Brillouin zone was proposed by Barišić /8/ from his treatment of the electron phonon interaction within the tight binding theory as due to the long range Coulomb interactions. A different theory by Horovitz et al./19/ also predict that the softest phonons of the $2k_F$ anomaly will be those with the shortest wave length. Below 200 K the spectra at both points are similar. Within our resolution a soft mode frequency cannot be defined and the spectra looks like a central peak superimposed upon a broad overdamped phonon component. Remarkably as soon as the central mode reflects the beginning of 3d-ordering the soft mode at both q-points seems to increase in frequency and the phonon spectrum becomes hard well above $T_o = 100$ K /10/. A hardening of the $2k_F$ phonons is explained within the mean field picture by the formation of an electronic gap. Fluctuations cause the formation of a pseudo gap instead of a real gap but with essentially the same properties /2/ . With   increasing coherence length and inter chain coupling the influence of the fluctuations is diminished /15/ and the pseudogap approximates more and more a real gap, and could be responsible for a hardening of the phonons at low temperatures   above T_o. Dietrich /20/ who has calculated the temperature dependence of the elastic structure factor of the central mode including inter chain coupling also finds that the observed hardening would be in agreement with his theory but a self consistent calculation of the dynamical structure factor $S[\omega, 2k_F]$ is still missing.

4. CONCLUSIONS

It has been shown that the experimental techniques of X-ray and thermal neutron scattering provide a lot of information about the metal insulator transition in linear conductors. The enhanced Kohn anomaly which has been detected in KCP at room temperature proves 1d-metallic properties at higher temperatures. But additionally a very low frequency peak was found at the same wave vector which was explained by the assumption of critical fluctuations. Thus at room temperature we do not observe a real metallic state but the system already performs fluctuations around an equilibrium position which is the isolating state at lower temperatures. The fluctuations will cause the existence of a pseudo gap which is responsible for a reduced electron mobility and a hardening of the soft mode frequencies above 100 K.

The temperature dependent intensity of the central component has been measured. It was found that the transition to the Peierls state where the fluctuations are frozen in, is caused by a 3d-ordering of the distortions impressed on the single chains. Finally it was found that this 3d-transition does not lead to a real long range ordered structure. It leads to domains which are fairly large in chain direction [ξ > 170 Å] but which are small perpendicular to that direction [ξ = 33 Å].

The X-ray experiments where performed at Orsay, and the neutron scattering experiments partly at the Karlsruhe FR2 reactor and the Grenoble ILL - high flux reactor.

REFERENCES

/1/ For a review see:
W. Gläser, Festkörperprobleme XIV, ed. H.J. Queisser, p. 205, Pergamon / Vieweg [1974]

/2/ M.J. Rice and S. Strässler, Solid State Commun. 13, 1931 [1973]

/3/ P.A. Lee, T.M. Rice and P.W. Anderson, Phys. Rev. Lett. 31, 462 [1973]

/4/ R. Comès, M. Lambert and H.R. Zeller, Phys. Status Solidi [b] 58, 578 [1973]

/5/ R. Comès, M. Lambert, H. Launois and H.R. Zeller, Phys. Rev. 38, 571 [1973]

/6/ R. Comès [to be published in the Proc. of the German Phys. Soc. Conf. on "One Dimensional Conductors", University of Saarbrücken, 10 - 12 July [1974]

/7/ K. Krogmann, Angew. Chem. Int. Edit. 8, 35 [1969]

/8/ H.P. Geserich, H.D. Hansen, K. Krogmann and P. Stampfl, Phys. Stat. Solidi [a] 10, 537 [1972]

/9/ For a review see: H.R. Zeller, Festkörperprobleme XIII, ed. H.J. Queisser, p. 31, Pergamon / Vieweg

/10/ B. Renker, H. Rietschel, L. Pintschovius, W. Gläser, P. Bruesch, D. Kuse and M.J. Rice, Phys. Rev. Lett. 30, 1144 [1973]

/11/ B. Renker, L. Pintschovius, W. Gläser, H. Rietschel and R. Comès [to be published in the Proc. of the German Phys. Soc. Conf. on "One Dimensional Conductors", University of Saarbrücken, 10 - 12 July [1974]

/12/ H. Rietschel, Solid State Commun. 14, 699 [1974]

/13/ H.R. Zeller, this issue

/14/ B. Renker, L. Pintschovius, W. Gläser, H. Rietschel, R. Comès, L. Liebert and W. Drexel, Phys. Rev. Lett. 32, 836 [1974]

/15/ D.J. Scalapino, M. Sears and P.A. Ferrell, Phys. Rev. B6, 3409 [1972]

/16/   M.J. Rice and S. Strässler, Sol. State. Commun. 13, 1389
         [1973]
/17/   H. Niedoba, H. Launois, D. Brinkmann, H.R. Zeller, Phys.
         Stat. Solidi [b] 58, 309 [1973]
/18/   Barišić, Ann. Phys. 7, 23 [1972]
/19/   B. Horovitz, M. Weger and H. Gutfreund, Phys. Rev. B9, 1246
         [1974]
/20/   W. Dietrich [to be published in the Proc. of the German Phys.
         Soc. Conf. on "One Dimensional Conductors", University of
         Saarbrücken, 10 - 12 July [1974]

NMR STUDIES OF THE PEIERLS TRANSITION

IN ONE DIMENSIONAL METALS

Huguette LAUNOIS - Halina NIEDOBA

Laboratoire de Physique des Solides

Université Paris-Sud - 91405 ORSAY - France

Nuclear magnetic resonance, using the nuclei as probes of their local environment provides information on the electronic and atomic properties : susceptibility of the electron gas, and electric field gradient at the nucleus. Both the static and low frequency dynamic properties can be studied respectively by C.W. or transient NMR experiments.

This ability of NMR is used here to study the Peierls transition of the pseudo 1-D metallic system $K_2 Pt (CN)_4 Br_3 3H_2O$ called KCP in the following. In the present state of this work mainly electronic properties have been investigated in this system, and we particularly insist here on the effect on the electronic system of the critical fluctuations at the Peierls transition. The only direct insight on the atomic properties of KCP in this paper consists of a study of the positions and movements of the water molecules. The advantage of a microscopic simultaneous measurement of both electronic and atomic properties with a single probe is better illustrated in a recent NMR study of a "spin-Peierls" transition : the dimerization of a linear Heisenberg chain in doped VO_2 (1) . These data being published in detail, we will not present them in these notes.

I.INTRODUCTION : NMR in 1 - D METALS (2)

A.Measurement of the electronic properties

Nuclear spins I = 1/2, having no quadrupolar moment, do not interact with the electric field gradients. In a metal the interaction of a spin 1/2 with the conduction electrons plays a dominant role in the NMR spectra. We will only consider this simple case here for the measurement of the electronic properties.

In an external magnetic field H_o the Zeeman energy levels of a nuclear spin are split, and a radiofrequency field H_1 can induce transitions between these levels for the Larmor frequency $\omega_o = \gamma \hbar H_o$. In a metal the nuclear spin interact with the conduction electrons. The <u>static contribution</u> of the interaction will produce a shift of the resonance to a field H.Taking the symbol and sign definition used by physicists for the Knight shift due to s electrons in metals, we define this shift by $K = (H_o - H)/H_o$. The <u>dynamic contribution</u> of the interaction corresponds to its ability to induce transitions between the Zeeman energy levels, and is proportional to its spectral density at the nuclear frequency $J(\omega_o)$.   It leads to energy exchanges between the nuclear spins and the electronic system, which is in good thermal contact with the lattice, in a transient NMR experiment it corresponds to the spin-lattice relaxation rate $(T_1)^{-1}$.

a) <u>s electrons in a metal in the absence of electron - electron interactions</u> : The contact interaction between the s electrons and the nucleus shifts the resonance line proportionally to the spin susceptibility of s electrons χ_s, itself proportional to the s density of states at the Fermi-level $n_s(E_F)$. The strength of the interaction is represented by an hyperfine field $(H_{hf})_s$, and $K_s = c (H_{hf})_s \chi_s$. In this expression the constant c is $0.895 \times 10^{-4} A\Omega$ when H_{hf} is expressed in Oe per spin and χ in emu/mole. A is the Avogadro number and Ω the atomic volume.

The dynamic effect of the contact interaction leads to a relaxation rate $(T_1)^{-1}_s$ given by the well-known Korringa relation :

$$(T_1)^{-1}_s = \frac{4\pi \, k_B}{h} \left(\frac{\gamma_n}{\gamma_e}\right)^2 K_s^2 T$$

In presence of electron-electron interaction another formulation will be used:

$$(T_1)^{-1} = \frac{k_B T}{2 \mu_B^2 \hbar} (H_{hf})^2 \sum_q \frac{Im\chi(q,\omega_o)}{\omega_o}$$

where $Im\chi$ represents the imaginary part of the susceptibility, and ω_o the Larmor frequency. For free electrons this formula reduces to the Korringa relation.

b) <u>d electrons in a metal in the absence of electron -electron interactions</u> : The d electron density at the nucleus being zero, the contact interaction don't give any direct hyperfine field. But through the polarisation of the core s shells by outer d spins a d hyperfine field is produced, giving a d Knight shift $K_d = c (H_{hf})_d \chi_d$. The d hyperfine fields are negative, and much smaller than the s hyperfine fields.

Similarly to the case of s electrons the spin lattice relaxation rate $(T_1)^{-1}_d$ is related to the Knight shift K_d by a Korringa relation, the only difference beeing a reduction factor q of the rate. q is related to the orbital degeneracy of the wave function, and varies between 0.2 and 0.4.

c) <u>Effect of electron-electron interactions</u>: In a 3-D metal the Coulomb intra-atomic interaction U leads to an enhancement of the static susceptibility given by $\chi = \chi_0 \left[1 - U \chi_0 \right]^{-1}$ where χ_0 is the static susceptibility without interactions. The Knight shift is still given by $K = c \, H_{hf} \chi$, and the Korringa relation remains.

In a 1 - D metal, the divergency of the susceptibility at $q = 2k_F$ is reflected by the relaxation rate :

$$T_1^{-1} = \frac{k_B T}{2 \, \mu_B \, \hbar^2} \, (H_{hf})^2 \, \sum_q \frac{\text{Im} \, \chi \, (q, \omega_0)}{\omega_0}$$

As pointed out by Ehrenfreund et al. (2c) for a 1 - D system Im χ (q, ω_0) will have appreciable values only for $q = 2 \, k_F$ at low temperature ($k_B T \ll E_F$) Using the usual RPA approximation in tight binding at $0°K$ they write

$$\text{Im}\chi \, (q, \omega) = \frac{\text{Im} \, \chi_0(q, \omega)}{\left| 1 - U \, \text{Re} \, \chi_0(q, \omega) \right|^2}$$

They obtain an enhancement of $(T_1)^{-1}$ given by $\left[1 - U \, \chi_0(q = 2 \, k_f) \right]^{-2}$ whereas the static susceptibility is increased by only $\left[1 - U \, \chi_0 \, (q = o) \right]^{-1}$. The Korringa relaxation rate is then increased by a factor

$$\eta_{1D} = \left[1 - U\chi_0 \, (q = 2 \, k_F) \right]^{-2} \left[1 - U\chi_0(q = 0) \right]^2$$

d) <u>Orbital contribution in a partially filled d band</u> : In a partially filled d band the dominant orbital magnetism is the Van-Vleck paramagnetism : the coupling between the orbital moment L and the magnetic field produces an admixture of wave functions of different symmetry, dequenching the orbital moment in second order : for example in the atomic limit the coupling between a filled fundamental state $|o\rangle$ and empty excited states $|n\rangle$ of different symmetry gives

$$\chi_{\alpha\alpha} = 2 \, \mu_B^2 \quad \frac{\sum_n \left| < o \, | \, \sum_j L_{j\alpha} \, | \, n > \right|^2}{E_n - E_o}$$
$$\alpha = x, y, z$$

To this orbital susceptibility corresponds a Van-Vleck contribution to the shift :

$$K_{vv} = (H_{hf})_{vv} \, \chi_{vv} \quad \text{with} \quad (H_{hf})_{vv} = 2 < r^{-3} >$$

The Van-Vleck shift is known in insulating materials as the paramagnetic chemical shift, with a correct choice of gauge for the potential vector. The paramagnetic chemical shift has the opposite sign convention : $\sigma_p = - K_{vv}$

B. Dipolar interaction between nuclear spins

Let us first consider 2 nuclear spins separated by a distance a . It is the case of protons in a water molecule. The dipolar interaction yields a dedoubling of the NMR line, known as the "Pake doublet". The two lines are separated by

$$\delta\omega = \frac{3}{4} \quad \frac{h \gamma_n^2}{a^3} \quad (1 - 3 \cos^2\theta)$$

where θ is the angle between the external magnetic field and the axis joining the nuclei.

For a high number of interacting nuclear spins, the dipolar interaction result in a broadening of the resonance line typically of the order of a few KHz.

An effect of dipolar interaction on T_1 has to involve a modulation of the interaction by atomic movements. Except in the water molecule the dipolar interaction is too small to produce important relaxation rates.

C. Measurement of electric field gradients

For spins $I > 1/2$, the nucleus possesses a quadrupolar moment. Its coupling to the local electric field gradient produces static effects which are a measurement of the local electric field gradient. Quadrupolar effects are very sensitive to the local symetry, and an extremely small structure variation or disorder, not accessible to X-ray analysis, can be detected. An example is given in (1).

Dynamic effects : the modulation of the quadrupolar coupling by low frequency phonons leads to enormous relaxation rates, as observed in ferroelectrics. As an insight into the low frequency dynamics this method is complementary to neutron diffraction studies.

II . THE PEIERLS TRANSITION IN KCP

A general presentation of KCP is done in the preceding papers of H.R.Zeller and B Renker, and the most significant features are collected in Figure I . We want here to point out that

1 . Apart from the EPR measurements of Mehran and Scott, which show the appearance of d_{z^2} conduction electrons around 70 K, nothing is known about the magnetic properties of the electron gas, and the density of states.

2 . The low frequency dynamics of the atomic spectrum is unknown : in particular the central peak observed in neutron scattering is probably static below 100 K, but dynamic above 200 K.

3 . The observation of only short range 3 - D ordering even at low temperature is an evidence of the presence of disorder in the system. But nothing else is known on the disorder, and it could

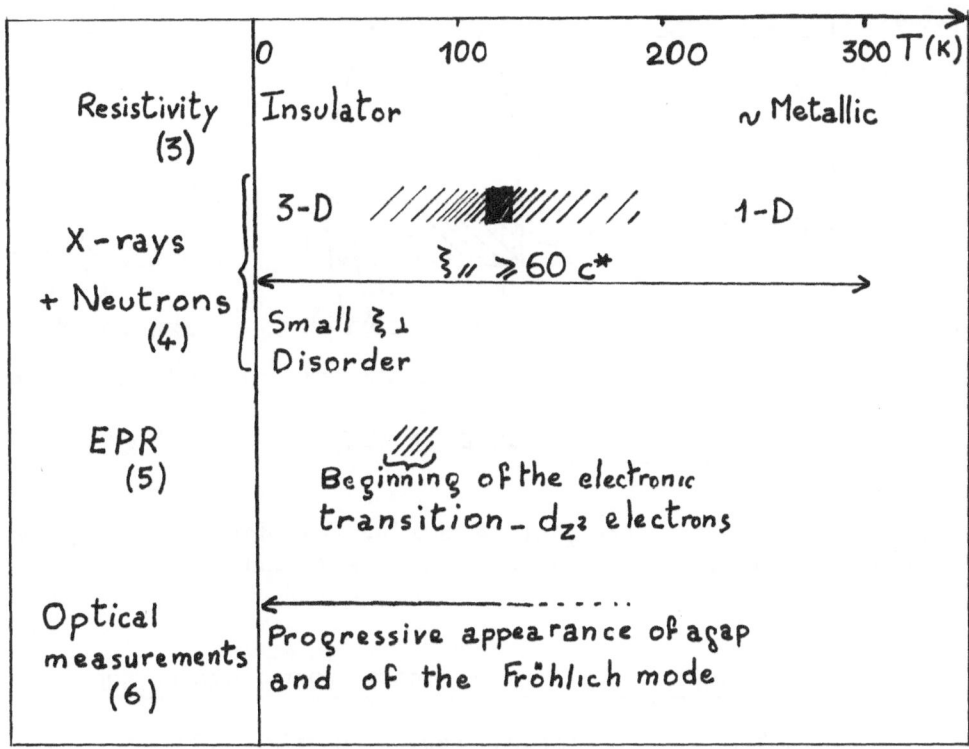

<u>Figure 1</u> : Properties of KCP in the transition temperature range.

severely influence the transition region.

The first point can get an answer from NMR study of ^{195}Pt, which is directly on the electronic chains. Having a spin 1/2, Pt cannot help to clarify points 2 and 3. Nuclei between the chains can be used for that : (i) pure NQR can be attempted on Br or N, but will probably be too sensitive to the disorder present in the system, making the experiment difficult or impossible. (ii) K can be used, but is at the limit of the sensitivity. (iii) Deuterium in a deuterated sample present the advantage of a good signal, but the big inconvenient that its dynamics will be influenced by the movement of the water molecules. To know these movements a preliminary study of the water molecules is necessary, even if it presents only a small interest for the Peierls transition.

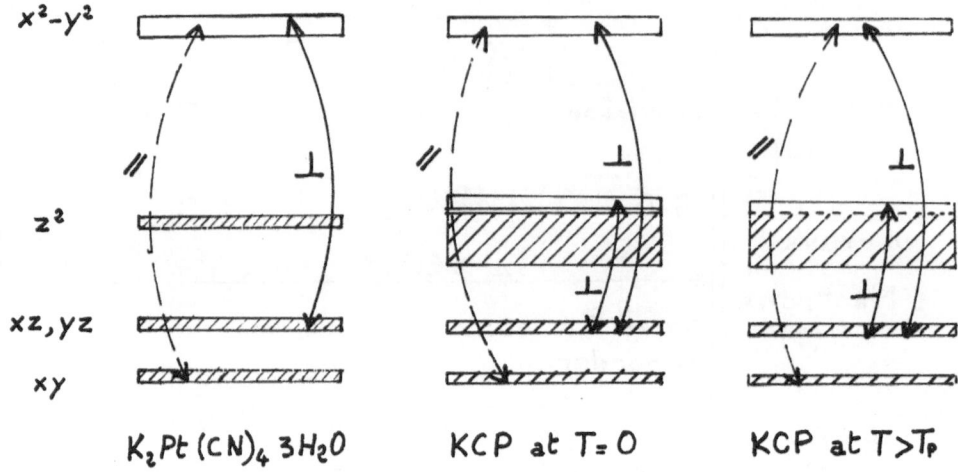

Figure 2 :   Schematic band picture showing the admixture of levels
at the origin of Van-Vleck components of the // and ⊥ shift.

III. ^{195}Pt NMR STUDY OF THE ELECTRONIC PROPERTIES OF KCP

A. The low temperature insulating state (7)

        The Knight shift is temperature independant between 1 and 77°K,
and when the magnetic field is perpendicular and parallel to the
chains, respectively equal to $K_\perp$ = - 0.56% and K   = - 0.33%. The
reference compound used here, as usually for Pt-NMR, is H_2PtCl_6
which presents itself an important orbital shift (~ 0.6 to 0.7%)(8)
The negative sign is an artefact and $K_\perp \sim$ 0. 1%, $K_{//} \sim$ 0.3%.
        The spin lattice relaxation rate $T_1^{-1}$is very small at liquid
helium temperature ($\sim 10^{-3}$ s $^{-1}$), is temperature and field dependent, and certainly due to relaxation by magnetic impurities
present in the system. This indicate that no elementary excitation
of the electronic gas is available to provide a relaxation mechanism.
Added to the temperature independant shift it shows that the fundamental state is a non magnetic band insulator. It corresponds to
the 3 - D Peierls distorted state.
        We will comment here on the values obtained for the shift,
in order to see wether the orbital contributions vary in the transition region. From a systematic study of Pt complexes, Rupp and
Keller (9) deduced that $K_\perp$ depends very slightly on the ligand and
is of the order of - 0.6% (which means roughly no orbital contributions to $K_\perp$),and that $K_{//}$ decreases when the strength of the equatorial ligand increases, from + 0.56% in K_2PtCl_4 to - 0.33% in
BaPt $(CN)_4$ 4 H_2O. In the insulating compound MPt $(CN)_4$ nH_2O with

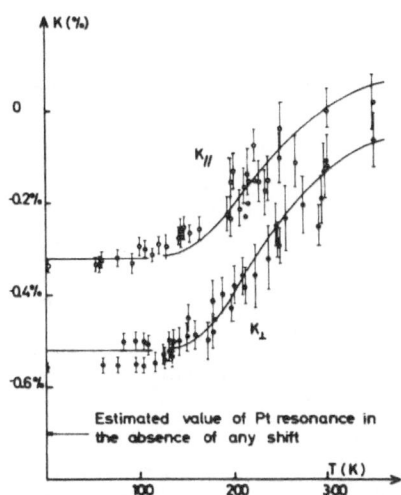

<u>Figure 3</u> : ^{195}Pt Knight shift in KCP, in the $//$ and $\perp$ directions.
The solid lines represent a fit to the susceptibility calculation
of Lee, Rice, Anderson (13).

M = K, Mg, Sr, Ba the values obtained for $K_{//}$ (-0.33 to - 0. 38%)
and $K_\perp$ (-0.55 to - 0.61%) were not very sensitive to the Pt-Pt
distance.

The Van-Vleck shift observed in KCP is exactly the same as in
the compound without Bromine, which is a priori very surprising,
at least for $K_\perp$: a crude scheme of the band levels is shown in figure 2. The $//$ orbital component comes from the coupling of the xy
level with the high empty $x^2- y^2$ level, and no additional component
appears when the z^2 band is partially empty. Obtaining the same value
for both cases is therefore quite reasonable. No variation will occur due to the opening or disappearance of the Peierls gap. The $\perp$
component presents one extra contribution in KCP, due to the coupling of the xz or yz band to the empty states of the z^2 band. The
experimental observation of the same value for KCP and $K_2Pt(CN)_4 3H_2O$
has then to correspond to the fact that the orbital hyperfine field
$< r^{-3} >$ is very small, due probably to covalency with CN groups (10)
This explains quite well that approximately the same very small value has been found for all planar complexes. In this context, the
small change due to the Peierls transition is not expected to affect the NMR orbital components.

B. <u>Pt NMR data in the transition region</u> (11)

When the temperature increases, resistivity data show that KCP
undergoes a transition to a metallic state. Knight shift measurements give us the static spin susceptibility of the electron gas,

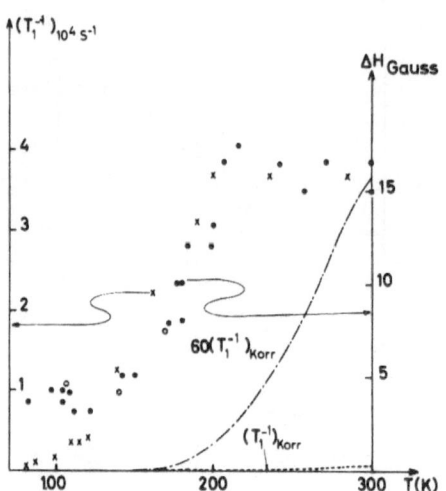

<u>Figure 4:</u> [195] Pt relaxation rate and line width versus temperature.

and spin lattice relaxation measurements some knowledge of the dynamic susceptibility.

The temperature variation of the Knight shift is shown in Figure 3. Above 120 K the shift increases continuously in the studied temperature range, limited by the decomposition of the sample; 120 K is the temperature below which 3 – D ordering takes place. Having ruled out in § III.A.an orbital contribution to this variation, we can write

$$K(T) - K(0) = K_{spin} = c(H_{hf}) \chi \sim n(E_F)$$

In a z^2 band, the hyperfine field due to d electrons is negative, and the contribution to the Knight shift should be negative. The experimental observation of a positive shift is then a proof of the presence of s admixture in the d wave functions. In transition metals a 2 bands picture is generally used to describe the presence of s and d character of the conduction electrons, leading to the following decomposition of the spin susceptibility and Knight shift

$$\chi_{spin} = \chi_s + \chi_d$$
$$K_{spin} = K_s + K_d = c(H_{hf})_s \chi_s + c(H_{hf})_d \chi_d$$

The susceptibility data existing at the present time (12) exhibit a low temperature increase for $\chi_{\parallel}$ and not for $\chi_{\perp}$, due to magnetic defects which have a very anisotropic g factor ($g_{\perp} = 0$). In addition they are limited to temperatures smaller than 230 K. We can only deduce a very rough value of the magnitude of χ spin from these measurements, and we use for that $\chi_{\perp}$ which is not affected by

magnetic impurities : $\chi_{spin} \sim 2 \times 10^{-6}$ emu/mole at 230 K. We have
then $\chi_d < 2 \times 10^{-6}$ emu/mole. Using the value of the d hyperfine
field of Pt metal, it yields $K_d < 0.05\%$. At that temperature we
have the experimental result $K(230) - K(0) \sim 0.2\%$ and it is also equal to $K_s + K_d$. This gives $K_s < 0.25\%$, which corresponds to a
s contribution to the susceptibility $\chi_s \sim 0.7 \times 10^{-6}$ emu/mole, if
we use the s hyperfine field of Pt metal. We get then a s admixture
of the order of 25 %.

    As long as the s admixture is not perturbed by the Peierls
distorsion, which would seem very unreasonable as the Peierls gap
corresponds to a very small energy range, the temperature variation
of K (T) - K (0) should be proportional to the static susceptibility and to the density of states at the Fermi level. We can see on
Figure 3 that no saturation is observed at room temperature. The
metallic state is not reached at 300 K . This was not evident,
a maximum in the conductivity occuring around 250 K in the best
samples. Out of any precise model, it shows that the mean field
Peierls temperature is higher than 300 K.

    The spin lattice relaxation rate T_1^{-1} is extremely high
for T > 100 K, and   it determines the line width for T > 150 K.
In fact the Pt spin lattice relaxation was observed not to be exponential up to 150 K, and the points of figure 4 represent a
"mean" T_1 which corresponds to the relaxation of Pt nuclei within
a factor of 3. Above 150 K the signal to noise ratio was too poor
to check the exponentiality of the relaxation.

    In a simple s metal, without any interaction between con-
duction electrons, the relaxation rate would be given by the Korringa relation, and vary as $n^2 (E_F) T$. The Korringa relaxation rate
$(T_1^{-1})_{Korr}$ deduced from the temperature variations of the Knight
shift $K (T) - K (0)$ is shown in Figure 2. The experimental relaxation rate is strongly enhanced by a factor η that we define as

$$\eta = \frac{T_1^{-1}}{(T_1^{-1})_{Korr}}$$

    The temperature variation of η is given in Figure 5.
Below 160 K, as K (T) - K (0) becomes very small, η is difficult to
define inside the experimental error bars, and we have not presented values of η in this region. Very careful measurements of K and
T_1 on the same sample would be necessary to go further in the low
temperature range. It is clear from figure 5 that η increases very
rapidly when temperature decreases. We will establish below that
this cannot be attributed to metallic effects like s - d mixing or
electron-electron interactions in a 1-D metal, and show that such
an effect is expected from critical fluctuations at the transition.

    The effect of s-d mixing in a 2 bands picture, as used
above to determine the s admixture would lead to

$$(T_1^{-1}) = (T_1^{-1})_s + (T_1^{-1})_d$$

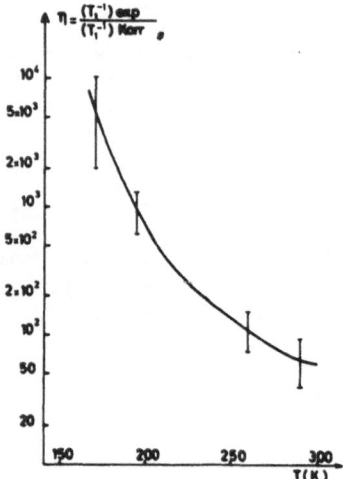

<u>Figure 5</u> : The enhancement η of the relaxation rate versus tempera-
ture in KCP. η is the ratio of the experimental "mean" T_1^{-1} as
defined in the text, to the Korringa relaxation rate deduced from
Knight shift measurements on 9 different samples.

where the s and d relaxation times are related to the s and d spin
components of the Knight shift by the Korringa relation, with
a reduction factor q for d electrons (see part I). Using the separation of s and d effects made earlier, one get

$$(T_1^{-1}) \sim (T_1^{-1})_{Korr} \times 1.5$$

The Korringa value deduced from K(T) - K (O) is only 1.5 time
smaller than the relaxation rate obtained in a 2 - band picture.
This model in addition is quite unrealistic in our case, and gives
an upper limit of the enhancement due to s - d mixing.
<u>   The effect of electron-electron interactions in a 1 - D metal</u>,
in the absence of critical fluctuations, is given in part I :

$$\eta_{1-D} = \left[1 - U\chi_0 \ (q = 2 \ K_F \ )\right]^{-2} \left[1 - U\chi_0(q = 0)\right]^2$$

In a 1 - D metal $\chi_0 (q = 2 \ K_F) \ / \ \chi_0 \ (\ q = 0)$ is the Lindhart function
which diverges when T → O. We are discussing here the T_1^{-1} values in
a temperature range (150 K - 300 K) where $E_F/K_B \ T$ varies roughly
from 50 to 100 (taking $E_F \sim$ 1e V) and the ratio
$\chi_0 \ (q = 2 \ k_F) \ / \ \chi \ (q = 0)$is then of the order of 5 and varies
slowly with T. When T decreases from 300 K to 150 K, an eventual
enhancement due to these electron - electron correlation effects
would decrease very rapidly with T, because $\chi \ (q = 2 \ K_F)/\chi(q = 0)$
remains roughly constant, but $\chi \ (q = 0)$ which is proportional to
the Knight shift K (T) - K (O) decreases rapidly.If we suppose for
example that at room temperature η = η_{1-D} = 60, which corresponds

to U χ_o (2 k_F) $\sim$ 0.9 and to an enhancement of the static susceptibility
of 1.2, we get for example at 200 K a value of $\eta_{1-D} \sim$ 2.

All the above statement supposes no peculiar low frequency variation of the dynamic susceptibility. We have demonstrated that in
this picture even if the room temperature enhancement was due to
electron - electron interactions, nearly no enhancement would be
present for T < 200 K, when the density of states is very low.

But at the opposite if critical fluctuations lead to low frequency effects on the dynamic susceptibility 1 - D enhancement
should be important. The above formula for η_{1-D} deduced from RPA,
can no more be used and one has to come back to the formulation

$$(T_1)^{-1} = \frac{k_B T}{2 \mu_B \hbar^2} (H_{hf})^2 \sum_q \frac{Im \chi (q,\omega_o)}{\omega_o}$$

C. Critical fluctuations at the Peierls transition

The enhancement of T_1^{-1}, and its temperature dependency cannot
be explained in the absence of low frequency effects on the susceptibility in the transition region. In addition we know from the
X ray and neutron data that the behaviour of KCP is one dimensional
above 200 K, and from the Knight shift that the metallic state is
not reached at room temperature. In a one dimensional metal, critical fluctuations are expected below the mean field Peierls transition. Their static effect have been estimated by Lee, Rice and Anderson. In figure 3 we present a fit of our experimental data to their
calculation of the static susceptibility, using a mean field Peierls
temperature of 625 K, and arbitrary vertical scales. The apparent
very satisfactory agreement between theory and experiment should
not be taken too seriously, in view of the experimental error bars,
due to the difficulties of measurement in presence of a very short
T_1. In addition we have used here 2 adjustable parameters. The
Lee-Rice Anderson curve of n (E_F) used here corresponds to a real
order parameter, but the curve they obtain with a complex order parameter is very similar, and could probably not be distinguished
within the error bars.

The dynamical properties have not been estimated, and in the
absence of a calculation of the dynamical susceptibility χ (q,ω)
we will just try to understand if a high enhancement of the relaxation rate can be expected from the critical fluctuations in the
case of a Peierls transition (no electron-electron interaction). A
priori the Peierls transition in 1 - D is an effect restricted to
the freezing of charge density waves, without any correlated effect
of spin density waves, and it is difficult to see how the dynamical
susceptibility can be affected, or in other words how the spin couples to the charge density waves.

We would like to suggest here a very crude model leading to
an enhancement of T_1^{-1}. Taking an instantaneous picture of a 1 - D
chain, we can visualize it in space : some regions present a charge

density wave, and their extension is essentially the correlation length $\xi_{//}$ of the critical fluctuations. They are separated by metallic regions. They fluctuate in time and space; but for frequencies much higher than the frequencies of the critical fluctuations, they can be treated as static. In particular conduction electrons, moving with the Fermi velocity ($\sim 10^{-16} - 10^{-17}$ in a usual metal) can be considered as localised in the metallic regions. The rough energy quantification of the partially localized states would then lead to a longer life-time of the non-compensated spin states, leading to a longer correlation time for the electrons.

This picture of metallic regions fluctuating slowly (at the electron scale) in time and space is very similar to the picture obtained in the case of a liquid semiconductor, with the two differences that here (i) the insulating regions are due to charge density waves, and not to a localization induced by disorder (ii) the temperature variation comes from the slowing down of critical fluctuations, and not from the slowing down of atomic motions in a liquid. In the case of liquid semiconductors, an enormous enhancement of T_1^{-1} have been observed by W.W.Warren (14) near the liquid - solid transition when the electronic wave functions are weakly localized. It has been attributed to an increase of the time spent by an electron on a nucleus, and related to the transport properties.

The analogy of the Peierls fluctuations with liquid semiconductors indicate that en enhancement of T_1^{-1} is expected. The over simplified description given above is no more valid when the correlation length of the critical fluctuations becomes too high. We would like to point out that in this mechanism the $2 k_F$ instability plays a role in creating the metallic regions, but that the dynamical effect which increases $\chi (\omega_o)$ is at $q = 0$. It is not clear that a more direct coupling of the spins to the charge density waves should not be expected and lead to enhancements of $\chi (\omega_o)$ at $2 k_F$.

D. <u>Is a more complicated picture of the transition necessary ?</u>

In the preceeding section we have tried to understand the properties of KCP by the 1 - D fluctuations at the Peierls transition. We have then neglected the effect of 3 D coupling and static disorder.

From the neutron data between 200 K and 50 K one can consider that the dimensionality changes continuously from 1 to 3 dimensions (cross over region) : the intensity of the central peak at (π/a, η/a, $2 k_F$) increases in this range. The effect of 3 - D coupling on the critical fluctuations has been discussed by Rice and Strassler (15). If we believe our fit of the Knight shift to the Lee Rice Anderson susceptibility, we can then conclude that the 3 - D coupling does not affect the curve of the density of states at the Fermi-level, and that its main importance is to couple the

chains when the correlation length becomes very long as argued in
(13). But this conclusion is not very significant, as based on a
quantity which is going to zero and can no more be measured with
precision. And the dynamical susceptibility being unknown even in
the simplest case, we cannot get any conclusion from our T_1^{-1} value.
The situation of neutron experiments in this respect is not better
(16). The importance of static disorder in 1-D metals had been pointed out by many authors (17). In KCP it is clear that the fundamental state is a 3 - D Peierls insulator, and that disorder plays
at most a secondary role at the transition. In particular could it
be at the origin of the increase of T_1, that we have described here
as due to a dynamical disorder intrinsic to the critical fluctuations ? There is no answer to this question inside the experimental data collected at the present time.

E. Conclusion

Our ^{195}Pt NMR study have shown that (i) the density of states
at the Fermi level begins to increase above 120 K, and is not saturated at 300 K, indicating a mean field Peierls temperature
T_p > 300 K. Using the calculations of Lee, Rice, Anderson to fit our
data we obtain $T_p \sim$ 600 K. (ii) The relaxation rate is anomalously
large ; the enhancement relative to a Korringa mechanism increases
when T decreases, and reaches a few orders of magnitude between
150 and 200 K. We have shown that in the absence of fluctuations
there was no explanation of this temperature dependency. We have
proposed an origin of increase of T_1^{-1} in the presence of 1 - D
critical fluctuations, by analogy with liquid semiconductors. This
mechanism would not necessary be the only one accompanying the
critical fluctuations.
We have not been able to exclude neither the effects on the
fluctuations of a cross-over between 3 - D and 1 - D behaviour,
nor an influence of static disorder. Any progress in this direction
goes through experimental studies of similar systems in which the
relative influence of 1 - D fluctuations, 3 - D coupling, and disorder could be different.

III. POSITION AND MOVEMENTS

OF THE WATER MOLECULES IN KCP

It is not obvious that the water plays a role in the Peierls
transition in KCP. The interest of studying the water is nevertheless the following :
1 - The movement of the water molecules will influence the quadrupolar relaxation of the different nuclei (Br, D in deuterated samples, K) . As a preliminary step to the study of the low frequency
dynamics of the phonon spectrum, it is indeed necessary to study

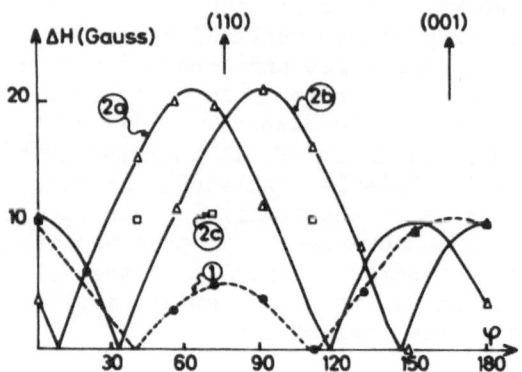

Figure 6 : The angular dependency of the Pake doublet at 130 K in the (1̄10) plane. Water 2 a and 2 b are in this plane and make respective angles of + 14° and - 14° with the [110] direction. Water 2 c is equivalent to these water, but situated in the perpendicular plane (110). Water 1 are along the [100], [010] directions.

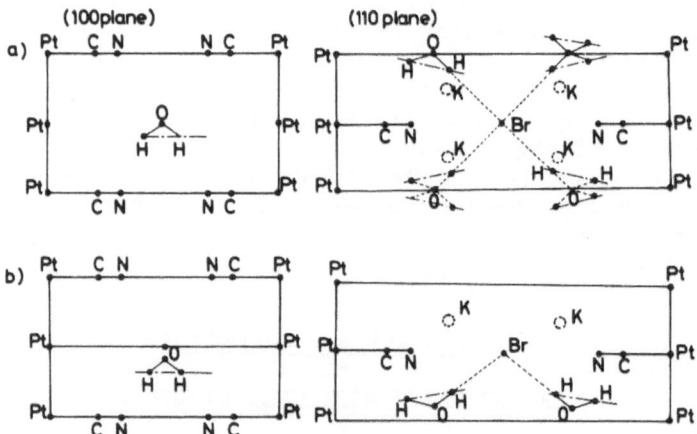

Figure 7 : Tentative positions of the water molecule in a) the structure of Krogmann (20) b) the new structure proposed for KCP (19).

these movements.

2 - The importance of the water content on the transport properties
has been pointed out by many authors. It has been suggested recently that the electric dipole moments of the molecules could screen
the effects of Br disorder (18). The experiments accompanying this
suggestion were done on powders, and absolutely unable to distinguish the two types of existing water molecules, leading to inexact
conclusions as will be pointed out by our work on single crystals.

3 - A new structure of KCP has been recently proposed (19). It presents far less disorder than the structure of Krogmann (20) : the
potassium atoms are no more disordered, but occupy the 4 positions
given by Krogmann at z = 3/4c. We will try to situate the water
molecule in these two structures and to understand its movements.

A. Position of the frozen water molecule

The dipolar interactions between the 2 protons of the water
molecule leads to a dedoubling of the NMR line. The two lines of
the doublet (called the Pake doublet) are separated by

$$\Delta H = \frac{3 \mu}{r^3} \quad 3 \cos^2\delta \cos^2 (\varphi - \varphi_o) - 1$$

where δ is the angle made by the proton-proton vector and the plane
of rotation of the crystal, and $(\varphi - \varphi_o)$ the angle of the external
magnetic field H_o and the projection of the proton-proton vector
on the plane of rotation.

The angular dependency of ΔH yields the direction of the proton-
proton vector. An example is given in Figure 6.

Our results are the following for the "freezed" water molecules
(figure 7).

- Water 1 is in the (100) plane and has its H-H vector parallel
to the [100] or [010] direction. The most probable position is
the first of these two, with the OH vector directed toward the CN
group. They cannot be distinguished because of the four fold symmetry axis.

- Water 2 is in the (110) plane, where its H-H vector makes an angle of 14 ± 1° with the [110] direction.

Because of the symmetry again it is a priori impossible to distinguish between this position and a position of the water molecule
perpendicular to the plane. But the (110) plane is a mirror plane,
and has to contain this water molecule.

This gives a position of the molecule in which one OH points to
the Br, and will probably be H bonded to this Br. The other OH is
directed to the CN group, and less tightly bonded.

B. Movement of the water

Water movements in solids leads to a narrowing of the Pake
doublet. The angular dependency gives the axis of rotation of the
molecule (21). Some results are already known on these movements

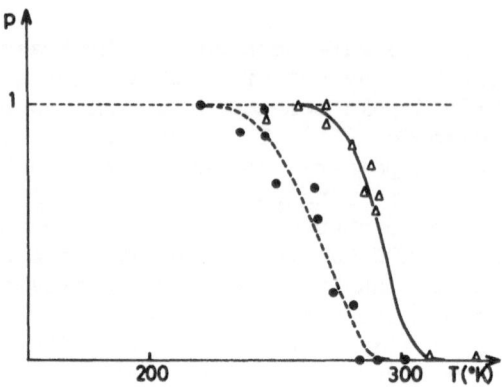

Figure 8 : Proportion of frozen water versus temperature. Triangles
refer to water 1, and full circles to water 2.

in KCP from the room temperature studies of R.Brugger and D.Brink-
mann (22a) . Our room temperature data agree with their data, corrected for the false orientation of their crystal (22b).
The movement of the 2 types of waters is the following :
- Water 1 is rotating freely, above 310 K.
- Water 2, above 290° K, presents a partially free rotation which
has been presented by Brugger and Brinkman as due to the rotation
around 2 perpendicular axis R_1 // $[00\bar{1}]$ and $R_2 \perp R_1$. Knowing now
that water 2 is the water molecule which is H bonded to the Br
atom, it is perhaps more convenient to consider this movement as
tunneling between the equivalent positions of the water, as ahead,
observed in hydrated solids (23). The new structure of KCP would
then give only a tunneling between 4 positions, which is equivalent to a rotation around a $[001]$ axis. The partially free movement is difficult to understand in this structure.
     The proportion p of frozen water is given in figure 8 for
water 1 and 2. p has been obtained from the intensity of the Pake
doublet of the frozen molecule versus temperature. A significant
portion of moving water 2 is obtained above 230 K, and it is only
at 290 K that all the water 2 molecules are moving. The fact that
the movement does not occur at the same temperature for all those
molecules indicate that disorder influence this movement.
Water 1 rotates freely above 290 K ± 10 K.
     The experiments done on powders (18) were not able to distin-
guish these types of waters and in fact correspond to p (water 1)
+ 2 p (water 2) . The interpretation of this sum as due to a distribution of tumbling times τ_c of slightly different water sites
is not valid.

Acknowledgments : We are very grateful to R.Comès, J.Friedel, T.M.Rice, W.W.Warren, and H.R.Zeller for stimulating discussions and to C.Weyl - Bertinotti for performing the orientation of one of our crystals.

REFERENCES

1 - J.P.Pouget, H.Launois, T.M. Rice, P.Dernier, A.Gossard, G.Villeneuve and P.Hagenmüller, Phys.Rev.B $\underline{10}$, 1801 (1974)

2 - a) A.Abragam, Principles of Nuclear Magnetism (Clarendon Oxford, 1961)
    b) J.Winter, Magnetic Resonance in Metals (Clarendon Oxford, 1971)
    c) E.Ehrenfreund, E.F. Rybaczewski, A.F. Garito and A.J.Heeger, Phys. Rev. Lett $\underline{28}$ 873 (1972)

3 - D.Kuse and H.R.Zeller, Phys.Rev.Lett. $\underline{27}$, 1060 (1971)
    H.R.Zeller and A.Beck, J.Phys.Chem.Solids $\underline{35}$, 77 (1974)

4 - R.Comès, M.Lambert, H.Launois, and H.R.Zeller, Phys.Rev. $\underline{B\ 8}$, 571 (1973).
    B.Renker, H.Rietschel, L.Pintschovius, W.Gläser, R.Comès, L. Liébert, W.Drexel Phys.Rev.Lett. $\underline{32}$, 836 (1974)

5 - F.Mehran and B.A. Scott, Phys.Rev.Letters, $\underline{31}$, 1347 (1973)

6 - H.R.Zeller, these proceedings

7 - H.Niedoba, H.Launois, D.Brinkmann, R.Brugger, and H.R.Zeller, Phys.Stat.Sol.$\underline{58}$, 309 (1973)

8 - L.E.Drain, J.Phys.Chem.Solids $\underline{24}$, 379 (1962)

9 - H.J.Keller and H.H. Rupp, Z.Naturf.$\underline{26\ a}$, 785 (1971)
    H.H. Rupp, Z.Naturf.$\underline{26\ a}$, 1937 (1971)

10- Similar effects have been observed in Co $(CN)_6$: R.E.Walstedt, J.H. Wernick and V.Jaccarino Phys.Rev.$\underline{162}$, 301 (1967)

11- H.Niedoba, H.Launois, D.Brinkmann, H.U.Keller to appear in J.Physique Lettres (1974)

12- A.Menth and M.J.Rice, Solid state Comm.$\underline{11}$, 1025 (1972)

13- P.A.Lee, T.M.Rice and P.W. Anderson, Phys.Rev.Lett.$\underline{31}$,462(1973)

14- W.W.Warren, Jr Phys.Rev. $\underline{B\ 3}$, 3708 (1971); Phys.Rev. B 6

2522 (1972) ; J.Non Crys.Solids 8 - 10 , 241 (1972)
W.W.Warren, Jr and G.F.Biennert, Amorphous and Liquid Semiconductors, ed.Taylor and Francis, London (1974) p 1047.

15- M.J.Rice and S.Strässler, Sol.State Comm.13, 1389 (1973)

16- A central peak is expected in the neutron spectrum in one dimensional effects even in RPA : B.Horowitz, H.Gutfreund and M.Weger, Sol.State Comm.11, 1361 (1972)
S.Barisic, A.Bjels and K.Saub, Sol.State Comm. 13, 1119 (1973)

17- I.F.Schegolev, Phys.Stat.Sol.(a) 12, 9 (1972)
A.N.Bloch, R.B.Weisman and C.Varma, Phys.Rev.Letters 28 753 (1972).

18- M.A.Butler and H.J. Guggenheim, Phys.Rev.B 10 1778 (1974)

19- J.M.Williams, J.L.Petersen, H.M.Gerdes and S.W.Peterson, to be published.
H.J.Deiseroth and H.Schulz, to be published.

20- K.Krogmann and H.D.Hausen, Z.Anog.Chem.358, 67 (1968).

21- H.S.Gutowsky and G.E.Pake, J.Chem.Phys.18 162 (1950).

22- a - R.Brugger and D.Brinkmann, Solid State Comm.13, 889 (1973)
b - D.Brinkmann and H.U.Keller, to be published.

23- V.F.Izlev, N.I.Zavarzina, S.P.Cabuda. Soviet Physics Crystallography 13 705 (1969).

STRUCTURAL ASPECTS OF ONE-DIMENSIONAL METALS

Klaus Krogmann

Institut für Anorganische Chemie der Universität

D 75 Karlsruhe, Germany (West)

Structural Elements

If one chemist tries to define regions of stability for a
certain type of compounds, this is always appealing to other
chemists to prove by synthesis that he is wrong. For that purpose,
and for still other chemists who are simply interested in a
rationalized representation of observed facts together with short
range extrapolations, I shall do the job above mentioned.

Several conditions must be met in order to obtain crystals
with metallic properties in only one direction :
1. they have to contain linear rows of atoms or molecules with
   distances which are small enough to allow the formation of
   energy states delocalized in one dimension through the crystal
   (1-D bands);
2. the distances between these rows must be great enough to
   provent a similar delocalisation in the other two dimensions;
3. the 1-D bands have to be incompletely filled.

As the metallic bond between free atoms can not be restricted
to one dimension, one way towards 1-D metal construction starts
with introducing stable saturated bonds in two dimensions at a
suitable "center", thereby blocking these two dimension against
further interactions. The "center" may be a special atom or a less
well defined region in a more complex unit. Examples are the
square planar complexes (TCMP = tetra coordinated metal plane),
and TCNQ (= 7,7,8,8-Tetracyanoquinodimethane) compounds,
respectively. The next step might be done if the "centers" are able
to combine and to form metallic bonds in the third and last
dimension left for them.

A second way may lead to the direct synthesis of a "linear" chain of atoms with strong covalent bonds, on which a metallic bond is superimposed. The question remains, then, whether further insulation is needed by proper side groups or a suitable matrix to establish strictly 1-D behaviour. On the other hand, "linearity" does not necessarily mean collinear bonds between atoms. (SN)$_x$ is perhaps an example for this second way. In the following, I shall discuss square planar complexes only, as other lectures at this symposium will cover the other fields.

Square planar complexes or TCMP are formed by
1. central atoms with d^8 configuration in a strong ligand field : Ni(+2) in [Ni(CN)$_4$]$^{2-}$, Ir(+1) in Ir(CO)$_2$acac (acac = acetylacetonate);
2. d^4 configurations : Cr(+2) in [Cr(acetate)$_2$]$_2$ · 2 H$_2$O, Re(+3) in [Re$_2$Cl$_8$]$^{2-}$;
3. some main group elements : [JCl$_4$]$^-$, XeF$_4$;
4. ligands chelating a central atom in one plane, like the phthalocyanine anion (Fig. 1a).
Here, the planar ligand forces the central atom in a planar surrounding, whereas in the other three cases, the special electron

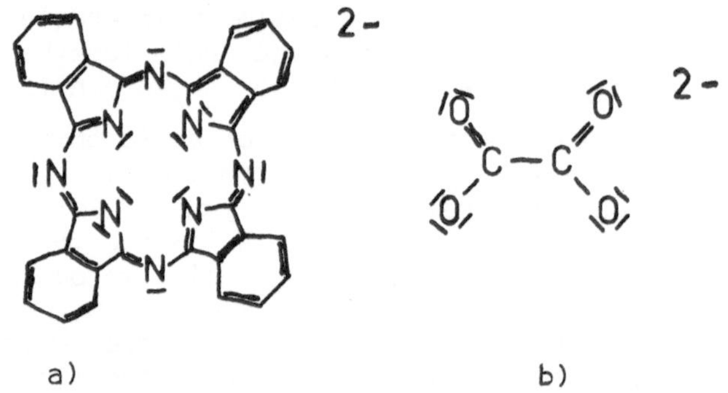

a)   b)

Fig. 1

Structure of planar ligand :

a) Phthalocyanine anion ;   b) oxalate .

configuration is responsible for the deviation from the otherwise electrostatically favoured tetrahedral coordination of four ligands. Two lone pairs can be assumed to be localized mainly outside the plane in case 3 (Fig. 2a). The orbitals and their occupation in cases 2 and 1 are depicted in Fig. 3. The degeneracy of the upper d- orbitals in an octahedral complex is removed, if

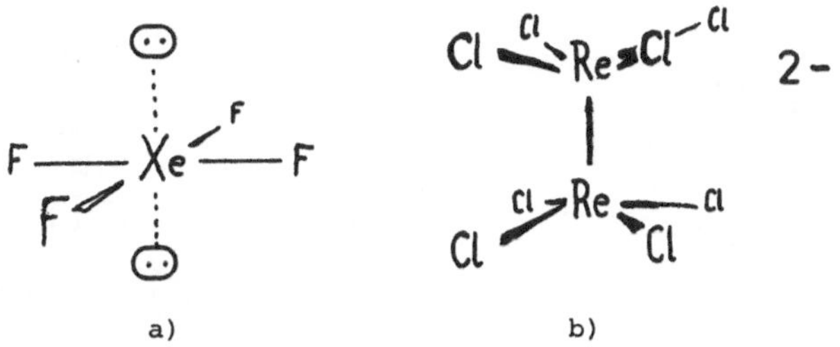

Fig. 2

Structure of XeF_4 (a) and $Re_2Cl_8^{2-}$ (b) .

two trans ligands are pulled out to form a square planar complex (Fig. 3a and 3b). With four d electrons (Fig. 3c), dimers are formed (Fig. 2b), in which the central atoms are connected by rather strong, localized bonds.

The d^8 configuration (Fig. 3d) stabilizes monomer square planar complexes, if the ligand field is strong enough to allow spin pairing. These units are found in the so called "columnar structures" which contain columns of TCMP stacked in a staggered manner with the central atoms forming a linear row. This structural arrangement is related to a weak bonding interaction which becomes active if other lattice requirements, e.g. size of other atoms, do not overrule them (1). If the metal-metal distance is closer than 3.3 Å , the crystals show strongly anisotropic properties and unusual colours. These structures meet the first two conditions for 1-D metals mentioned in the beginning, and there are many of them, with different central atoms : Rh, Ir, Ni, Pd, Pt, Cu, Au; and also with many different ligands like halide ions, NH_3 and some organic derivatives, oximes, nitriles, isonitriles, cyanide, oxalate (Fig. 1b), acetylacetonate (acac) etc.

An easy recipe for the synthesis of a 1-D metal should be to take one of these columnar compounds and to oxidize it partially in order to obtain an incompletely filled band (condition 3). But of all that manifold, only a few central atoms and ligands are really able to form 1-D metals. In table I is given a list of accessible d^8 configurations with its oxidation numbers, which change with group according to the atomic number. Only the underlined elements Pt and Ir are found in 1-D metal structures. This may be rationalized by looking at the side groups in oxidation states 0 and +3 first. Au(+3) has a higher effective charge than Pt(+2) and, therefore a more contracted d shell, which is

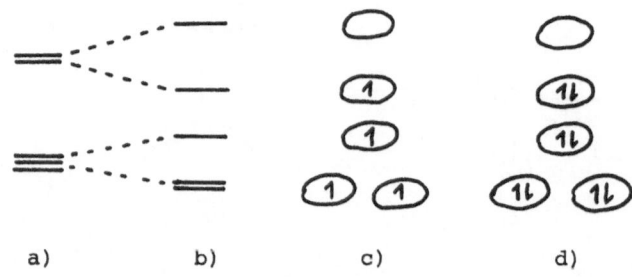

Fig. 3
Electron configuration in square planar fields :
a) d-orbitals in octahedral, and b) square planar
fields; c) d^4, d) d^8 configuration.

unfavourable for the metal-metal interaction. This is the more
true for Ag(+3) and Cu(+3), since the radial part of the 4d and
3d wave functions has smaller values than 5d at some distance
from the nucleus. On the other hand, the zero valent states of Fe,
Ru, and Os prefer the coordination number five in their compounds,
because the effective radius is so much larger than that of Pt(+2).
This is demonstrated by the well known molecule of $Fe(CO)_5$, iron
pentacarbonyl. The carbonyl ligands are at the corner of a
trigonal bipyramid, a geometry which prevents close contact of
adjacent complex centers.

There is still a preference for coordination number five in
the chemistry of Ir(+1) and Rh(+1), but there are also square
planar complexes. Little is known about Co(+1), which is a highly
unstable state. Rh(+1), however, might be regarded as a possible
1-D metal center, though there are still no well established
examples.

Table I.
d^8 central atoms and their oxidation number

$3d^8$	Fe(0)	Co(+1)	Ni(+2)	Cu(+3)
$4d^8$	Ru(0)	Rh(+1)	Pd(+2)	Ag(+3)
$5d^8$	Os(0)	Ir(+1)	Pt(+2)	Au(+3)

Ni(+2) and Pd(+2) form many square planar complexes, even
columnar structures, isomorphous with the corresponding Pt
compounds. If Pt, however, is found in columns with distances of

less than 3.2 Å, then the lighter elements either do not crystallize in the same lattice or at least with longer distances. The tendency of Ni and Pd towards special metal-metal interaction seems to be much less pronounced than for Pt. One may understand this again as a result of the difference between 5d and other d electrons.

In the 1-D metal systems, the distances of the central atoms are still smaller, about 2.9 Å or less. This requires small VAN DER WAALS radii not only for the co-ordinating atoms but also for the other parts of the ligand system, if it is not monoatomic. This rules out Br^-, J^-, S^{2-}, but also NH_3 because of its hydrogen atoms standing out of the plane. Only two Cl^- are allowed (at least until now), together with two CO ligands. On the other hand, the small F^- ligand, a "hard" base, is unwilling to combine with the "soft" LEWIS acids Pt(+2) and Ir(+1). Multiatomic ligands like phthalocyanine also lead to too much interligand repulsion, even if the atoms themselves are small.

As the incompletely filled band is usually achieved by partial oxidation, the square planar complexes must be oxidizable (e.g. Au(+3) is not!), but the "completely" oxidized unit, e.g. a Pt(+4) complex, should not be too stable.

After all these strict requirements, we are not too surprised to find so relatively few 1-D metals of this type.

Structures

The substances best known and most studied by both physicists and chemists are K_2 [Pt(CN)$_4$] $Br_{0.30}$ × $3H_2O$ and the analogous chloride. Their structure contains closely stacked [Pt(CN)$_4$] units with distances of 2.88 Å between the central atoms (2).The complex columns are negatively charged by 2 - (DPO) per unit, where DPO is the degree of partial oxidation. In the structure above mentioned the decrease of negative charge after partial oxidation is balanced by additional halide ions, which may be regarded as acceptor atoms for (DPO) electrons from every Pt atom. The structure offers suitable lattice sites for the halide, which are occupied in about 60 % of the unit cells.

Also anionic chains of complexes exist in the dioxalato platinates, which form 1-D metals with many different cations. They prefer a defect of cations for the charge counting. The structure of $Mg_{0.82}$ [Pt(C$_2$O$_4$)$_2$] · 6 H_2O was studied in detail by single crystal methods (3), but the metal periods of many others are known by single crystal or powder data, which are about 2.85 Å or less (1,3,4). The DPO ranges for these Pt compounds from 0.30 to 0.40, only a few phases deviate to lower numbers, none to higher.

The first Ir compound recognized as a 1-D metal was what we claimed to be Ir(CO)$_{2.9}$ Cl$_{1.1}$ (5), what others first described as Ir(CO)$_3$Cl (6). Preliminary reports of single crystal structure

determinations (5, 7) reveal only the stacking mode of $Ir(CO)_3Cl$
units, with a certain randomness in the orientation of the $Ir - Cl$
bond. This makes it impossible to decide by x-rays whether the
system contains $Ir(CO)_2Cl_2$ units or not. Recent MÖSSBAUER spectra
(8, 9, 10) reveal only one type of Ir atoms, indicating that
either the correct formulation is $[Ir(CO)_3Cl] Cl_{0.1}$, or our analysis
of DPO is incorrect, and the 1-D metal character is established by
overlapping filled and unfilled bands.

Meanwhile, we found a number of Ir compounds containing
$[Ir(CO)_2Cl_2]$ anionic units with cation defect structures (11),
which are listad in Table II. Just as the other substances, they
had been prepared and described previously (12, 13), but their
unusual composition lead to wrong proposals for the structure. The
DPO varies much more than it was experienced for Pt compounds,
whereas the distance remains equal. The exact composition of some
of the salts is yet unknown. Although we had only disturbed single
crystals or powders, the short distance is an indication for their
1-D metal character.

Powder diagrams of pure 1-D metal type compounds (Fig.4) may
quite often be indexed without great difficulties. Neglecting weak
lines, which may belong to the superstructure, the pattern of lines

Table II.
columnar structures with Ir

compound	oxidation number	distance ($\overset{o}{A}$)
$Ir(CO)_{2.93} Cl_{1.07}$	+1.07	2.85
$(H_3O)_{0.38}[Ir(CO)_2Cl_2]$	+1.62	2.86
$K_{0.58}$ "	+1.42	2.86
$Cs_{0.50}$ "	+1.50	2.86
$(N(me)_4)_{0.55}$ "	+1.45	2.86
$(As(ph)_4)_{0.62}$"	+1.38	2.86
Li_x "	?	2.86
Mg_x "	?	2.86
Ba_x "	?	2.86
$[Ir(CO)_2Cl_2][Ir(CO)_2acetat]$	+1.50	2.78

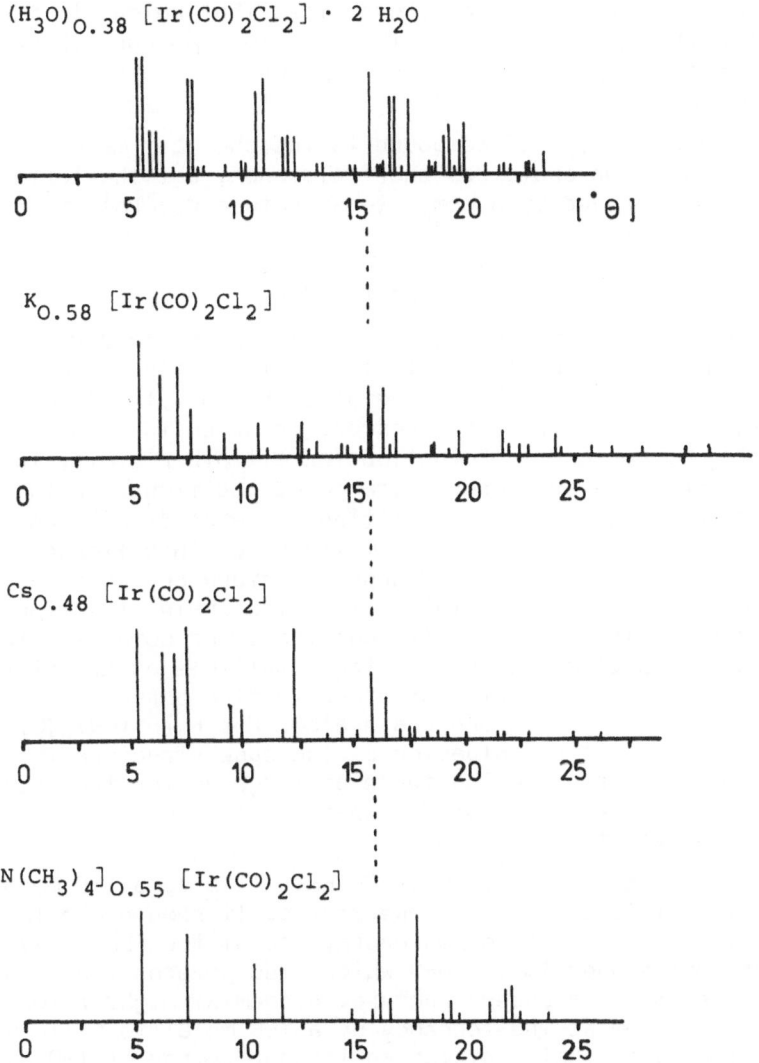

Fig. 4
GUINIER powder diagrams of Ir 1-D metal phases.
(Dotted line indicates, where the "third dimension" begins.)

up to θ = 15 degrees is produced by a two-dimensional reciprocal lattice, which might be guessed easily. The next step is to find indices for the remaining strong lines that belong to the "third dimension", the reciprocal lattice vector c_M^x corresponding to the heavy atom subcell constant c_M, which is also the metal-metal distance.

A special type of compound is the one last mentioned in Table II with an anomalous short distance, but the formulation has still to be proved by a complete structure determination.

Superstructures

The periodicity of heavy atoms in the chain direction was named the metal subcell constant c_M, since this decides, together with composition, on the 1-D metal character. The "real" cell constant must of course be greater, in order to allow a reasonable stacking manner and coordination for all other atoms. The ligands of adjacent complexes are in staggered positions. In the simple case of $Pt(CN)_4$ columns, a twisting angle of 45 degrees for the following complex leads to minimization of interligand repulsions and, there fore, to a ligand superstructure c_L, which has double the periodicity of the metal cell in the chain direction :
c_L = 2 · c_M. This may be different for other complex columns, but we should expect c_L to be an integer multiple of c_M. Similar considerations might hold for other lattice partners present in integer proportions to the chain atom. For $K_2[Pt(CN)_4]Cl_{0.32}$ · $3H_2O$ ("KCP"), a recent reevaluation of the superstructure intensities lead to a potassium period twice as large as for the previonsly assumed randomness of four K^+ on eightfold positions (14), and to a less symmetric space group.

According to a theorem of PEIERLS (15), a 1-D metal is unstable, at least at low temperatures. It should create a lattice distortion, that produces an energy gap at the FERMI level. The latter corresponds to the DPO which also governs the composition, the former means a superstructure. A chemist might reformulate the PEIERLS theorem as the tendency of a 1-D metal to build a superstructure, that contains an integer number of 1-D electron pairs in its unit cell. As (DPO/2) is the number of 1-D defect electron pairs per metal atom, we met the above condition in periods c_P which are integer multiples of (2/DPO) x c_M ; e.g. for the KCP (chloride) we should get c_P = (2n/0.32) x c_M = 6.26 n x c_M (n = 1, 2, ...) ; the oxalate complex mentioned above leads to c_P = (2n/0.36) · c_M = 5.56 n x c_M. In all these cases the PEIERLS distorted cell cannot be a small integer multiple of the metal cell. Either the theorem is obeyed only approximately, or the PEIERLS cell is "aperiodic" with respect to c_M.

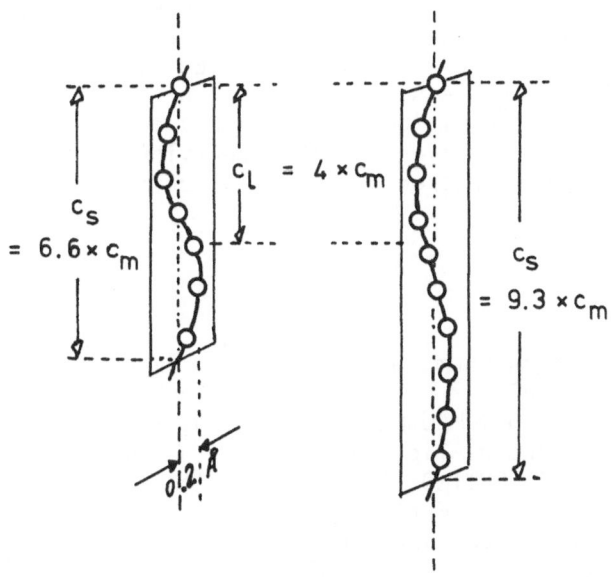

Fig. 5

"Aperiodic" superstructures in $K_{1.64}$ $[Pt(C_2O_4)_2] \cdot 4 H_2O$,
heavy atom positions only; a) Phase A_α , b) Phase A_β .

Studies of diffuse x-ray (16) and inelastic neutron
scattering (17) showed effects exactly matching n = 1 in the
above formula, that can be discussed in terms of a soft lattice
vibration becoming a static distortion at low temperatures. Which
atoms take part in these PEIERLS shifts needs still to be proved.
The heavy atoms, of course, but the ligands also ? The scattering
experiments allow an estimation of 0.01 Å for the PEIERLS shift.
Why does not the big halide ion impose a more ordered arrangement ?

Recently, we found examples of still other "aperiodic"
superstructures in terms of c_M, which are static because they
cause BRAGG reflexions, and which have no simple relation to the
DPO. Exept one Ir compound, these belong to the dioxalato platinates,
which tend to display a manifold of different phases with the same
cation, as was already reported for the Mg salts (3). This is due,
in part, to different stacking sequences for the ligands; water and
cation distribution may also participate in the play. But
"aperiodic" structures are found in at least two phases of the
potassium salt, in which "aperiodic" shifts of the complexes up to
0.2 Å occur, which are perpendicular to the chain direction. As
a result, the Pt atoms form a sinusoidal line instead of a straight

one in these two phases (not in all, nota bene), the period of which is again no integer multiple of c_M, namely c_S = 6.6 x c_M in one, c_S = 9.3 x c_M in the other case (Fig. 5). But it also does not coincide with c_P or c_L . A similar case with a helix of heavy atoms appear to be present in $[Ir(CO)_2Cl_2]$ $[Ir(CO)_2acetate]$. We are unable to give an explanation for this behaviour. All these systems try to avoid strictly ordered crystals, which may favour the 1-D metal state.

References

1. K.KROGMANN, Angew.Chem. 81, 10 (1969); Angew.Chem.Internat. Edit. 8, 35 (1969).
2. K.KROGMANN, and H.D.HAUSEN, Z.Anorg.Allg.Chem., 358, 67 (1968).
3. K.KROGMANN, Z.Anorg.Allg.Chem. 358, 97 (1968).
4. A.BERTINOTTI, C.BERTINOTTI, and G.JEHANNO, Compt.Rend.Acad.Sci.(Paris) B, 278, 45 (1974).
5. K.KROGMANN, W.BINDER, and H.D.HAUSEN, Angew.Chem. 80, 844 (1968); Angew.Chem.Internat.Edit., 7, 812 (1968).
6. W.HIEBER, H.LAGALLY, and H.MAYR, Z.Anorg.Allg.Chem. 246, 138 (1941).
7. L.F.DAHL, and H.VAHRENKAMP, unpublished results.
8. A.P.GINSBERG, R.L.COHEN, F.DISALVO, and K.W.WEST, to be published.
9. H.J.KELLER et al., to be published.
10. K.FRÖHLICH, K.KROGMANN, and P.STAMPFL, to be published.
11. H.J.ZIELKE, Dissertation Karlsruhe 1973.
12. L.MALATESTA, and F.CANZIANI, J.Inorg.Nucl.Chem., 19, 81 (1961).
13. M.J.CLEARE, and W.P.GRIFFITH, J.Chem.Sec.A., 2788 (1970).
14. H.J.DEISEROTH, and H.SCHULZ, Conference on "One-Dimensional Conductors" in Saarbrücken 1974, Lecture Notes in Physics, Springer 1974.
15. R.E.PEIERLS, "Quantum Theory of Solids", p. 108 Clarendon Press, Oxford (1955).
16. R.COMES, M.LAMBERT, H.LAUNOIS, and H.R.ZELLER Physical Review B 8, 571 (1973).
17. B.RENKER, H.RIETSCHEL, L.PINTSCHOVIUS, W.GLÄSER, P.BRÜESCH, D.KUSE, and M.J.RICE, Phys.Rev.Lett. 30, 1144 (1973).

ELECTRICAL PROPERTIES OF INTEGRAL OXIDATION STATE METAL COMPLEXES WITH COLUMNAR STRUCTURES

A. E. Underhill

School of Physical and Molecular Sciences
University College of North Wales
Bangor, Caerns, Great Britain

Metal complexes which possess a columnar structure (see Figure 1) and contain an infinite number of directly interacting metal atoms arranged in a linear chain throughout the crystal lattice can be divided into two types.

Type A. Those complexes in which the metal atoms are in an integral oxidation state (e.g. $K_2Pt(CN)_4,3H_2O$).

Type B. Those complexes in which the metal atoms are partially oxidised and therefore in a non-integral oxidation state (e.g. $K_2Pt(CN)_4Br_{0.3},3H_2O$).

This chapter will give an introductory account of the general structural and electrical conduction properties of the Type A compounds. A review dealing mainly with Type A compounds has been published recently.[1]

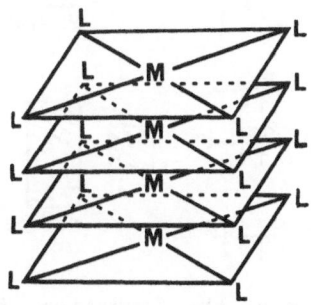

Figure 1. Idealised metal complex with columnar structure

   To form a metal atom chain compound with a columnar structure,
a co-planar monomer complex is required. The most common co-planar
stereochemistry is that of a four-coordinate square co-planar complex
involving a metal ion with a d^8 configuration. The d-block elements
which possess an oxidation state with a d^8 configuration are shown
below.

$$Co^I \quad Ni^{II} \quad Cu^{III}$$
$$Rh^I \quad Pd^{II} \quad Ag^{III}$$
$$Ir^I \quad Pt^{II} \quad Au^{III}$$

A dashed line has been drawn around those ions which form metal atom
chain compounds. It is observed that the tendency to form square
co-planar complexes, and hence the possibility of metal atom chain
compounds, increases down the group Ni, Pd and Pt.

   The relative energies of the orbitals of a metal atom situated
at the centre of a square co-planar array of ligands (D_{4h} symmetry)
is shown in Figure 2. The actual order of the d orbitals will vary
with the ligands involved. In the solid state where the co-planar
units are stacked above one another along the z-axis of the crystal
there will be an interaction between adjacent metal atoms and this
will give rise to spectral bands which are not present in the isolated
molecules in solution. The interaction in the solid state can be
described either in terms of an electrostatic interaction or in

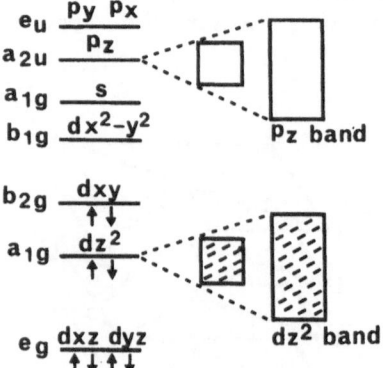

Figure 2.   Representation of the formation of a band structure in
            d^8 metal-atom chain compounds and the effect of
            decreasing inter-metallic distance

terms of metal-metal-orbital overlap. In the latter case the p_z and d_z2 orbitals (or a combination of these orbitals with s orbitals) are considered to overlap with the corresponding orbitals on adjacent metal atoms. Such an overlap will give rise to a d_z2 band which will be completely occupied and to an empty p_z band at higher energies. Calculations have indicated however, that an interaction of this type can, at the most, only be weakly bonding in the ground state.[2]

It is convenient to divide integral oxidation state metal atom chain compounds into four categories depending on the type of molecule or ion comprising the chain.

(a) Neutral molecule chain, e.g. $Rh(CO)_2acac$
    where acac = $CH_3C-CH=C-CH_3$
    $\quad\quad\quad\quad\quad\quad\quad O\quad\quad O$

(b) Anion chains, e.g. $K_2[Pt(CN)_4],3H_2O$

(c) Cation chains, e.g. $[Pt(NH_3)_4]Cl_2$

(d) Anion-Cation chains, e.g. Magnus's green salt $[Pt(NH_3)_4]^{2+}$ $[PtCl_4]^{2-}$.

Although categories (a)-(c) are of the more general type, a considerable number of conduction studies have been performed on examples of category (d). In this chapter attention will be focussed mainly on the more general categories (a) and (b).

There is some evidence that several of these categories of integral oxidation state compounds are capable of partial oxidation to form a non-integral oxidation state metal atom chain compound. However these have only been extensively characterised for the compounds derived from the tetracyanoplatinum (II) and dioxalatoplatinum(II) anions (see Table 1).[3]

Although from the electrical conduction viewpoint the Type A compounds can be considered as a 1-dimensional strand of interacting metal atoms it is worthwhile attempting to relate the differences observed between one compound and another to differences in their molecular structure. Indeed the successful application of this type of material may only come about by 'molecular engineering' to produce the desired electronic properties. In this type of system the conduction properties are going to depend to a greater or lesser extent on the following structural features:

(a) the metal-metal distance in the chain
(b) the metal atom involved in the chain
(c) the ligands attached to the metal
(d) the ions and molecules in the lattice between the chains.

TABLE 1.   Relationship between Integral Oxidation State compounds
           (IOS) and Non-Integral Oxidation State compounds (NIOS)

Type of unit in chain	IOS	NIOS
Neutral molecule	$Ni(dpg)_2$	$Ni(dpg)_2I$ (?)
Anion	$K_2Pt(CN)_4,3H_2O$	$K_2Pt(CN)_4Br_{0.30},3H_2O$
	$MgPt(C_2O_4)_2,7H_2O$	$Mg_{0.82}Pt(C_2O_4)_2,xH_2O$
Anion-Cation	$Pt(NH_3)_4.PtCl_4$	$Pt(NH_3)_4.PtCl_4$
	(dpg = diphenylglyoxime)	– acid adducts

Because the review is concerned with anisotropic systems the
conductivity data discussed will be restricted to that obtained
from single crystal measurements.

ANION CHAIN COMPOUNDS

In many respects this type of system is probably one of
best in which to study the effect of structural modifications on
the conduction properties. Isomorphous compounds containing
different central metal atoms can be obtained and the metal–metal
distance can be varied by changing the cation or the degree of
hydration. In addition, single crystals of this type of complex can
be readily obtained, but dehydration effects may present measurement
problems.

The tetracyano complexes of nickel(II), palladium(II) and
platinum(II) are probably the most investigated series of compounds
possessing a columnar stacked structure. The crystal structure of
$Mg[Pt(CN)_4],7H_2O$ has been reported and shown to consist of a body-
centred tetragonal unit cell with successive $[Pt(CN)_4]^{2-}$ units in
a staggered configuration along the c axis.[4] The Pt–Pt distance was
shown to be 3.155Å.Moreau-Colin has shown the tetracyano complexes
of these metals having the same cation exhibit isotypism and examples
of this are shown in Table 2.[5] It should be noted that in these
compounds the metal–metal distance = C/2 and therefore the Pt–Pt
distance is significantly less than the Ni–Ni or Pd–Pd distance.
This suggests considerably greater metal–metal interaction in the
platinum complexes.

TABLE 2. Isotypism in the tetracyanides of Ni(II), Pd(II) and Pt(II).[5]

		a(Å)	b(Å)	c(Å)
BaNi(CN)$_4$,4H$_2$O	monoclinic	12.07	13.61	6.72
BaPd(CN)$_4$,4H$_2$O	monoclinic	11.98	13.83	6.73
BaPt(CN)$_4$,4H$_2$O	monoclinic	11.89	14.08	6.54
CaNi(CN)$_4$,5H$_2$O	orthorhombic	17.13	18.76	6.77
CaPd(CN)$_4$ 5H$_2$O	orthorhombic	17.33	19.29	6.84
CaPt(CN)$_4$,5H$_2$O	orthorhombic	17.36	19.30	6.72

TABLE 3. Tetracyanoplatinum(II) Compounds.[1]

Compound	Pt-Pt distance (Å)	Colour
Sr[Pt(CN)$_4$],3H$_2$O	3.09	violet
Mg[Pt(CN)$_4$],7H$_2$O	3.15	red
Ba[Pt(CN)$_4$],4H$_2$O	3.32	yellow-green
Mg[Pt(CN)$_4$],4.5H$_2$O	3.36	yellow
Rb[Pt(CN)$_4$],1.5H$_2$O	3.39	green
Sr[Pt(CN)$_4$],5H$_2$O	3.60	colourless

In these columnar stacked structures the metal-metal distance varies with both the cation and the degree of hydration as illustrated in Table 3. This table also indicates the very striking colour changes observed on changing the cation and the origin of those colours has interested chemists for a long time. The spectrum of the free [Pt(CN)$_4$]$^{2-}$ ion in aqueous solution contains no bands below 3000Å. In the solid state however, new bands appear at lower energies and these bands are strongly polarised. It has shown that the position of the lowest energy band polarised parallel to the metal atom chain varies linearly with the Pt-Pt separation and a similar relationship has been observed for the bands in the fluorescence spectrum.[5] In the tetracyanopalladium(II) complexes a similar relationship is observed but for a given metal-metal separation the bands occur at higher frequencies for the palladium complex. This indicates that for a given intermetallic distance there is greater interaction between the platinum atoms, presumably due to the larger size of the 5d orbitals on platinum compared with the 4d orbitals on palladium. The origin of these new low energy bands in the solid state was originally attributed to transitions between the filled d_z2 band and the empty p_z band (see Figure 2). More recently these efforts have been explained on the basis of a Davydov coupling of intramolecular transitions without the need to invoke electron exchange in the ground state.[6]

An investigation of the electrical conduction properties of this type of complex has been started but only preliminary results are available.[7] The room temperature dc conductivities for a series of tetracyanoplatinum(II) complexes are shown in Table 4. It can be seen that the conductivity is strongly dependent on the metal-metal distance. Conductivity results obtained at microwave frequencies suggest that the conductivity along the metal atom chain direction is about 100 x greater than that perpendicular to the chain direction. Further work is in progress on these systems including an investigation of the rôle of Pt(IV) impurities on the conductivity in this type of system. This has been shown to be very important in $Pt(NH_3)_4 \cdot PtC\ell$.

TABLE 4. Room temperature dc conductivity in metal atom chain direction.[7]

Compound	$\sigma(\Omega^{-1}cm^{-1})$	Pt-Pt distance (Å)
$Li_2Pt(CN)_4, gH_2O$	$4x10^{-4} - 3x10^{-3}$	3.18
$(NH_4)_2Pt(CN)_4, nH_2O$	$4x10^{-7} - 1x10^{-6}$	3.25
$BaPt(CN)_4 4H_2O$	$8x10^{-7} - 5x10^{-6}$	3.27, 3.32
$K_2Pt(CN)_4, 3H_2O$	$5x10^{-8} - 1x10^{-7}$	3.50
$SrPt(CN)_4, 5H_2O$	$6x10^{-9} - 4x10^{-8}$	3.57
$(NEt_4)_2Pt(CN)_4, nH_2O$	$1x10^{-8} - 6x10^{-8}$	-

NEUTRAL MOLECULE CHAIN COMPOUNDS

A wide range of metal atom chain compounds containing neutral molecules is known and there is currently a great deal of work in progress on this type of system.

Collman and his co-workers made a detailed study of the conduction properties of $M(CO)_2$acac complexes (where M = Rh or Ir) in the 1960's.[8] The structures of both compounds contain a columnar stack of $M(CO)_2$acac molecules with alternate molecules staggered by 180°.[8] The metal-metal distance of the complex of the third row d-block element iridium is again less than that of the corresponding second row element, rhodium (see Table 5). Table 5 also shows that the iridium complex has a room temperature dc conductivity 10^6 times greater than that of the rhodium complex and that the activation energy for conduction in the iridium complex is much less than that of the rhodium compound. Mixed crystals of the two complexes were also investigated and found to have conductivities in between those of the two parent compounds, but with an activation energy equal to that of the iridium compound. Thermoelectric effects suggest that negative electrons are the predominant charge carriers. Flash mobility experiments indicate a carrier mobility of $\sim 1 cm^2 v^{-1} s^{-1}$ suggesting a hopping mechanism for conduction but this does not agree with the observation of identical activation energies both parallel and perpendicular to the metal atom chain direction. The compounds were also shown to be photoconductors but with a low photon yield. The effect of pressure on the conductivity of $Ir(CO)_2$ acac has also been studied and it was found that the conductivity of a compressed polycrystalline sample could be increased to $10^{-1} \Omega^{-1} cm^{-1}$.[9]

Some of the best known examples of complexes containing infinite chains of interacting metal atoms occur in crystals of planar complexes which contain two anionic planar organic ligands. Bisdimethylglyoximatonickel(II) (See Figure 3) is an example of this

TABLE 5. Dicarbonylacetylacetonato complexes of Rh(I) and Ir(I).[8]

Compound	M-M dist ($\overset{\circ}{A}$)	σ_{11}[+] ($\Omega^{-1} cm^{-1}$)	$\sigma_{11} : \sigma_{\perp}$	ΔE (ev)
$Rh(CO)_2$acac	3.26	10^{-11}	>100	0.32-0.44
$Ir(CO)_2$acac	3.21	10^{-5}	>500	0.25-0.27

[+]σ_{11} = specific conductivity in the metal atom chain direction at room temperature

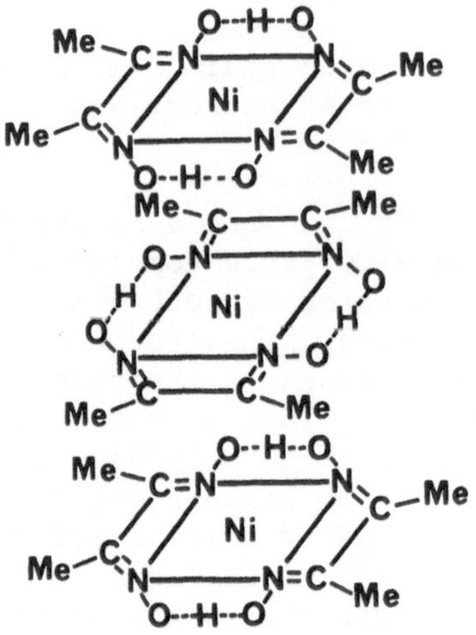

Figure 3.   Structure of bisdimethylglyoximatonickel(II)

type of compound with a Ni-Ni distance of 3.245Å. The monomer units
are stacked above one another with successive molecules staggered
by 90°. This enables the methyl groups, which are the thickest part
of the molecule, to interlock and hence allows the closest possible
approach of the nickel atoms.[10] Many complexes of this type have
been studied and for those compounds which have short metal-metal
distances solid state spectroscopic effects have been observed which
are similar to those already described for the tetracyano complexes.[1,6]
Some electrical conduction studies have been made on this type of
complex but they all appear to be very poorly conducting semiconduc-
tors along the metal atom chain direction   with $\sigma < 10^9 \ \Omega^{-1} cm^{-1}$.[11]

ANION-CATION CHAIN COMPOUNDS

     The electrical conduction properties of Magnus's green salt,
$[Pt(NH_3)_4]^{2+}[PtCl_4]^{2-}$ have been extensively studied by several
workers (see Table 6).[8,9,11,12] Of particular interest is the fact
that replacement of one of the platinum complex ions by the analagous
palladium complex reduces the conductivity in the metal atom chain
direction by at least a factor of 10^3.[11] By studying the effect of
pressure the conductivity has been shown to be very sensitive to the
Pt-Pt separation.[9] Mehran and Scott have shown the presence of
Pt(IV) impurities in concentration of >200 ppm in all the samples
of $[Pt(NH_3)_4]^{2+}[PtCl_4]^{2-}$ they examined and found that the highest
conductivity was associated with the crystals with the highest
Pt(IV) content.[13]

TABLE 6. Anion-cation compounds.[8,9,11,12]

Compound	σ_{11}(RT) $(\Omega^{-1}cm^{-1})$	ΔE (ev)
$[Pt(NH_3)_4]^{2+}[PtCl_4]^{2-}$	$10^{-5}- 10^{-2}$	0.2-0.4
$[Pt(NH_3)_4]^{2+}[PdCl_4]^{2-}$	10^{-9}	-
$[Pd(NH_3)_4]^{2+}[PtCl_4]^{2-}$	10^{-8}	-

GENERAL CONCLUSIONS

It can be seen from the foregoing discussion that there are a wide range of metal atom chain compounds which contain the metal atom in an integral oxidation state. In addition it has been shown that by varying the metal atoms, ligands, and in some cases, the other molecules and ions present in the lattice, large differences can be found in the extent of the metal-metal interactions.

Although the electrical conduction properties of this type of compound have not received the same detailed attention as $K_2Pt(CN)_4Br_{0.3}.3H_2O$ nevertheless a general relationship between the electrical conduction properties and molecular structure has emerged.

(a) All the integral oxidation state metal atom chain compounds examined so far exhibit semiconducting and not metallic properties· over the temperature range studied with activation energies in the range 0.1-1.0 ev.

(b) The conductivity in the metal atom chain direction is very dependent on the metal-metal separation as evidenced from the work on the $[Pt(CN)_4]^{2-}$ complexes and the pressure studies on $Ir(CO)_2acac$ and Magnus's green salt.

(c) The conductivity falls by many orders of magnitude when a third row d-block element is replaced with the corresponding second row element for the same intermetallic distance. This is shown by the results on $M(CO)_2acac$ and the anion-cation chain compounds.

(d) Anisotropy of conduction of >500 have been reported for this type of system but more accurate determinations using methods such as those of Montgomery are needed to confirm these results.[14]

(e)   The results obtained for M(CO)$_2$ acac suggest that the
      activation energy for conduction is lower for third row
      d-block elements than for second row elements but more
      work is needed to confirm this as a general trend.

(f)   The presence of small quantities of metal atoms in higher
      oxidation states may be of great importance in determining
      the magnitude of the conductivity.

Much more work is required on these systems before their
conduction properties can be fully understood in terms of a detailed
mechanism.   It is clear however that by a suitable selection of the
central metal atom and ligand systems it is possible to make anisotropic semiconductors which will possess a dc conductivity in most
ranges from 10^{-12} to $10^{-2}\Omega^{-1}cm^{-1}$.

REFERENCES

1.   T. W. Thomas and A. E. Underhill, Chem. Soc. Rev., 1972, **1**, 99
     and references therein.

2.   L. L. Ingraham, Acta Chem. Scand., 1966, **20**, 283;
     L. V. Interrante and R. P. Messmer, Inorg. Chem., 1971, **10**, 1174.

3.   K. Krogmann, Angew. Chem. Internat. Edn., 1969, **8**, 35.

4.   R. M. Bozorth and L. Pauling, Phys. Rev., 1932, **39**, 537.

5.   M. L. Moreau-Colin, Structure Bonding 1972, **10**, 167 and
     references therein.

6.   P. Day, Inorg. Chim. Acta Rev., 1969, **3**, 81.

7.   J. H. O'Neil and A. E. Underhill, unpublished results.

8.   J. P. Collman, L. Slirkin, L. F. Ballard, L. K. Monteith,
     C. G. Pitt in "International Symposium on Decomposition of
     Organometallic Compounds to Refractory Ceramics, Metals and
     Metal Alloys", K. S. Mazdiyasni, Ed. University of Dayton
     Research Institute, Dayton, Ohio 1968, p. 269; L. K. Monteith,
     L. F. Ballard and C. G. Pitt, Solid State Commun., 1968, **6**, 301.

9.   L. V. Interrante and F. P. Bundy, Inorg. Chem., 1971, **10**, 1169.

10. L. E. Godycki and R. E. Rundle, Acta. Cryst. 1953, **6**, 487
     D. E. Williams, G. Wohlauer and R. E. Rundle, J. Amer. Chem.
     Soc., 1959, **81**, 755.

11. P. S. Gomm, T. W. Thomas and A. E. Underhill, J. Chem. Soc. (A), 1971, 2154.

12. L. V. Interrante, J. Chem. Soc., Chem. Commun. 1972, 6, 302.

13. F. Mehran and B. A. Scott, Phys. Rev. Lett., 1973, 31, 99.

14. H. C. Montgomery, J. Appl. Phys., 1971, 42, 2971.

(Acknowledgement is made to the Chemical Society for permission to reproduce Figures 1, 2 and 3.)

ELECTRON TRANSPORT IN SOME ONE-DIMENSIONAL METAL

COMPLEX SEMICONDUCTORS

L. V. Interrante

General Electric Corporate Research and Development

P. O. Box 8, Schenectady, New York 12301

INTRODUCTION

The subject of one dimensional electronic interactions in transition metal complexes has been treated in several other papers in this volume, with particular emphasis on systems which exhibit "metallic" properties. The main purpose of this paper is to develop the argument that transition metal complex semiconductors are also of significant scientific and technological interest and to encourage further work on such materials.

A general survey of the integral oxidation state transition metal complexes with pseudo-one dimensional electrical properties has been provided in the preceding paper by Underhill. We will discuss here two specific systems of this type in more detail in order to illustrate the potential capabilities of this class of materials.

The systems to be discussed are representatives of two structural types of one dimensional transition metal complex solids, the "metal-chain complexes" (Figure 1a) in which square-planar complex units assume a columnar stacking arrangement, and the "ligand-bridged" chain systems where a chain structure is generated by the sharing of ligand atoms on adjacent complex units (Figure 1b). The specific examples chosen, Magnus' green salt (MGS), $Pt(NH_3)_4PtCl_4$, and the $M(A)_nX_3$ (M=Pd, Pt; A=NH_3, ethylenediamine, etc.; X=Cl, Br, I), complexes reflect, in part, the directions of our past[1,2] and current research; however, these examples also serve to illustrate rather different types of electron transport behavior in transition metal complex systems and are, at the present time, the most studied and best characterized

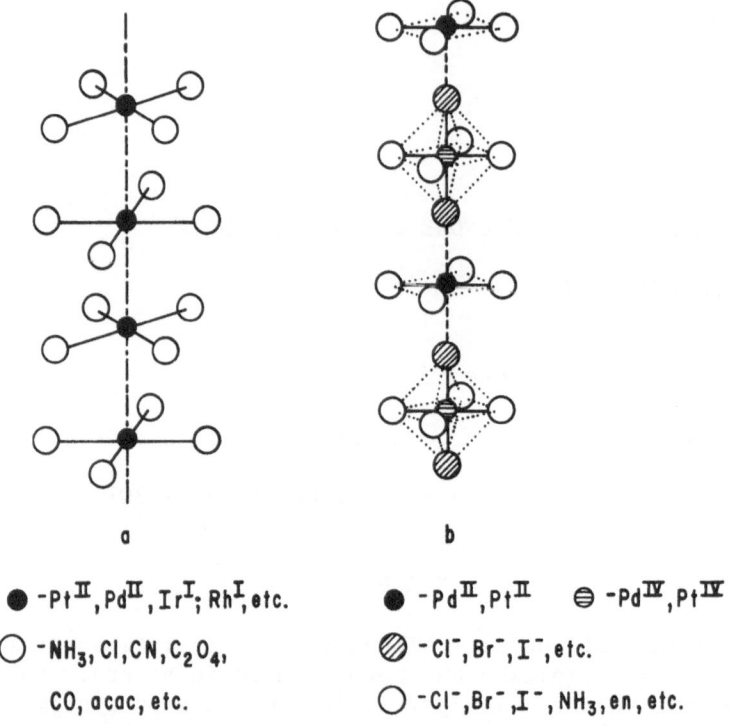

a   b

● $-Pt^{II},Pd^{II},Ir^{I},Rh^{I}$,etc.   ● $-Pd^{II},Pt^{II}$ ⊖ $-Pd^{IV},Pt^{IV}$

○ $-NH_3,Cl,CN,C_2O_4$,   ▨ $-Cl^-,Br^-,I^-$,etc.

      CO,acac,etc.   ○ $-Cl^-,Br^-,I^-,NH_3$,en,etc.

Figure 1 – Illustration of the chain structure in two types of one dimensional transition metal complex solids: a. the "metal-chain" complexes and, b. the halide-bridged, $M(A)_nX_3$ compounds.

examples of one dimensional, metal complex semiconductors.

MAGNUS' GREEN SALT (MGS)

The one dimensional Pt chains in MGS result from the alternate stacking of $Pt(NH_3)_4^{2+}$ and $PtCl_4^{2-}$ units in a tetragonal unit cell.[3]   The metal-metal separation within the chains is one-half of the $\underline{c}$ repeat distance or 3.25Å.   This compound has been of interest for many years for its unusual solid state absorption spectrum and, in particular, its green color which was quite unexpected since the component $PtCl_4^{2-}$ and $Pt(NH_3)_4^{2+}$ ions in aqueous solution and in salts such as K_2PtCl_4 and $Pt(NH_3)_4Cl_2$ are, respectively, pink and colorless.   Subsequent spectral studies of the solid have identified the source of the green color as a "window" in the absorption spectrum at ~ 20,000 cm^{-1} arising from a red shift and intensification of largely intramolecular "d→d" transitions on the $PtCl_4^{2-}$ ion.[4]

These spectral shifts have been explained without resort to the concept of extended band states using the theory of Frenkel excitons in molecular crystals;[4] however, recent work on the UV reflectance spectra of mixed Pd – Pt Magnus' salts[5] suggests that a purely "molecular" description of this type does not provide a complete explanation for the spectral properties of MGS.

In an early study of the MGS absorption spectrum by Miller,[6] attention was focused on an apparent strong, broad absorption band centered at 6000 cm^{-1} in the near infrared which could not be readily assigned to an intramolecular transition. It was suggested that this could be intermolecular in character and a qualitative energy band description was proposed to account for its occurence (Figure 2). This band model considered the extended interaction of the "d_{z^2}" and "p_z"-like orbitals on the Pt II atoms in the one dimensional chain, producing filled "d_{z^2}" and empty "p_z" bands in the solid. The near infrared absorption was attributed to transitions connecting the uppermost d_{z^2} and the lowest p_z bands, which placed the "band gap" at ~ 0.6 eV.

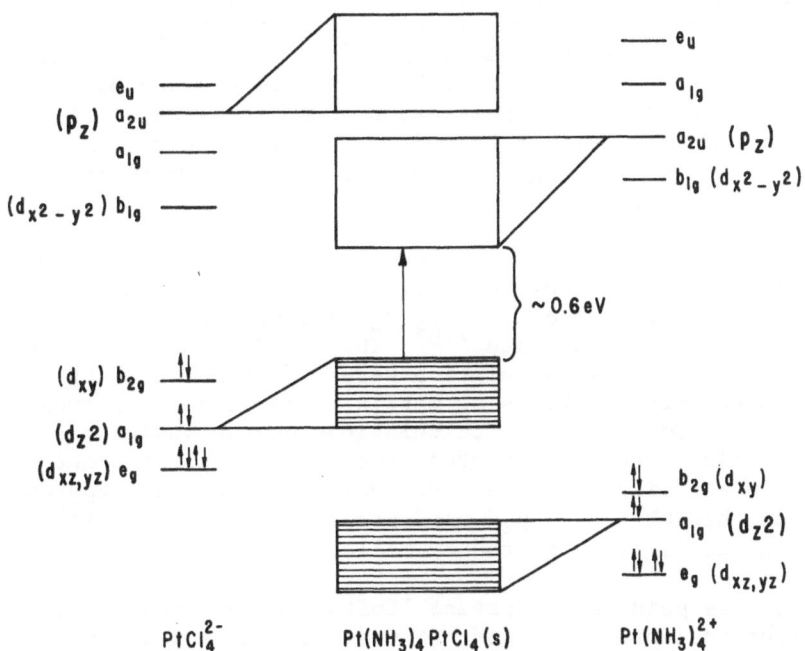

Figure 2 – Qualitative band model for MGS proposed by Miller (Reference 6).

The possibility of semiconduction and photoconduction in this system was anticipated by Miller and later confirmed by Collman and co-workers[7] who reported the first electrical property studies on single crystals of MGS. The close correspondence in the energy for the onset of photoconductivity, twice the thermal activation energy for conductivity, and the "near infrared absorption band" reported by Miller, provided support for the energy band model in Figure 2 and the identification of MGS as an intrinsic semiconductor. Later conductivity measurements on MGS[8] confirmed the the observations of Collman, et al and indicated a conductivity anisotropy of 100/1 for this material, with the highest conductivity along the metal chain direction in the solid.

Subsequent spectral studies on MGS[9] and more quantitative band structure calculations,[1b] on the other hand, began to raise some question about the relationship of these electrical properties to the simple band picture of Miller. In particular, qualitative observations on single crystals in the vicinity of 6000 cm^{-1} indicated a much weaker absorption band than that reported by Miller, and a possible vibrational origin for this band was suggested.[9] This suggestion was later confirmed by detailed spectral studies in the near infrared on deuterated and undeuterated crystals.[1c] All of the weak bands within 12,000 and 4,000 cm^{-1} were identified as overtones and combinations of the N – H stretching modes of the $Pt(NH_3)_4{}^{2+}$ unit and an upper limit of ~ 0.2 liter/mole·cm was set for the maximum intensity of any electronic absorption bands in this region. The previous observation of an apparent strong, broad absorption band at ~ 6000 cm^{-1} was attributed to experimental difficulties arising from the use of polycrystalline samples in the earlier studies.

The first semiquantitative band structure calculations on MGS were carried out in 1970 using a tight binding approach, considering only the interactions of the $d_z{}^2$ and p_z orbitals on adjacent Pt atoms in the chain.[1b] The off-diagonal and diagonal matrix elements in this calculation were approximated, within the extended Hückel framework, using matrix elements derived from semiempirical extended Hückel calculations on the individual complex units. The results suggested that bands of significant width are indeed likely at the internuclear separations appropriate to MGS, but that the "$d_z - p_z$" band gap is probably more like 4.5eV rather than the <1 eV suggested by Miller.

Up to this point an important limitation to detailed studies of the electrical properties of MGS was the availability of suitable single crystal samples. The virtual insolubility of this compound in common solvents and its infusibility precluded the application of most of the usual crystal growing methods. Single crystals could be grown by slow diffusion in aqueous solution using glass frits as diffusion barriers, but the crystals prepared in this manner were generally of poor quality and of small cross section. The development of a silica gel method for the preparation of MGS crystals allowed for the first time the preparation of high quality single crystals with large cross section (up to 0.7 mm).[1d]

Several samples of these crystals were examined by X-ray precession techniques to verify their single crystal character and to establish the relationship between the gross crystal morphology and the crystallographic directions. After screeing for visible defects under a microscope with polarized light, these crystals were mounted for electrical measurements using fine gold wire leads and silver paint contacts. Initial three-probe experiments using guard electrodes indicated that surface conduction was not an important factor and that the measured conductivity was indeed a bulk property. D. C. conductivity measurements were then carried out on samples from different crystal growth preparations using the usual four-probe, voltage-current method with the current leads and the voltage probes along one direction of the crystal. In addition several crystals were wired in the four-electrode configuration described by Montgomery[10,11] to measure directly the conductivity anisotropy.

A typical log σ vs 1/T plot obtained in the four-probe studies is illustrated in Figure 3. The linearity of this plot over the wide temperature range studied (120 - 350 K) suggests that we are dealing with one thermally activated conduction process in this temperature regime. Above ~ 80°C irreversible changes in resistance were observed, probably arising from thermal degradation of the sample or the contacts.

Measurement of the conductivity anisotropy were carried out both by four-probe measurements on individual samples and by the Montgomery method[10,11] and gave values in the range 18→25/1 for the ratio of the c (metal chain direction) and a axis conductivities. This is somewhat less than the anisotropy previously reported by Gomm et al[8] but is still basically consistent with the suggestion that the conductivity in MGS is largely a one dimensional process.

A detailed study of the conductivity as a function of frequency has not been carried out as yet; however, preliminary two electrode A. C. measurements at room temperature on a single crystal sample indicated essentially no change from the D. C. conductivity value with A. C. frequencies up to at least 1000 hertz.[11]

In addition to these conductivity studies, measurements of the thermopower were carried out on several crystals as a function of temperature[11] using an electrode arrangement similar to that described by Chaikin.[12] The results (Figure 4) indicate p-type

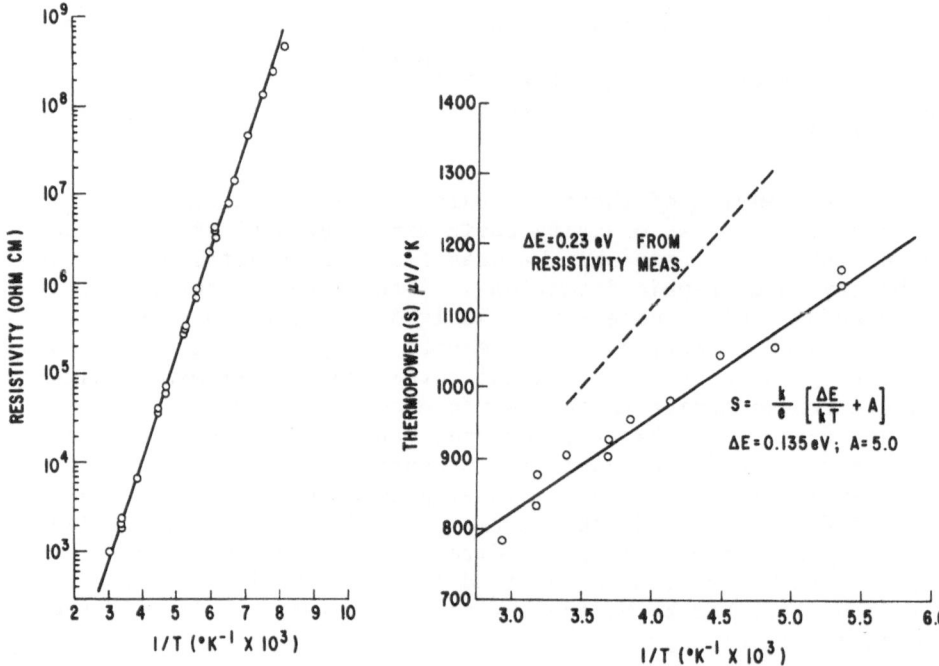

Figure 3 – Representative log σ vs 1/T plot for a MGS crystal wired for four-probe D. C. measurements along the c axis.

Figure 4 – The thermopower of MGS as a function of 1/T. The slope of the log σ vs 1/T curve for this same crystal is included for purposes of comparison.

conductivity behavior with thermopower values in the range 800 to 1200 μV/°K. The large magnitude of the thermopower and its essentially linear dependence on 1/T is consistent with previous observations on semiconductors and with the expression, S = k/e[ΔE/kT + A], derived by Fritzsche for a semiconductor in which only one band is involved in the conduction process.[13a] As is illustrated in Figure 4 the "ΔE" obtained from this S vs 1/T plot is substantially lower than the conductivity activation energy measured for the same crystal. This could occur if the mobility in MGS was exponentially dependent on temperature, as in a hopping-type conductivity situation;[13] however, in view of the variety of factors which can contribute to the thermopower in a semiconductor[13] this can hardly be considered as good evidence for such a conduction mechanism.

The results of the four-probe D. C. conductivity measurements on 19 different crystals of MGS obtained from 4 crystal growth preparations are summarized in Table I. These data reveal several

TABLE I.   D. C. Conductivity Data for MGS Crystals[a]

Preparation	Number of Samples	c-Axis Conductivity at 25°C $(Ohm^{-1}cm^{-1})$	Thermal Activation Energy (eV)
A	6	$4.7(\pm2.4) \times 10^{-5}$	$0.352(\pm.033)$
B	3	$8.5(\pm4.2) \times 10^{-4}$	$0.257(\pm.025)$
C	4	$1.3(\pm0.3) \times 10^{-3}$	$0.255(\pm.020)$
D	6	$7.5(\pm1.6) \times 10^{-3}$	$0.212(\pm.012)$

[a]Average values with average deviations in parentheses.

interesting features regarding the conduction process in MGS.   In
particular, whereas the room temperature conductivies of crystals
obtained from the same preparation are in reasonably good agreement, the variation from one preparation to the next is as much as
three orders of magnitude. A marked variation in the thermal
activation energy for conductivity is also apparent between preparations with, in general, a lower thermal activation energy for
the samples of higher conductivity. Careful examination of these
crystals under a microscope and by X-ray precession methods
revealed no apparent structural or morphological differences among
the samples from different preparations. Microanalytical determinations indicated small amounts (ppm level) of other transition
elements (primarily Fe, Cu, and Pd) but not obvious variations in
composition from one preparation to the next.

By means of additional experiments carried out using the same
$Pt(NH_3)_4^{2+}$ and $PtCl_4^{2-}$ solutions, it was shown that certain solutions consistently gave crystals of higher conductivity, suggesting
that differences in the composition of these solutions were responsible for the observed conductivity variations. On the basis of
these findings and the previously mentioned arguments against the
intrinsic band model, it was suggested that the conductivity in MGS
was impurity dominated.[1c,d]

The first definitive information regarding the nature of these
impurities was provided by EPR measurements on MGS carried out on
both single crystal and powder samples prepared under different
experimental conditions.[14] An axially symmetric resonance was
observed in the single crystal samples which was identified,on the
basis of the g-values and the observed hyperfine pattern, as
arising from an electron in a d_z^2-like state extending over several
Pt atoms. Spectrophotometric analysis of the solutions used to

prepare MGS and doping experiments suggested further that the source of these unpaired electrons were Pt^{IV} complexes present in the solutions. These Pt^{IV} complexes apparently induce a partial oxidation of the Pt^{II} ions in the solid which is compensated by the addition of negative ions (presumably halides), either at interstitial sites or as a replacement for the neutral NH_3 groups. The existence of Pt^{IV} species in the nominal Pt^{II} complexes used to prepare MGS was found to be a general occurrence, presumably due to air oxidation or to incomplete reduction of the Pt^{IV} starting material in the preparation of the Pt^{II} complexes.

EPR studies on some of the same crystals employed in the previously mentioned conductivity studies revealed a general correspondence between the intensity of the EPR resonance and the conductivity of the crystals, thus establishing a relationship between the impurity induced conductivity effects and the presence of "Pt ^{III} -like" states in the solid.

Using the available information regarding the nature of the "impurity" states in MGS and the results of the semiquantitative band structure calculations, a modified "band model" for the MGS electronic structure can be constructed which appears to account for much of the available data. This model (Figure 5) is similar to the intrinsic band model proposed by Miller, in that "d_z^2" and "p_z" bands of significant width are postulated; however, here the $d_z - p_z$ band gap is on the order of 4.5 eV and we have introduced acceptor levels, corresponding to the Pt ^{III} -like states localized in the vicinity of the charge compensating defects, ~ 0.6 eV from the top of the the d_z^2 band.

Intrinsic MGS in this context would be a good insulator and the observed conductivity is attributed entirely to the presence of the "impurity" states. Thermal excitation of electrons from the d_z^2 band into these acceptor levels should lead to hole conduction in the d_z^2 bands, consistent with the observed "p-type" conductivity behavior. Also the essentially frequency independent conductivity and the weak photoconductivity observed in the near infrared is understandable on this basis, as is the lack of any strong absorption bands in this region, considering the relatively low concentration of impurity states present (200 ppm "Pt ^{III}" sites estimated by EPR[14]). It is also possible to account for the apparently smaller thermal activation energies observed for the crystals of higher conductivity by considering the likelihood of impurity banding in these presumably more highly doped samples.

Although this extrinsic band model can be used to account for the available information concerning the electrical properties of MGS, especially in view of the thermopower data obtained, the alternative possibility that conventional band theory is not appropriate to MGS and that a more localized description of the

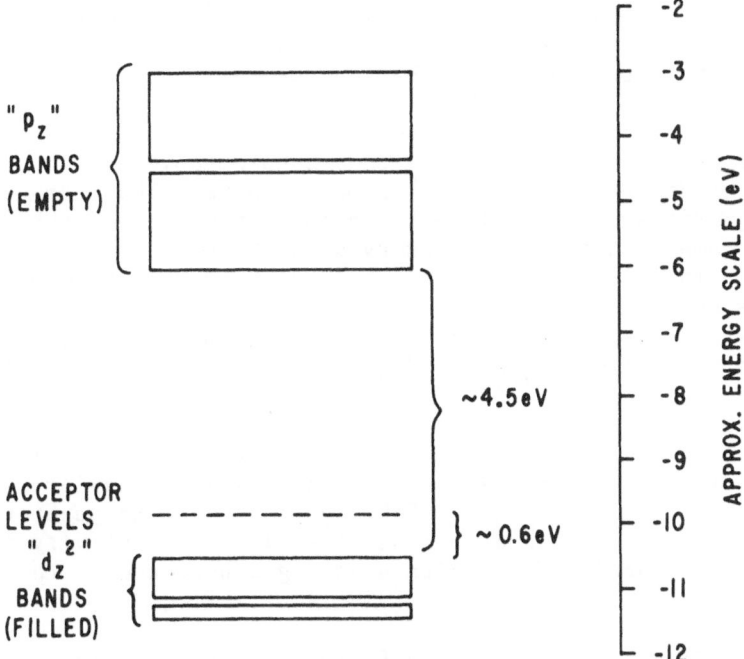

Figure 5 - Extrinsic band model description for the electronic
structure of MGS. The acceptor levels correspond to localized
"Pt^{III}-like" states, each containing one unpaired electron.

electronic structure is required clearly cannot be discounted at
this point. A definite answer to this question must await further
studies on this system, including a direct determination of the
charge carrier mobility.

THE $M(A)_n X_3$ (M = Pd, Pt; A = NH$_3$, en; X = Cl, Br, I) COMPLEXES

The linear chains in the $M(A)_n X_3$ complexes arise from the
alternate arrangement of planar and octahedral complex units in
the solid and consist of metal atoms bridged by halide ions.[15]
Initially, these complexes were thought to contain Pd and Pt in
the unusual +3 formal oxidation state;[16] however, later studies
showed that a "mixed-valence" description, involving the more
typical +2 and +4 oxidation states was more appropriate.[17]
These valences are locked into alternate metal sites along the
linear chain by the position of the "bridging" halide ions which
are located at approximately one-third of the distance between
M^{IV} and M^{II} sites.[15] Aside from the asymmetry in the position

of these halide ions the coordination environment about the metal
atom in these two sites is essentially identical, involving
typically two nitrogen and two halide donors.

Consistent with this "mixed-valence" description the properties of these complexes reflect, to a considerable extent, the
presence of discrete M^{II} and M^{IV} complex units. Thus, the solid
is diamagnetic and spectral features characteristic of the Pt^{II}
and Pt^{IV} complex species are observed in the solid state absorption
spectrum. In addition, a strong visible absorption band polarized
along the chain axis direction is typically observed in these
spectra which has been attributed to an intermolecular transition
involving the filled and empty "d_z^2"-like orbitals on the M^{II} and
M^{IV} ions, respectively. [17]

Previous electrical conductivity measurements on these complexes indicated highly anisotropic conductivity behavior with
the highest conductivity ($\lesssim 10^{-8} ohm^{-1} cm^{-1}$ at room temperature) in
the direction of the linear chains. [18] This conductivity is
thermally activated and exhibits a strong dependence on the nature
of M, A and X.

As is illustrated in Figure 6, the conductivity is enormously
enhanced at high pressures with values at ~ 100 Kbar pressure as
high as 2.0 $ohm^{-1} cm^{-1}$. Substantial changes in the optical absorption spectra are also observed, consistent with shifts in the
intermolecular $M^{II} d_z^2 \to M^{IV} d_z^2$ charge transfer transition to
lower energies. [2a]

These observations can be explained with the aid of the
schematic energy level diagram in Figure 7, which shows the anticipated changes in the M d_z^2 and halide p_z energy levels as the
M^{II}... halide distance decreases under pressure. In particular, a
broadening of the energy bands due to the increased $M^{II} d_z^2$ -
halide p_z orbital overlap, along with a general merging of the two
"d_z^2" bands as the environment about the metal ions becomes more
nearly equivalent, is expected. At the limit of equal metal -
halide distances, this should ultimately lead to a transition to a
system with half-filled, one dimensional metallic band.

The results of X-ray diffraction measurements up to ~ 60 Kbar
pressure and measurements of the thermal activation energy for
conductivity with pressure [2a] suggest that this "transition" does
not occur within the pressure limits of the experiment, although,
as is evidenced by the spectral and conductivity measurements, a
substantial movement along this reaction coordinate apparently is
effected.

In addition to demonstrating the extraordinary sensitivity
of the solid state properties of these complexes to the interatomic

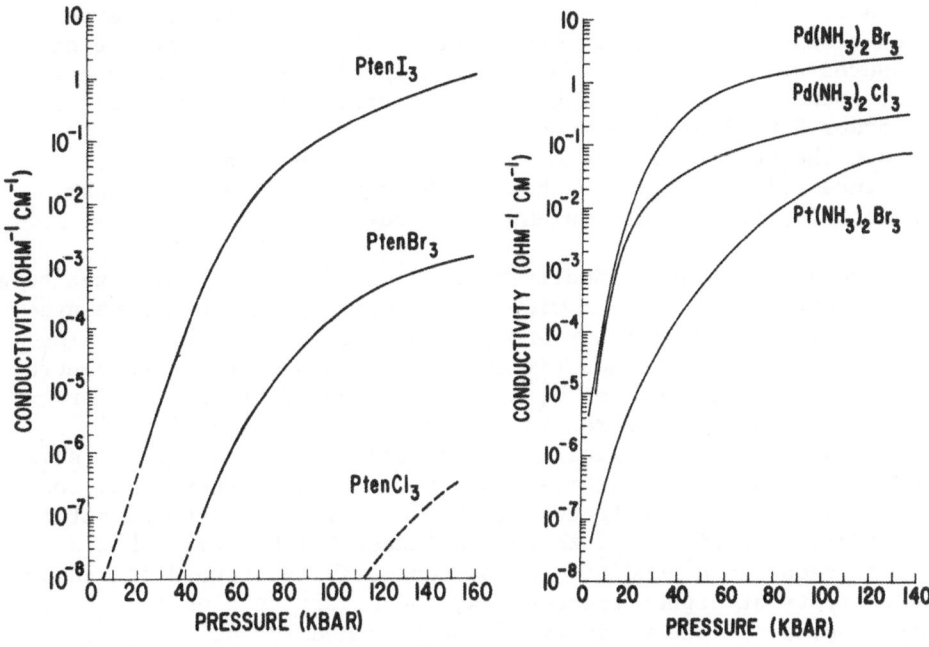

Figure 6 - The conductivity of the $M(A)_nX_3$ complexes as a function of pressure (reproduced with permission from Reference 2a).

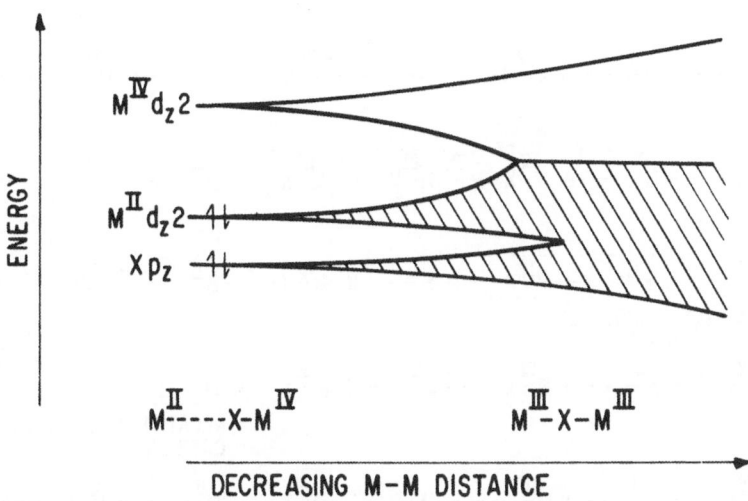

Figure 7 - Anticipated changes in the $M(A)_nX_3$ electronic structure as the $M^{II} - M^{IV}$ distance decreases under pressure (reproduced with permission from Reference 2a).

separations within the linear chains, these high pressure experiments also reveal a strong dependance of the conductivity at high pressures on M, A and X. The general trends indicated by these measurements are toward higher conductivities at any pressure in the order X = Cl<Br<I; A = en<NH_3 and M = Pt<Pd. These results support the tentative conclusions regarding the conductivity-structure relationships reached on the basis of the preliminary conductivity studies at ambient pressures.[18]

Recently, these conductivity studies of $M(A)_nX_3$ complexes have been extended to include a wider range of complexes and some preliminary A. C. conductivity and photoconductivity work.[2b] The results of the D. C. conductivity measurements support the general conductivity dependence on M, A and X indicated by the earlier work and the high pressure studies, and show further that the thermal activation energies for conductivity decrease in the same order that the conductivities increase. Low frequency A. C. conductivity and D. C. photoconductivity measurements carried out on single crystals of one of these complexes, $Pd(NH_3)_2Cl_3$, show a strong frequency dependence of the conductivity, toward increased conductivity at higher frequencies, and a very weak photoconductivity response in the vicinity of the presumed $Pd^{II}d_z^2 \rightarrow Pd^{IV}d_z^2$ transition in the visible region. These results do not support the application of a conventional intrinsic band model for the description of the electronic structure of these materials but suggest instead that a more localized description, involving perhaps a hopping motion of the charge carriers might be more appropriate. One such hopping model is illustrated in Figure 8.

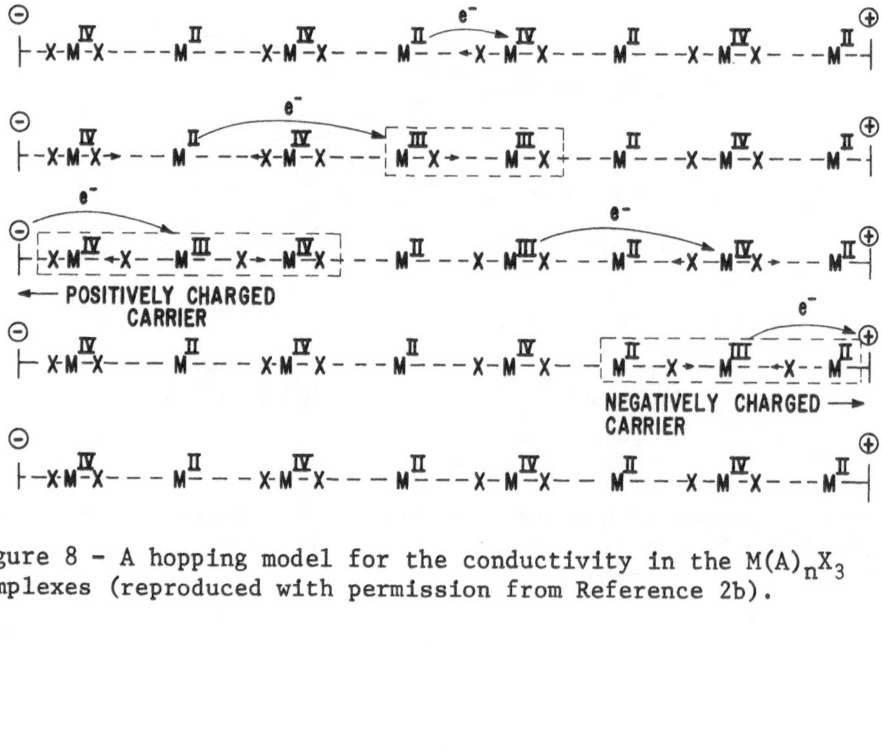

Figure 8 - A hopping model for the conductivity in the $M(A)_nX_3$ complexes (reproduced with permission from Reference 2b).

The essential feature of this model is that carrier motion is a phonon assisted process involving a combined electron and ion transfer along the one dimensional chains. The "carriers" are envisioned as a cluster of three metal ions with a formally tri-valent metal ion at the center. The carrier "moves" by a sequence of electron hops between next nearest neighbor metal ions which is accompanied by the transfer of halide ions between adjacent metals. The halide motion serves to make the metal ion sites more nearly equivalent thus lowering the activation barrier for electron hopping.

In addition to explaining the frequency dependence of the conductivity (which is a general characteristic of low mobility solids) and the very weak photoconductivity response (photoexcita-tion in this model does _not_ produce carriers directly), this mechanism also provides an explanation for the conductivity and activation energy dependence on M and X. In particular, since electronic motion in this model is facilitated by halide ion motion the activation energy for the conduction process should reflect the M – X stretching force constant. Indeed, comparison of the M – X stretching frequencies for the corresponding MX_6^{2-} complexes[19] reveals a general decrease in the stretching frequency with M and X in the same order as the observed decrease in the thermal activation energy for conductivity.

In addition to the "intrinsic" process for generating charge carriers which is suggested in this model, as in MGS such carriers could also arise from impurities or a slight off-stoichiometry such as is shown in Figure 9. In this alternative mechanism for the conduction process, the source of the carriers are M^{III} ions which replace a M^{II} at a chain terminus, with an additional halide ion added for charge compensation.

Once created, the carriers move in exactly the same manner as in the "intrinsic" process and their mobility is subject to the same considerations as in the previous model; the only real difference is that the origin of the carriers are M^{III} ions which are present in the solid in its electronic ground state.

The observation at 4K of an EPR resonance in single crystals of one of these complexes $[Pd(NH_3)_2Cl_3]$[20] with g values essentially identical to those previously attributed to a $Pd^{III}Cl_5^{2-}$ species in a radiation damaged K_2PdCl_4 crystal[21] lends support to the supposition that a mechanism such as that in Figure 9 may be responsible for the conductivity of the $M(A)_nX_3$ complexes.

CONCLUSION

More work on both MGS and the $M(A)_nX_3$ complexes must be done

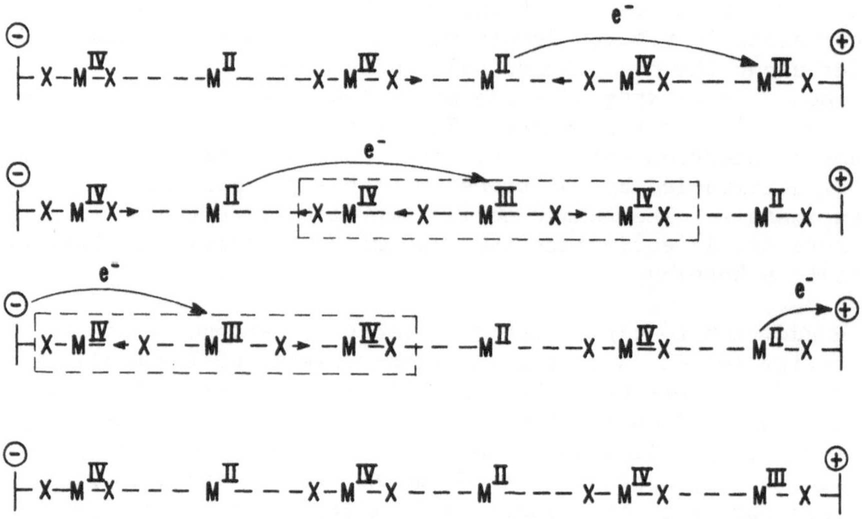

Figure 9 - A possible "impurity" mechanism for the conductivity in the $M(A)_nX_3$ complexes (reproduced with permission from Reference 2b).

in order to determine whether the particular models for electron transport suggested here or various other alternatives will ultimately provide the best description.

However, one thing that can be stated with some confidence at this point is that in transition metal complex semiconductors impurity effects can be equally as important in determing the electrical properties as they are in more conventional types of semiconductor materials, and that in future work on such systems the assumption of intrinsic behavior must be applied with considerable caution. On the positive side one need only recall the enormous practical advantage that has resulted from the characterization and control of impurity effects in covalent semiconductors to recognize the considerable potential here. The large enhancement in the conductivity of MGS (up to 10^{-2} ohm^{-1}cm^{-1} at room temperature) resulting from the introduction of impurities provides a good illustration of this potential.

ACKNOWLEDGMENT

This work was supported, in part, by the Air Force Office of Scientific Research (AFSC), United States Air Force, under contract

F - 44620 - 71 - C - 0129. The contributions of Drs. F. P. Bundy, H. R. Hart, Jr., R. P. Messmer and Mr. K. W. Browall to the work described herein are also gratefully acknowledged.

REFERENCES AND FOOTNOTES

1. a. L. V. Interrante and F. P. Bundy, Inorg. Chem. 10, 1169 (1971);

   b. L. V. Interrante and R. P. Messmer, Inorg. Chem. 10, 1174 (1971);

   c. E. Fishman and L. V. Interrante, Inorg. Chem. 11, 1722 (1972);

   d. L. V. Interrante, J.C.S. Chem. Commun. 302 (1972).

2. a. L. V. Interrante, F. P. Bundy and K. W. Browall, Inorg. Chem. 13, 1158 (1974);

   b. L. V. Interrante and K. W. Browall, Inorg. Chem. 13, 1162 (1974).

3.   M. Atoji, J. W. Richardson, and R. E. Rundle, J. Amer. Chem. Soc. 79, 3017 (1957).

4.   D. S. Martin, Jr., R. M. Rush, R. F. Kroening, P. F. Fanwick, Inorg. Chem. 12, 301 (1973); D. S. Martin, Jr., ACS Symposium Series, "Extended Interactions between Metal Ions in Transition Metal Complexes," L. V. Interrante, ed., American Chemical Society (1974), in press.
      P. Day, Inorg Chim. Acta Rev. 3, 81 (1969); ACS Symposium Series, "Extended Interactions between Metal Ions in Transition Metal Complexes," American Chemical Society, L. V. Interrante, ed., (1974) in press

5.   B. G. Anex, S. I. Foster, A. F. Fucaloro, Chem. Phys. Lett. 18, 126 (1973); B. G. Anex, ACS Symposium Series, "Extended Interactions between Metal Ions in Transition Metal Complexes," American Chemical Society, L. V. Interrante, ed., (1974) in press.

6.   J. R. Miller, J. Chem. Soc. 713 (1965).

7.   J. P. Collman, L. F. Ballard, L. K. Monteith, C. G. Pitt and L. Slifkin, in "International Symposium on Decomposition of Organo-Metallic Compounds in Refractory Ceramics, Metals and Metal Alloys," K. S. Mazdiyasni, ed., Dayton, Ohio (1968) pp. 269-283.

8.   P. S. Gomm, T. W. Thomas, and A. E. Underhill, J. Chem. Soc. (A) 2154 (1971).

9.   P. Day, A. F. Orchard, A. J. Thomson and A. J. P. Williams, J. Chem Phys. 43, 3763 (1965).

10.   H. C. Montgomery, J. Appl. Phys. 42, 2971 (1971).

11.   We are indebted to H. R. Hart, Jr. and W. R. Giard of the GE Corp. Res. and Dev. for carrying out these measurements.

12.   P. M. Chaikin, J. F. Kwak, T. E. Jones, A. F. Garito, and H. J. Heeger, Phys. Rev. Letters 31, 601 (1973).

13. a. H. Fritzsche, Solid State Commun. <u>9</u>, 1813 (1971);
    b. F. Gutmann and L. E. Lyons, "Organic Semiconductors,"
       John Wiley and Sons, Inc., New York (1967), pp. 72-79,
       512-514.
14.   F. Mehran and B. A. Scott, Phys. Rev. Letters <u>31</u>, 99 (1973);
       B. A. Scott, F. Mehran, B. D. Silverman and M. A. Ratner,
       ACS Symposium Series, "Extended Interactions between Metal
       Ions in Transition Metal Complexes," L. V. Interrante, ed.,
       (1974) in press.
15.   J. Wallen, C. Brosset, and N. Vannerberg, Ark. Kemi. <u>36</u>,
       541 (1962); T. D. Ryan and R. E. Rundle, J. Amer. Chem. Soc.
       <u>83</u>, 2814 (1961); B. M. Craven and D. Hall, Acta Crystallogr.
       <u>14</u>, 475 (1961).
16.   H. D. K. Drew and H. J. Tress, J. Chem Soc. 1244 (1935).
17.   M. B. Robin and P. Day, Advan. Inorg. Radiochem. <u>10</u>, 247
       (1967) and references therein.
18.   T. W. Thomas and A. E. Underhill, J. Chem. Soc. (A) 512
       (1971).
19.   K. Nakamoto, "Infrared Spectra of Inorganic and Coordination
       Compounds," John Wiley and Sons, Inc., New York (1963),
       p. 119.
20.   Private communication from Dr. G. Watkins, General Electric
       Corp. Res. and Dev.
21.   T. Krigas and M. T. Rogers, J. Chem. Phys. <u>54</u>, 4769 (1971).

PREPARATIVE ASPECTS OF ONE-DIMENSIONAL TRANSITION METAL COORDINATION COMPOUNDS

H. J. Keller

Anorganisch-Chemisches Institut der Universität
Heidelberg, D-6900 Heidelberg 1,
Im Neuenheimer Feld 270, GFR

ABSTRACT

The results of a chemical and physical investigation of known linear chain metal compounds with direct metal-to-metal contacts can be successfully used as a guideline in the directed synthesis of new linear chain transition metal complexes with strong intermolecular interactions. One-dimensional metallic behaviour of crystallized four-coordinate planar transition metal compounds with 8 d electrons is expected if small and strongly π electron-accepting equatorial ligands like carbon monoxide are used. I.r., ^{1}H-n.m.r., ^{195}Pt-n.m.r., e.s.r., u.v. and ^{193}Ir-Mößbauer spectra show that increasing bulkiness and increasing electron donating properties of the (mostly organic) ligands considerably decrease the strength of the intermolecular metal interactions in all compounds. Planar transition metal complexes with less than 8 d electrons per metal seem to be far better suited for the formation of 1d metals. But because of the strong Lewis acid activity of the individual molecules in solution, the axial positions in the coordination sphere are blocked by donating solvent molecules. The self-association reaction necessary for the building of linear chains is prevented for this reason. Preparative procedures to overcome these problems are proposed. One of the "mixed valence" solids obtained contains a linear I_3^- chain in addition to the linear metal chain.

I.   INTRODUCTION

Much theoretical and experimental work on cooperative states in one-dimensional (1d) or low-dimensional
solids has been published during the last few years and
excellent reviews are available [1-4]. The wide field
of these investigations might be divided into two major
regimes, one of which is that of magnetic exchange phenomena while the other is concerned mainly with metal-
like ("metallic") interactions leading at best to 1d
metals. Low-dimensional "magnetic" solids like for example 1d-antiferromagnets were studied on numerous transition metal compounds containing almost all known transition elements as central metal ions (many papers appear
every year describing new compounds of this type).

The investigations on 1d metallic solids are, on
the other hand, mainly restricted to only two types of
compounds which have been known for a long time. These
are:
   a) the charge transfer salts of the "TCNQ"
      type belonging to the realm of organic
      chemistry, and

   b) the mixed valence platinum complexes as
      typical inorganic coordination compounds.

In fact only a few 1d metallic TCNQ compounds[4] and
only one "metallic" transition metal complex[1] have
been studied thoroughly so far.   There are only a few
other known species like $(SN)_x$ which look promising in
respect to 1d metallic solid state behaviour[5, 6]. The
question therefore arises as to whether distinct guidelines for chemists could be set up which will help in
an effort of synthesizing new compounds with 1d metallic
properties or at least with considerable electron delocalization along the direction of the columnar stacks.
This problem was recently discussed by Heeger and Garito[7] with respect to the highly conducting organic
charge transfer salts of the "TCNQ" type and will be
outlined in the following section using planar four-
coordinate transition metal complexes as building blocks.

All these solids have in common a so-called "columnar structure" which allows considerable intermolecular
interactions[8, 9]. The packing in such a columnar structure can best be envisioned as a closely packed array of
discs or flat cylinders reminiscent of a roll of coins
(fig. 1).

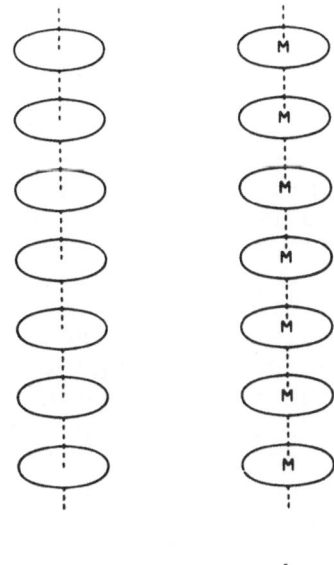

Figure 1 -

Schematic drawing of organic charge transfer salt (a) and planar transition metal complex (b) columnar stacks with molecular planes perpendicular to the stacking direction.

a   b

Most of the known <u>planar</u> transition metal complexes are four-coordinate (there are only a few three-coordinate exceptions) and contain d^8 metal ions. Compared to the numerous planar d^8 compounds only a few types of <u>planar</u> species with d^7 and d^9 ions have been isolated and only very few of them crystallize in columnar stacks (10, 11). The very stable planar, four-coordinate compounds of the ions Rh(I), Ir(I), Ni(II), Pd(II) and Pt(II) which crystallize in columnar stacks have therefore been chosen. The planned syntheses of such molecules should primarily aim at answering the following question: which kind of molecular parameters of the <u>individual</u> elements govern the properties of these sol<u>i</u>ds, and what special rules can be set up to prepare suitable compounds in a systematic way? In order to find such rules for the synthesis of tailor-cut molecules, aside from steric problems, the electronic structure of planar four-coordinate transition metal complexes has to be investigated in more detail.

II. STERIC CONSIDERATIONS

A quite fundamental difference between the intermolecular interactions in inorganic (transition metal) and organic linear chain compounds has to be considered if one tries to synthesize strongly interacting metal chains. The intermolecular contact in the organic charge transfer salts is brought about mainly by unpaired electrons in molecular π-functions which are delocalized to extend more or less all over the molecule, but the closed-shell linear chain metal complexes interact almost exclusively through electrons in d-functions which are centered at the metal ions. This gives an advantage to the metal complexes. In contrast to the rather concentrated π orbitals, made up of $2p_z$ carbon orbitals in the charge transfer salts, the 3d, 4d and 5d electrons of the metal ions are widespread and extend far to the neighbouring molecules. Their electron distribution is shown in figure 2. If one considers 'z' as the stacking direction, electrons in the d_{z^2} orbitals are more preferred for intermolecular interaction. For electrons in these orbitals an efficient overlap is guaranteed even if the metal atoms are not too close together. The same arguments hold if a small $(n+1)p_z$ contribution is taken into account.

But these highly directed and effective interactions through d_{z^2} of p_z electrons have some less agreeable consequences for solid linear chain metal compounds. These electrons are localized around the metal atoms of the chain, so a narrow hose of delocalized electrons runs along the interacting metal ions which does not extend into the ligands which behave like an isolating layer around the conducting spine. Some rather trivial conclusions concerning the bulk solid state properties of these compounds can be drawn from this simple picture.

First of all, the cooperative phenomena in the bulk linear chain metal complexes should depend very strongly on lattice distortions perpendicular to the chain. While in the organic salts a small slipping of molecules would only minimally effect the intermolecular interactions of the delocalized π electrons, a strong effect on the highly directional exchange in the metal chains is to be expected in this case. The bulk properties are therefore much more sensitive to lattice distortions in the transition metal chains.

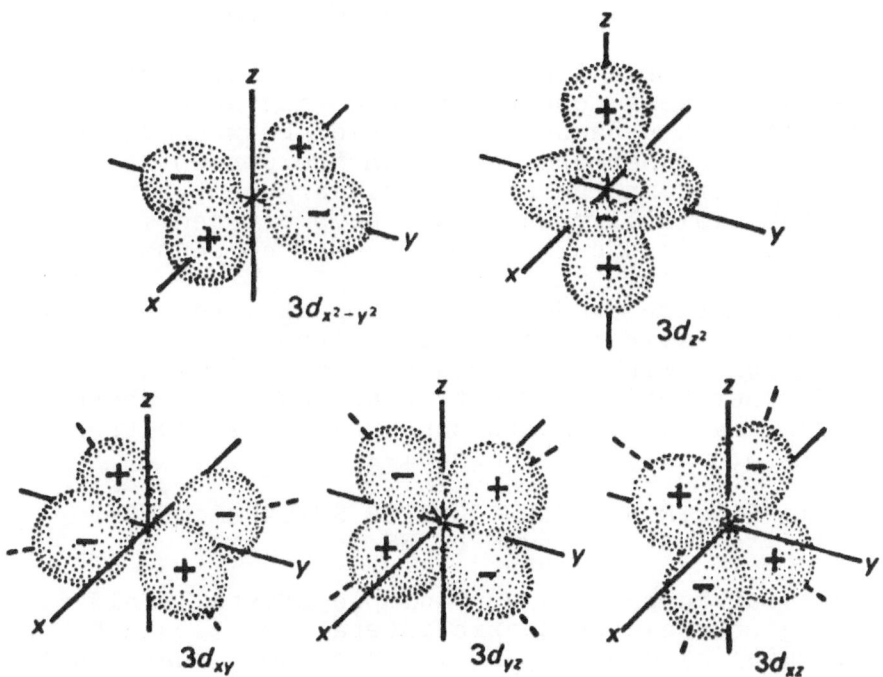

Figure 2 - Distribution of d electrons

Secondly, these properties should strongly depend
on the "thickness" of the isolating layer around the
metal chains made up by the ligands. Large organic ligands of about 20 - 30 Å in size isolate the metal
strands. Electrons of one strand will have no chance to
find another chain with which to interact unless a large
amount of activation energy is supplied to overcome
these barriers. The localization of electrons on one
chain can only be prevented if open-shell organic molecules are used as ligands (an inter-ligand interaction
is therefore added) or the lattice is made up of monoatomic linear chains without an isolating layer. $(SN)_x$
is a typical example for this latter type of compound.

Thirdly, large isolating organic groups attached
to the linear metal chains cause an increase in metal-
metal distances because of repulsion between the organic
groups and, furthermore, diminish the concentration of
charge carriers in the solid. This argument also predicts that small ligands should be preferred in synthesizing highly conducting linear metal compounds.

These rather crude conclusions at least hint at
some steric requirements for the ligands in order to prepare compounds with appropriate bulk solid state properties. Furthermore, the question arises as to whether
there are any requirements concerning the electronic
structure of individual molecules.

III. ELECTRONIC STRUCTURE OF SUITABLE METAL COMPLEXES

For our purposes a very simple crystal field model
including only electrostatic interactions could be used
as a working hypothesis, if it is assumed that the intermolecular contacts in the known linear chain metal
compounds with direct metal-metal bonds are brought
about mainly by the central metal's d electrons. In a
crystal field approach, the central metal ions of the
planar transition metal complexes residing in columnar
stacks are exposed to a tetragonal ligend field (fig.3)
exerted by the four "equatorial" ligands and the two
adjacent metal ions in the "axial" positions. The degree
of "tetragonal distortion" depends on the relative
strength of the axial and equatorial donor atoms. Equal
electrostatic interactions between electrons in d levels
on one hand and electrons of the axial and equatorial
ligands on the other hand result in a formally "octahedral" field with degenerate d_z2 and d_{x2-y2} levels. Since

the energetic position of the d_z2 level is most sensitive
to minor changes in "axial field strength", the energy
of d_z2 electrons in the solid compounds is difficult to
predict and quite controversial for this reason[12]. If
additional intermolecular overlap between d electrons
along the stacks is allowed, bands are obtained the
width of which depends upon the degree of overlap between the one electron d functions. The best intermolecular overlap is assumed between the d_z2 functions.

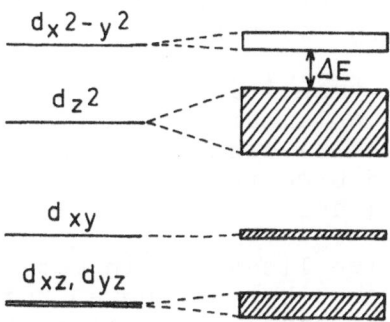

Figure 3 - Energy level diagram of d electrons
in a linear chain metal compound.
(Reproduced by permission of Zeitschrift
f. Naturforschung).

This model is in contrast to earlier proposed energy level diagrams[13] and more recent work[14] in which
appreciable "mixing in" of $(n+1)p_z$ states is suggested.
A small participation of $(n+1)s$ functions in the delocalization of d electrons with main quantum number n
should not be ruled out as well. In any case all the
proposed models have in common the assumption that empty
metal orbitals are in an energetically favourable position to participate in electron delocalization and that
partially filled conduction bands could result even in
pure d^8 compounds which have a filled d_z2 band in a
first approximation.

Nevertheless the diagram shown in figure 3 may help
in choosing appropriate ligands and central metal ions.
With central metal ions in a d^8 configuration and an
appreciable tetragonal distortion, the energy gap ΔE is
quite large, the d_z2 band is filled and the d_x2_-y2 band
is empty. A semiconducting behaviour with only weak intermolecular interactions is predicted. Strongly electron-accepting ligands which will decrease the energy

separation between the d_z2 and $d_{x^2-y^2}$ orbitals could
lead to a vanishing small energy gap in a formally
"octahedral" field, and 1d metallic d^8 compounds could
be obtained at best. The former case is that what is actually found in most known linear chain d^8 metal compounds. But the predicted weak interactions in d^8 complexes with large ΔE could be strengthened if it is possible to:

i)   diminish the energy gap between the d_z2 and
       $d_{x^2-y^2}$ bands by using strongly electron-
       accepting ligands;

ii)   oxidize transition metal complexes crystalli-
       zing in columnar stacks, so the filled d_z2
       band will be depopulated and a partially
       occupied band results (no matter how large
       ΔE is); and

iii)   synthesize linear chain metal complexes with
       less than 8 but an uneven number of d elec-
       trons per metal site, so partially filled
       bands below the d_z2 band arise.

These "electronic" effects have to be taken into
consideration in addition to the mentioned steric restrictions in the search for highly conducting metal
chain compounds.

IV. PREPARATION AND INVESTIGATION OF d^8 TRANSITION
     METAL COMPOUNDS

Most of the linear chain complexes with direct metal-metal interactions known up to now contain d^8 confi-
gurated central metal ions. The following types of complexes with directly connected linear transition metal
chains and 8 d electrons per central metal ion are known:

1) Tetrahalogenometal anions of the type $[MX_4]^{n-}$
   with X = Cl^-, Br^-, CN^- like $[Pt(CN)_4]^{2-}$ [13,15].

2) Complex cations of the type $[ML_4]^{n+}$ with L a
   monodentate neutral donor like NH_3 or isonitrile. The cation $[Rh(CNR)_4]^+$ is a simple example
   for this group of compounds [16, 17].

3) In mixing both types of complexes, stacks with
   alternating cations and anions are obtained
   (Magnus' Green Salt, e.g.) [9, 18].

4) Planar dicarbonylmetal(I) species containing the
   metals rhodium and iridium of stoichiometry
   $Ir(CO)_2Cl \cdot L$ and $Ir(CO)_2 \cdot L-L$, where L = monodentate neutral organic base and L - L = mononega-
   tive bidendate organic ligand. To this group
   belong the numerous derivatives of (pentan-2,4-
   dionato)dicarbonyliridium(I) and rhodium(I)
   (19, 20, 21) and, in our opinion, $Ir(CO)_3Cl$ prepared first by Hieber[22].

5) Bis(α,β-diondioximato)metal(II) complexes[9].

6) Bis(maleodinitriledithiolato)metal(II) compounds
   [23].

7) Bis(chelato)metal(II) compounds with bidendate
   mononegative or dinegative ligands like bis-
   (N-methylsalicylaldiminato)-nickel(II)[24] and
   -copper(II)[25] as well as potassiumbis(oxalato)
   platinate(II)[13].

8) Metalphthalocyanines[26].

The solid state properties of these known linear
chain metal compounds depend very strongly on the "steric" and "electronic" parameters of the individual molecules in the stacks and can therefore be varied systematically by chemical synthesis as shown by some examples
in the following.

IV.1. TETRAHALOGENOMETALLATE(II)ANIONS, TETRAKIS(DONOR)METAL(II)CATIONS AND ALTERNATE STACKS WITH THESE IONS

There are several well-known examples of the variation of metal-metal interactions with the kind of ligand. The compounds K_2PtCl_4 and its derivatives may be
used as a simple example of the changes which can be
done on this type of complexes and what solid state consequences will be expected. The intermolecular interactions between the platinum atoms are almost zero in
solid K_2PtCl_4. A large energy gap ΔE causes solid state
properties to show no sign of collective electron behaviour. Therefore, the intermolecular platinum-platinum
distance is quite large (4, 13 Å) [27]. However, as
pointed out in the described crystal field model, the
intermetallic interactions can be enhanced heavily by
using a more electron-withdrawing ligand than chlorine,
especially π-acceptors (e.g. cyano groups).

In a simple energy scheme this results in a lowering of the energy of the $d_{x^2-y^2}$ band thereby diminishing the energy gap. Actually the Pt-Pt distances in $[Pt(CN)_4]^{2-}$ compounds are considerably decreased and some of the solid state properties of compounds of this type clearly show the effect of anisotropic intermolecular interactions[13, 15]. Comparison of different solid $M_2[Pt(CN)_4]$ or $M'[Pt(CN)_4]$ species points out that at least one more important parameter besides the "electronic" factor has to be considered: the steric requirements of the lattice elements. This is proved by the dependence of intermolecular metal interactions on the hydration sphere around the cations[13]

Furthermore, the interaction can be enhanced by adding electrostatic attraction between lattice elements instead of repulsion in the purely anionic stacks of $PtCl_4^{2-}$. Magnus' Green Salt with alternating $[Pt(NH_3)_4]^{2+}$ and $PtCl_4^{2-}$ units and Pt-Pt distances of 3.25 Å shows considerably enhanced intermolecular interactions as evidenced by the unusual color and a considerable conductivity[28]. The electron accepting properties of a ligand and its steric requirements on one hand and the degree of intermolecular electron exchange on the other have been varied systematically over a wide range, but in small steps, in a group of isonitrile derivatives of the type $[Pt(CNR)_4]$ $[PtCl_4]$ with R = aryl groups[29]. As expected on the basis of the simple model, electron-withdrawing groups had an effect of strengthening the metal-metal interactions and hence the cooperative properties while electron-donating substituents on the ligand weaken these interactions. This is easily shown by studying the optical properties of the compounds.

Since increasing intermolecular metal-metal interaction results in enhanced "red-shifted" absorption of the solid compound, complexes with strongly electron-accepting groups (decrease of ΔE) should absorb strongly in the red part of the visible spectrum and appear deep blue or violet therefore in the solid state. Complexes with donating isonitrile groups adsorb in the blue part of the visible spectrum and appear yellow for this reason. Bulky isonitrile ligands decrease the interactions between the metal ions further but this steric influence seems to be minor compared to the "electronic" effect. It is clear that complexes with bulky and donating ligands generally look yellow, while the ones with small accepting ligands look blue and show a typical linearly polarized copper-like metallic lustre[29]. Nevertheless

none of these solids can be called metallic because of
their low electrical d.c.-conductivity in the range of
10^{-8} Ω $-1.cm^{-1}$ (30). A detailed discussion of the optical properties of several alkylisonitrile derivatives
has appeared recently[31]. That steric requirements of
lattice elements other than the metal complex molecules
play an important part in determining the cooperative
properties of linear chain compounds can easily be demonstrated by synthesizing tetrakis(isonitrile)rhodium
(I) cations with counter ions of different size[17]. The
chloride derivative crystallizes in violet, lustrous
needles and the perchlorate in blue ones while tetrakis-
(isonitrile)rhodium tetraphenyloborate with the huge
counter ion tetraphenyloborate is a yellow solid which
shows no sign of intermetallic interactions.

IV.2.   DICARBONYL- AND TRICARBONYLIRIDIUM(I) COMPOUNDS

Similar systematic changes in solid state behaviour
can be brought about by a systematic variation of equatorial ligands in planar chelatodicarbonylrhodium(I) and
iridium(I) compounds. Some of them have been proven to
crystallize in columnar stacks[19] and the resulting
anisotropic physical properties of these solids were
previously the subject of intense investigations[32].
Complexes of this type, e.g. of stiochiometry $Ir(CO)_2 \cdot$
(L-L), can be obtained easily by starting with $Ir(CO)_3Cl$.
The synthesis proceeds especially well if the tricarbo-
nylchloroiridium(I) is reacted with the "Triton B" (tri-
phenylbenzylammonium hydroxide) salt of the ligand[20].

Some conclusions about the intermolecular interactions can again be drawn by looking at the optical absorption spectra (fig. 4). Strong metal interactions can
be identified by broad and intense absorption in the
long wavelength part of the spectrum. As shown by the
variation of these bands with different ligands, the interactions are diminished with increasing size of the
organic chelates. The rather high electric conductivity
of $Ir(CO)_3Cl$ is decreased drastically when large organic
dye bases are used as ligands[33]. In some cases even
modification changes are observed which indicate that
the metal bonds do not contribute very much to the stabilization of the columnar stacks so the crystal structure is determined by the packing of the large organic
ligand molecules. The optical absorption spectra of the
two modifications are quite different, and suggest that
one of the modifications crystallizes without essential

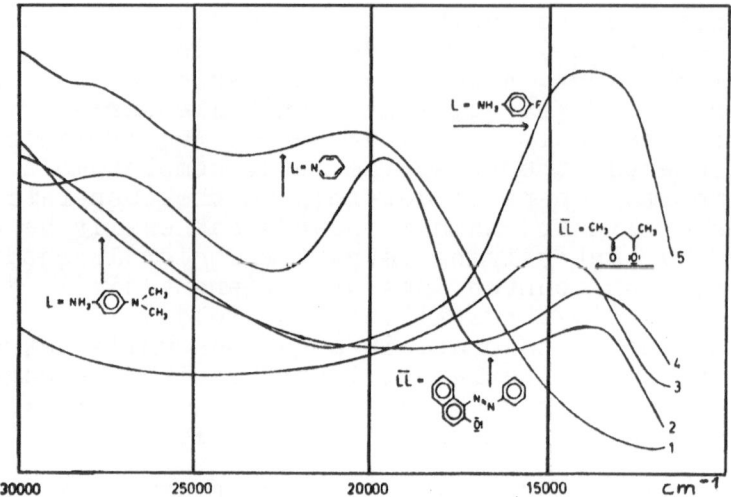

Figure 4 - Visible absorption spectra of $Ir(CO)_2L_2$
          compounds.

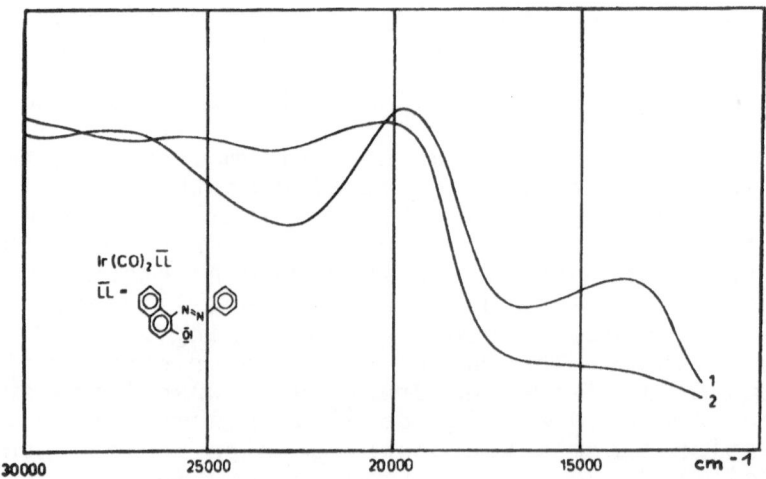

Figure 5 - Comparison of visible spectra of the "metallic"
          (1) and "nonmetallic" (2) modification of
          [1-(phenylazo)-naphtholato-(2)]dicarbonyliri-
          dium(I).

intermolecular interactions. One of them shows a broad
strong absorption in the region of 13.000 cm^{-1} which is
totally absent in the second modification[21](fig. 5).
The Mößbauer effect which works quite well with iridium
(^{193}Ir) can be used to identify intermolecular interactions as well. Isomer shifts and quadrupole splittings
depend strongly on the strength of the intermolecular
interactions along the metal chains[34, 35]. This is summarized in figure 6 which shows that a very consistent
list of shifts is obtained. Negative isomer shifts, relative to iridium metal as the standard, are found in
solids which give clear evidence of strong intermolecular
interactions in the solid state (optical spectra). The
compounds with only weak metal-metal interactions show
positive shifts and quite large quadrupole splittings.
The quadrupole splittings are smaller for the compounds
with strong intermolecular interactions growing larger
with decreasing influence of the axial ligands as predicted by the simple crystal field model.

Especially interesting are the data of those complexes which crystallize in two modifications (only one
of which contains columnar stacks) and the shifts of species with stoichiometry $Ir(CO)_2LCl$. As expected, the
isomer shift of a distinct iridium(I) complex are more
positive relative to the standard Ir metal for the modification without any sign of cooperative electron behaviour and are smaller for the deeply colored modification   crystallizing in columnar stacks. The results on $Ir(CO)_2LCl$-type compounds could be used to clarify the question of oxidation state of these species.
Very recently $Ir(CO)_3Cl$ was again claimed to be a
"mixed valence" solid[13, 36]. In our opinion the dependence of isomer shifts of $Ir(CO)_2LCl$-type compounds
rules out the existence of a mixed valence $Ir(CO)_{3-x}Cl_{1+x}$.
$Ir(CO)_3Cl$ is just an impressive example as to how the
cooperative behaviour of metal electrons can be enhanced
in linear chain metal compounds by using small and strongly accepting ligands as pointed out in section II. and
III.

IV.3. BIS(αß,-diondioximato)metal(II) COMPOUNDS

A similar relation between solid state properties
on one hand and steric and electronic ligand parameters
on the other hand can be found in the "dioximato" complex series. Numerous complexes of d^{8}ions with these ligands were described in the past[9, 37]. Especially

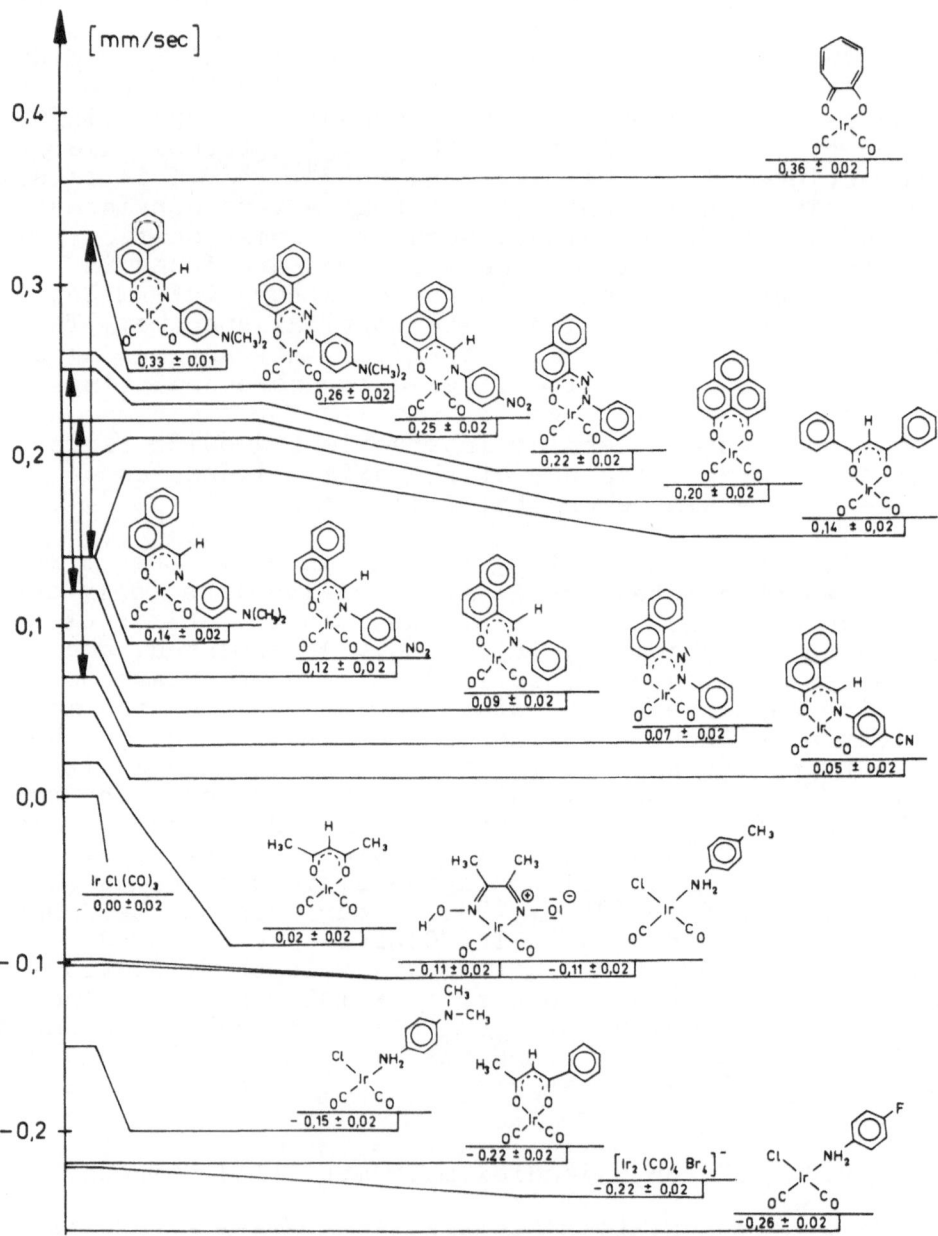

Figure 6 – ^{193}Ir-Mößbauer data (arrows connect two
different modifications of a complex)

strong interactions in these solids are found in the
Bis(1,2-benzoquinonedioximato)metal(II) species, the ligand of which is relatively small and is capable of accepting metal d electrons [38]. The enhanced delocalization of electrons along the metal chains in the benzoquinonedioximato compounds compared to other dioximato
metal(II) complexes can be proven by the decrease in
metal-metal distance, by the strong long wave length
optical absorption and by the higher d.c. conductivity.

The examples described in section IV prove that
even d^8 ions in linear metal chain complexes are capable
of strong interactions giving rise to 1d cooperative
electron behaviour if small and strongly electron accepting ligands are used. Ir(CO)$_3$Cl which is considered, in
our opinion, to be a pure d^8 compound is an impressive
example that the energy gap ΔE can be reduced considerably by using electron accepting groups and that the
d.c.-conductivity can be raised to almost metallic behaviour by using small ligands.

V LINEAR CHAIN "MIXED VALENCE" TRANSITION METAL COMPLEXES

Nevertheless the intermetallic interactions in all
of these d^8 transition metal complexes should be enhanced after oxidation. The results of oxidation reactions
on d^8 linear chain metal complexes were studied thoroughly for the tetrakis(cyano)platinum(II)complexes and some
oxalato derivatives. The first investigations were carried out more than 100 years ago and the results of all
of the work are well known and are summarized[13]. Sur-
prisingly only a few types of mixed valence complexes
with direct metal-metal contact are known[13, 39-41] although a considerable stabilization of these structures
should be predicted through intermolecular metal bonds.

One important question remains therefore to be answered: why does the oxidation of other square planar
platinum(II) compounds, crystallizing in columnar stacks,
not proceed to "mixed valence" one-dimensional metals?
This can be answered by investigating the association
reactions in solution which lead to linear chains. This
"piling up" reaction is essentially a donor-acceptor
association between a square planar platinum(II) donor
and a solvated platinum(IV) acceptor. The reaction is
prevented if solvent molecules occupy the "axial positions" in the strongly accepting platinum(IV) moiety
which cannot be removed by donating platinum(II) species.

A chain is obtained if the axially bound solvent molecules can be substituted by the donating platinum(II) species, that is, if the following reactions proceed to association:

$$[Cl \cdot Pt^{IV}(CN)_4 \cdot H_2O]^- + [H_2O \cdot Pt^{(II)}(CN)_4 \cdot H_2O]^{2-} \rightleftharpoons$$

$$[H_2O \cdot Pt^{(II)}(CN)_4 \cdot Pt^{IV}(CN)_4 \cdot H_2O]^{2-} + H_2O + Cl^-;$$

$$[H_2O \cdot Pt^{(II)}(CN)_4 \cdot Pt^{IV}(CN)_4 \cdot H_2O]^{2-} + [H_2O \cdot Pt(CN)_4 \cdot H_2O]^{2-} \rightleftharpoons$$

$$[H_2O \cdot Pt^{II}(CN)_4 \cdot Pt^{IV}(CN)_4 \cdot Pt^{II}(CN)_4 \cdot H_2O]^{4-} + 2H_2O$$

The reaction proceeds to association in aqueous solutions of Pt(II) and Pt(IV) species with cyanide and oxalato ions as ligands, but is prevented in the same solvent when using complexes of the type $PtCl_4^{2-}$ or $[Pt(NH_3)_4]^{2+}$ because either the reduced form of the oxidizing agent (mostly halogenide ions) or the solvent molecules are bonded very covalently onto the axial positions of the platinum(IV) acceptor. If this assumption is correct, the oxidation should proceed to "mixed valence" compounds in many other cases if the reaction can be carried out in non-donating and oxidizing solvents. If the medium is half concentrated sulfuric acid, the red K_2PtCl_4, the yellow $Pt(NH_2 \cdot CH_2 \cdot CH_2 \cdot NH_2)Cl_2$ and the green Magnus' Salt can be oxidized to deeply coloured compounds with a remarkable red metallic lustre[42, 43]. Chemical and physical evidence clearly proves that these compounds are mixed valence solids. Since the reaction is a heterogenous one and for this reason only powders could be obtained, this procedure is not appropriate in the synthesis of 1d metallic single crystals.

This problem could be solved by finding better suited solvents in the preparation of the above mentioned mixed valence solids or by starting with d^8 compounds which are soluble in non-donating organic solvents. Some of the linear chain bis(dioximato)metal compounds are of this type. "Mixed valence" solids derived from bis(1,2-diphenylglyoximato)-nickel(II) and palladium(II) have been known for a long time but were not recognized as strongly interacting metal chains[44, 45]. A structure very similar to the one found in the blue starch-iodine adducts with linear triiodide chains parallel to the metal complex matrix was proposed recently[46].

A similar solid was obtained recently by oxidizing bis(1,2-benzoquinonedioximate)palladium(II) and nickel(II) with iodine to mixed valence species which differ consi-

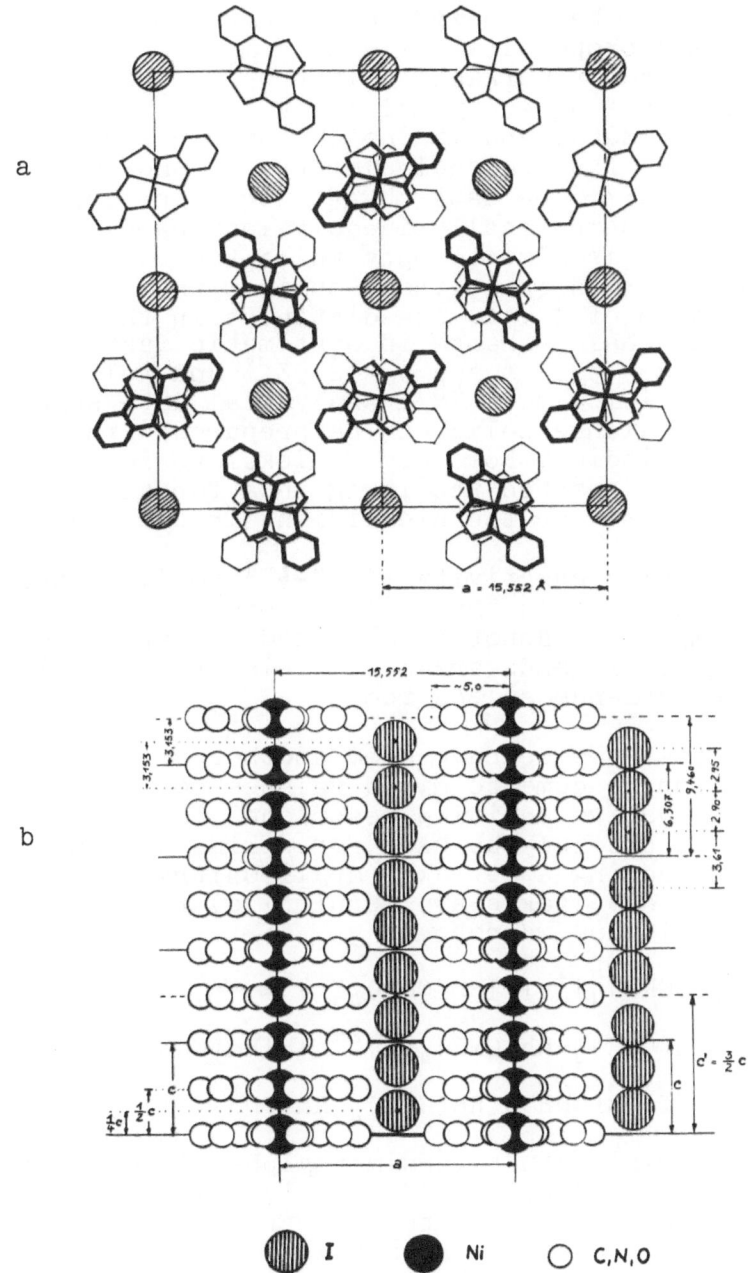

Figure 7 - Structure of oxidation product of Bis(1,2-
benzoquinonedioximato)nickel(II) with iodine.
(a) = projection parallel to the chains.
(b) = projection perpendicular to the chains.

derably from their unoxidized parent compounds in their
solid state properties[47]. The compounds crystallize in
needles with a remarkable anisotropic golden lustre. The
results of a single crystal X-ray analysis so for proves
that linear metal and iodine chains run parallel through
the lattice of tetragonal symmetry as shown schematically in figure 7. The data suggest that one type of the
two crystallographically inequivalent parallel channels
per unit cell is almost empty while the other type, hosts
tri-iodide units in a linear chain with a nickel to
iodine ratio of 3 : 1. The distances in the interrupted
tri-iodide chains change in systematic fashion. The spacings are a = 3.61 Å and b = 2.93 Å and follow an a b b
a b b repetition mode[48]. Analytical data suggest that
the mixed valence solid can be prepared with variing
amounts of iodine. Possibly a nickel to iodine ratio of
1 : 1 can be reached resulting in a formal oxidation
number of 2.33 for the nickel ions in the chain.

The low conductivity ($10^{-4} \Omega^{-1} \cdot cm^{-1}$) of these compounds again proves that highly conducting linear chain
metal complexes cannot be obtained if large and isolating organic ligands are used. This statement is valid
for mixed valence solids too.

Furthermore we have used another well-known type
of reaction to come to mixed valence solids: the so-
called oxidative addition reactions. This type of reaction could possibly be used for piling up linear metal
chains by adding a tetrakis(arylisonitrile)-rhodium(III)
species to a tetrakis(arylisonitrile) rhodium(I)-compound
in the following manner:

$$2[Rh(CNR)_4]^+ + [Rh(CNR)_4I_2]^+ \rightleftharpoons$$

$$[I-Rh(CNR)_4-Rh(CNR)_4 I]^{2+} + [Rh(CNR)_4]^+ \rightleftharpoons$$

$$[I-Rh(CNR)_4 Rh(CNR)_4 Rh(CNR)_4 I]^{3+}....$$

Upon mixing the 2,6-dimethylphenylisonitrile rho-
dium(I) and rhodium(III) derivatives in chloroform, a
sudden change of color, turning from a yellow-red to
dark blue is observed. From this solution a violet solid
with a striking linearly polarized metallic lustre can
be obtained. The preliminary spectroscopic data indicate
an iodide-bridged structure[17] as indicated schematically in figure 8. To my best knowledge this is the

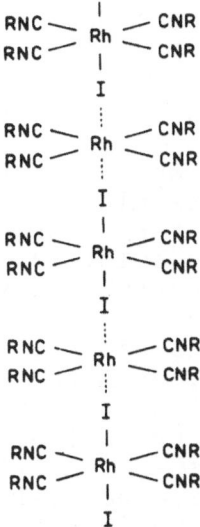

Figure 8 - Proposed structure of $[Rh(CNR)_4I]^+$.

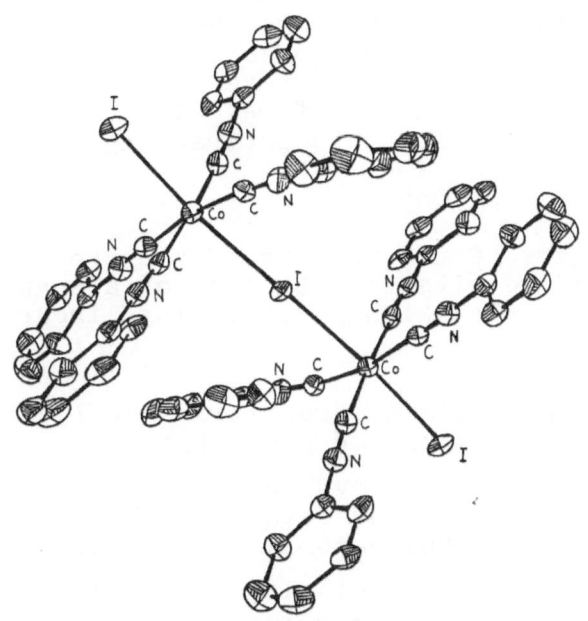

Figure 9 - Molecular structure of Di-iodotetrakis(phenyl-
          isocyanide)cobalt(II): a μ-iodo-bis[iodotetra-
          kis(phenylisocyanide)cobalt(II)] iodide. (Re-
          produced with permission from reference 48.)

first linear chain mixed valence solid with rhodium as
the central metal.

VI   COMPLEXES WITH d^7 CENTRAL METAL IONS

Finally, one result obtained in our search for linear chain compounds containing d^7 metal ions should be
mentioned. A typical and very stable ion with this electronic configuration is cobalt(II). The earlier report
of Malatesta that diiodotetrakis(arylisonitrile)cobalt(II)
compounds can be isolated in two crystalline modifications, one of which is diamagnetic and shows a remarkable metallic lustre[49], prompted us to determine the
exact structure of one of these compounds[50]. The molecular structure of the tetrakis(phenylisonitrile)co-
balt(II), shown in figure 9, is in fact μ-iodo-bis[iodo-
tetrakis(phenylisonitrile)cobalt(II)]iodide. The linearly polarized optical absorption of the solid corresponds
to a transition along the I - Co - I - Co - I heavy atom
chains, all of which run parallel to each other.

A linearly polarized "metallic" reflection (the
word "metallic" is used throughout this paper as a chemical description of the appearance of the solids and
not in the physical sense of a plasma edge in the optical
spectra) does by no means prove a strongly interacting
infinite metal chain therefore. It just gives a hint that
two or more metal ions are interacting electronically.

VII   CONCLUSIONS

A summary of the results of the synthetic work done
so far could be given in the form of a suggested procedure for preparing highly conducting linear chain metal
complexes:

    a) Prepare planar transition metal complexes
       with less than 8 d electrons.

    b) Insert only very small and highly π-accepting
       ligands without regard to the number of d elec-
       trons on the central metal ion.

    c) Use large organic ligands only if they are
       capable of intermolecular interactions of
       their own (planar organic open shell molecules
       as ligands).

d) Prepare solids which contain bundles of parallel metal chains which are surrounded by small multidentate ligands.

e) Add linear chains of interacting <u>atoms</u> or monoatomic ions to the metal complex <u>stacks</u>.

ACKNOWLEDGEMENTS

I would like to thank R. Aderjan, D. Baumann, J. Bolz, H. Breer, H. Endres, R. Lorentz, W. Moroni, M. Megnamisi-Bélombé, D. Nöthe, H.H. Rupp for their help in the preparative and spectroscopic work discussed in connection with this review.

This work was supported by Deutsche Forschungsgemeinschaft, Fonds der Chemischen Industrie and by Rheinisch-Westfälische Elektrizitätswerke, A.G.

REFERENCES

1. H. R. Zeller, Adv. Solid State Phys. <u>13</u>, 31 (1973)
2. Z. G. Soos, Ann. Rev. Phys. Chem. <u>25</u>, (1974)
3. E. B. Jagubskii, M. L. Khidekel, Russ. Chem. Rev. <u>41</u>, 1011 (1972)
4. A. F. Garito and A. J. Heeger, AIP Conf. Proc. <u>10</u>, 1476 (1973)
5. V. V. Walatka jr., M. M. Labes, J. H. Perlstein, Phys. Rev. Letters <u>31</u>, 1139 (1973)
6. R. Greene, NATO-ASI on "Low-dimensional cooperative phenomena" Starnberg, 1974
7. A. F. Garito, A. J. Heeger, Acc.Chem.Res. <u>7</u>, 232 (1974)
8. F. H. Herbstein, Perspect. Struct. Chem. <u>4</u>, 166 (1971)
9. T. W. Thomas, A. E. Underhill, Chem. Soc. Rev., <u>1</u>, 99 (1972)
10. U. Thewalt, Z. Naturforsch. <u>29 b</u>, 308 (1974)
11. Z. G. Soos, T. Z. Huang, J. S. Valentine and R. C. Hughes, Phys. Rev. <u>B 8</u>, 993 (1973)
12. L. V. Interrante, Symposium on "Extended interactions in linear chain compounds", Los Angeles, 1974
13. K. Krogmann, Angew. Chem. <u>81</u>, 10 (1969)
14. W. A. Little, NATO-ASI on "Low-dimensional cooperative phenomena" Starnberg, 1974
15. M. L. Moreau-Colin, Struct. and Bond. <u>10</u>, 167 (1972)
16. J. W. Dart, M. K. Lloyd, R. Mason, J. A. McCleverty, J. C. S. Dalton, <u>1973</u>, 2039

17. D. Baumann, Dissertation, Univ. Heidelberg, 1974
18. F. Mehran, B. A. Scott, Phys. Rev. Lett., 31, 99
    (1973)
19. N. A. Bailey, E. U. Coates, G. B. Robertson, F. Bo-
    nati, R. Ugo, Chem. Commun., 1967, 1041
20. R. Aderjan, H. Breer, H. J. Keller, H. H. Rupp, Z.
    Naturforsch. 28 b, 164 (1973)
21. R. Aderjan, H. J. Keller, Z. Naturforsch. 28 b,
    500 (1973)
22. W. Hieber, H. Lagally, A. Mayr, Z. Anorg. Allg.
    Chem. 246, 138 (1941)
23. R. E. Benson, U. S. Patent, 1966, 3-255-195
24. R. R. Bartkowski and B. Morosin, Phys. Rev. B 6,
    4209 (1972)
25. M. R. Fox and E. C. Lingafelter, Acta Cryst. 22,
    943 (1967)
26. T. S. Srivastava, J. L. Przybylinski and A. Nath,
    Inorg. Chem. 13, 1562 (1974)
27. G. E. Dickinson, J. Amer. Chem. Soc., 44, 2404
    (1922)
28. H. P. Fritz, H. J. Keller, Z. Naturforsch., 20 b,
    1145 (1965)
29. R. Lorentz, Dissertation, Univ. Heidelberg, 1974
30. H. H. Rupp, unpublished results
31. H. Isci, W. R. Mason, Inorg. Chem. 13, 1175 (1974)
32. K. L. Monteith, L. F. Ballard, C. G. Pitts, B. K.
    Klein, L. M. Slifkin, J. P. Collman, Solid State
    Comm., 6, 301 (1968)
33. R. Aderjan, H. J. Keller, H. H. Rupp, Z. Naturforsch
    29 a, (1974)
34. R. Aderjan, Dissertation, Univ. Heidelberg, 1973
35. R. Aderjan, H. J. Keller, H. H. Rupp, F. Wagner, U.
    Wagner, (to be published)
36. A. P. Ginsberg, R. L. Cohen, F. J. DiSalvo and K. W.
    West, J. Chem. Phys. 60, 2657 (1974)
37. C. V. Banks, D. W. Barnum, J. Amer. Chem. Soc., 80,
    3579 (1958)
38. H. J. Keller, M. Megnamisi-Bélombé, to be published
39. L. Malatesta, F. Canziani, J. Inorg. Nucl. Chem.,
    19, 81 (1961)
40. L. I. Buravov, R. N. Stepanova, M. L. Khidekel and
    I. F. Shchegolev, Dokl. Akad, Nauk. SSR, 203, 819
    (1972)
41. I. I. Chernaev and Z. M. Novozhenyuk, Russ. J. Inorg.
    Chem. 11, 1398 (1966)
42. R. D. Gillard and G. Wilkinson, J. Chem. Soc. 1964,
    2835

43. W. Gitzel, H. J. Keller, H. H. Rupp, K. Seibold, Z. Naturforsch., 27 b, 365 (1972)

44. M. Simek, Collect. Czech. Chem. Commun. 27, 337 (1962)

45. A. S. Foust and R. H. Soderberg, J. Amer. Chem. Soc. 89, 5507 (1967)

46. H. J. Keller, K. Seibold, J. Amer. Chem. Soc., 93, 1309 (1971)

47. H. Endres, H. J. Keller, M. Megnamisi-Bélombé, W. Moroni, D. Nöthe, Inorg. Nucl. Chem. Lett., 10, 467 (1974)

48. H. Endres, H. J. Keller, M. Megnamisi-Bélombé, W. Moroni, H. Pritzkow and J. Weiss, Acta Cryst.

49. L. Malatesta, "Isocyanide Complexes of Metals", J. Wiley, London, 1969

50. D. Baumann, H. Endres, H. J. Keller, J. Weiss, J.C.S. Chem. Comm., 1973, 853

APPENDIX

List of contributors to the discussion
sessions

G. Beni, Murray Hill

Polaronic effects in the
one-dimensional Hubbard
model

F. Borsa, Pavia

NMR study of electron spin
dynamics in one-dimensional
paramagnets

J. P. Boucher, Grenoble

EPR and nuclear relaxation
time in one-dimensional
systems
NON-MARKOVIAN interpretation

P. Chaikin, Los Angeles

Transport properties of
organic conductors

D. Davis, Palo Alto

Excitonic superconductivity
in a Pt-chain system;
numerical calculations

P. Delhaes, Gradignan

Electronic and thermal properties of some conductive
TCNQ salts

W. Dieterich, Grenoble

Theory of lattice softening
and spin-susceptibility in
A15-compounds

J. P. Farges, Nice

Electrical properties of
the highly anisotropic
organic semiconductor TEA
$TCNQ_2$

R. Greene, San José

Transport and thermal properties of Polysulfurnitri-
de $[(SN)_x]$

339

W. E. Hatfield, Chapel Hill   Magnetic studies on the nearly two-dimensional HEISENBERG system diethylenetriammoniomchlorocuprate(II)

D. Heitkamp, Jülich   Magnetic susceptibility of 1d-metals

L.V.Interrante, Schenectady   Solid state properties of some charge transfer derivatives of Tetrathiofulvalene with Bis-dithiolene metal complexes

R. Liebmann, Hamburg   Phonon dispersion of a single chain of Pt atoms

C. Mavroyannis, Ottawa   Resonant coherent pairing of excitons in molecular crystals

H. Morawitz, San José   Orientational PEIERLS transition in organic conductors

W.H.G. Müller, Orsay   Pressure dependance of the metal-semiconductor transition of $K_2Pt(CN)_4Br_{0.3} \cdot 3H_2O$

M. Nechtschein, Grenoble   Nuclear relaxation and exiton motion in semiconductor salts of TCNQ

J. O'Neill, Bangor   Electrical conduction properties of various phases of the non-intregral oxidation state metal atom chain compounds $Mg_{0.82}Pt(C_2O_4)_2 \cdot yH_2O$

A. Otto, München   Reflection spectroscopy on monoclinic organic molecular crystals for the example anthracene

L.Pintschovius, Karlsruhe   On the temperature dependance of the phonon anomaly in the linear conductor $K_2Pt(CN)_4Br_{0.3} \cdot 3D_2O$

R. Ranvaud, Grenoble
Ultrasonic investigation of $K_2[Pt(CN)_4Br_{0,3}] \cdot 3H_2O$

D. Reinen, Marburg
Cooperative Jahn-Teller interactions in transition metal compounds

R. Schöllhorn, München
Problems in structure and stability of superconducting two-dimensional intercalated chalcogenides

H. Schuster, Saarbrücken
Influence of COULOMB interaction on properties of the PEIERLS phase

D. Watkins, Bangor
The attempted preparation and characterisation of new non-intregral oxidation state metal chain complexes

SUBJECT INDEX

Absorption spectra, 300,
302, 308
Acoustical mode, 208, 215
Activation energy, 51, 52,
83 -85, 224, 293 - 295,
304 - 307
Alloy, 11, 35, 106
Alternation parameter, 45
Angular dependence of
e.p.r. spectra, 48 ff.,
172 ff.
Anisotropy, 7, 48, 157, 165
Anisotropy of conductivity,
93 ff., 295 ff.
Anisotropy of exchange, 147,
171 ff.
Anisotropy of reflectivity,
97
Antiferromagnetism, 2, 16,
45 - 47, 68, 159, 184

Band edge, 113
Band structure, 12, 13, 25,
28, 90, 217 ff.
Band width, 90
BCS-theory, 29, 34, 36, 98
Bond length, 9, 17, 98, 118
Bragg scattering, 240 ff.
Brillouin zone, 154
Bronzes, 198, 203, 212

Carrier density, 97
Carrier mobility, 97, 293

Chain,
anion, 281, 289
axis, 48, 57, 163, 186
cation, 90, 289
compounds, 91, 183, 187,
288, 315
inequivalent, 160
linear, 23, 45, 66, 91,
156
of spins, 150
Charge density, 97, 114
Charge density wave, 1, 24-
32, 100, 216 - 231, 269
Charge transfer band, 69,
109, 203, 308
Charge transfer energy, 90
Charge transfer salts, 1,
9, 46 ff., 69, 89 - 93,
317
Chloranil, 50
Closed shell molecules, 9,
193
Clusters, 3, 47, 172 ff.,
198
Coherence length, 252, 256
Collective mode, 99 ff.
Collision frequency, 52, 54,
83, 85
Columnar stacks, 90, 281,
293, 316
Columnar structure, 279,
287, 315
Conduction band, 215, 242
Conduction electrons, 97,
259
Conduction properties, 89,
289